Applied mathematics for advanced level
(The Mechanics of Particles and Rigid Bodies)

Butterworths Publications On Related Subjects

Pure mathematics for advanced level
(Second Edition 1983)
B. D. Bunday and H. Mulholland

Fundamentals of statistics
H. Mulholland and C. R. Jones

Also by H. Mulholland

Calculus made simple
(W. H. Allen, 1976)

Guide to Eire's 3000-foot Mountains
(Mulholland Wirral, 1980)

Guide to Wales' 3000-foot Mountains
(Mulholland Wirral, 1982)

Guide to Lakeland's 3000-foot Mountains
(Mulholland Wirral, 1983)

Applied mathematics for advanced level
(The Mechanics of particles and rigid bodies)

H. Mulholland, MSc
Liverpool Polytechnic (retired)

J. H. G. Phillips, MSc, AFIMA
(Faculty of Engineering, Science and Mathematics),
Middlesex Polytechnic

BUTTERWORTHS
London Boston Durban Singapore Sydney Toronto Wellington

All rights reserved. No part of this publication may be reproduced or transmitted in any form or by any means, including photocopying and recording without the written permission of the copyright holder, application for which should be addressed to the publishers. Such written permission must also be obtained before any part of this publication is stored in a retrieval system of any nature.

This book is sold subject to the Standard Conditions of Sale of Net Books and may not be resold in the UK below the net price given by the Publishers in their current price list.

First published 1969
Reprinted (with revisions) 1970
Reprinted 1971
Reprinted (with revisions) 1973
Reprinted 1976, 1977, 1980, 1982
Second edition 1984

© **Butterworth and Co. (Publishers) Ltd 1984**

British Library Cataloguing in Publication Data

Mulholland, H. (Henry)
 Applied mathematics for Advanced Level.—2nd ed.
 1. Mathematics
 I. Title   II. Phillips, J. H. G.
 510   QA36

ISBN 0-408-01445-8

Printed and bound in Great Britain by
J. W. Arrowsmith Ltd, Bristol

Preface to the second edition

In this edition alterations have been made to the chapter on Vector Algebra, also the chapter on Projectiles has been entirely rewritten using a vector approach. This brings the latter chapter more into line with current practice and, indeed, more into line with the rest of the book. Various small changes have been made throughout to improve layout and clarity, and, occasionally, to bring the text up to date.

A large number of worked examples remains a feature of the book, and there are plenty of exercises for the student to develop the necessary skills. The text covers virtually all the requirements in mathematical mechanics (both "Statics" and "Dynamics") of the various examining boards. However, a candidate studying on his or her own is recommended to obtain a copy of the appropriate syllabus and copies of immediate past examination papers to check that all topics are covered.

We have included more examples from recent papers set by the Joint Matriculation Board (JMB), the Delegates of Local Examinations, Oxford (Oxford) and The University of London (London). Also, questions set by the Associated Examining Board (AEB) and the Cambridge Local Examinations Syndicate (C) have now been included. We are grateful to all these boards for granting us permission to use their questions and the abbreviations shown have been used to indicate the source of such questions.

HM
JHGP

Preface to the first edition

The book is intended as a class text in Applied Mathematics at Advanced Level GCE and covers practically all the requirements of the various examining boards in Applied Mathematics and Theoretical Mechanics.

It is not meant to be a "teach yourself" book, but it is hoped that the intelligent reader would be able to use it in this way. The book provides an introduction to the Mechanics of Particles (Chapters 4–15) and Rigid Bodies (Chapters 16–21), quite apart from its Advanced Level orientation. An introduction to vector algebra (Chapters 2 and 3) is a main feature of the book and this is used to unify and clarify the work.

The chapters have been arranged in a logical order which should help student and teacher alike to find their way about the book. This is not to imply, however, that the order in which the chapters are read should necessarily follow that in the book. One system that has worked well in practice is to study Chapters 1 and 2, then work through Chapters 4–11 (omitting Chapter 10)—doing the easy examples and some of the easier exercises, omitting formal proofs—then, when the student's mathematical techniques are sufficient, to return to Chapter 3 and rework thoroughly.

The exercises set at the ends of sections within the chapters are, for the most part, straightforward. All our readers should attempt these. Those at the ends of the chapters are often more testing and include questions from past papers set by the various examining boards.

Even with the above approach the contents of Chapter 3 (Vector Algebra) will probably need more than one reading. Hence some of the sections have been starred so that they may be omitted at a first reading of Part I.

Other points of interest are:

(*a*) The traditional division into Statics and Dynamics has been abandoned in favour of a division into the study of particles and rigid bodies. This allows a simple logical development with Statics being treated as a special case of Dynamics. However, for those who prefer the other approach, the appropriate chapters can easily be treated separately.

(*b*) The work in Chapter 16 (Motion of Systems of Particles) does not appear on examination syllabuses but is included to provide a framework for the subsequent work in Part II. Indeed, in the book as a whole, there are a number of topics which do not appear on all examination syllabuses. They are included in order that the book should be coherent and give a more or less complete account of the theoretical mechanics of a particle and rigid bodies at this level. If Chapter 16 is omitted, the teacher will have to supply some discussion of the factors affecting the motion of a rigid body.

(*c*) A chapter on quantities and their units is included, SI units being used throughout the book. British Standard signs, symbols and abbreviations have been used almost without exception.

We should like to express our thanks to the Joint Matriculation Board (JMB), the Delegates of Local Examinations, Oxford (Oxford), London University (London) and the Welsh Joint Education Committee (WJEC) for granting us permission to use questions from their examinations in this book. The abbreviations above have been used to indicate the source of such questions. Also some of the London University and Welsh Joint Education Committee questions have been "metricized"; such modified questions are indicated by a † and are the responsibility of the authors.

Finally, we should like to thank our publishers for the care and trouble they have taken over the general presentation of the text, and those of our colleagues who have helped with the checking of some of the questions.

HM
JHGP

Contents

Preface to the second edition   v
Preface to the first edition   vii

Part I—Particles   1

1 Quantities and units   3
    1.1 Basic quantities   3
    1.2 Systems of units   3
    1.3 Derived quantities   4
    1.4 Practical units   4

2 Vector quantities   6
    2.1 Scalars and vectors   6
    2.2 The angle between two vectors   7
    2.3 Compounding vectors   8
    2.4 Components of a vector   11
    2.5 Resultant of two or more vectors   13

3 Vector algebra   18
    3.1 Introduction   18
    3.2 Addition and subtraction of vectors   20
    3.3 Multiplication of a vector by a scalar   23
    3.4 Resolution of a vector   25
    3.5 Rectangular resolution of a vector   27
    3.6 Differentiation and integration of a vector   30
    3.7 Differentiation of a unit vector   34
    3.8 Scalar (or dot) product   35

CONTENTS

    3.9 Vector (or cross) product*   39
    3.10 Moments*   42
    3.11 Position vectors and geometrical applications*   46
    3.12 The vector equation of a straight line*   48

4 Speed and velocity   57
    4.1 Average speed   57
    4.2 Speed   58
    4.3 Velocity   62
    4.4 Relative velocity   64
    4.5 Angular speed   73
    4.6 Angular velocity as a vector quantity*   76

5 Acceleration   85
    5.1 Introduction   85
    5.2 Motion in a straight line   86
    5.3 Uniform acceleration   86
    5.4 Speed-time graphs   91
    5.5 Variable acceleration (in a straight line)   94
    5.6 Simple harmonic motion   99
    5.7 Angular acceleration   108

6 Force   115
    6.1 Momentum   115
    6.2 Force   115
    6.3 Force and acceleration   116

7 Physical laws   121
    7.1 Introduction   121
    7.2 Gravitation   121
    7.3 Hooke's law (for a spring or elastic thread)   122
    7.4 Friction   123
    7.5 Newton's third law of motion   124
    7.6 Suggested approach to problem solving   129

8 Motion in a straight line under constant forces   134
    8.1 General method   134

CONTENTS

    8.2 Compound problems   139
    8.3 Linked accelerations   143

9 Motion under forces causing simple harmonic motion   155
    9.1 Basic methods   155
    9.2 Further examples   161

10 Motion in a straight line under variable forces   175

11 Equilibrium of a particle   188
    11.1 Resolution of forces   188
    11.2 Three-force problems   191
    11.3 Harder examples   195

12 Work, power and energy   206
    12.1 Work done by a constant force   206
    12.2 Work done by a variable force   210
    12.3 Power   213
    12.4 Power and velocity   214
    12.5 Work and kinetic energy   221
    12.6 Conservative forces and potential energy   226

13 Impulse and impact   236
    13.1 Impulse of a constant force   236
    13.2 Impulse of a variable force   236
    13.3 Impulse and momentum   237
    13.4 Inelastic impacts   240
    13.5 Impulsive tensions in strings   244
    13.6 Direct impact of elastic bodies   249
    13.7 Oblique impact of elastic bodies   254

14 Projectiles   270
    14.1 Basic theory   270
    14.2 Advanced examples   275
    14.3 Projectiles and impact   282

CONTENTS

15 Circular motion 295
- 15.1 Introduction 295
- 15.2 The acceleration of a particle moving in a circle 295
- 15.3 Force and motion in a circle 298
- 15.4 Uniform motion in a circle 298
- 15.5 Motion in a vertical circle 301

Part II—Rigid bodies 313

16 Motion of a system of particles 315
- 16.1 Centre of mass 315
- 16.2 Motion of the centre of mass 317
- 16.3 Motion about a fixed point 320
- 16.4 Motion relative to the centre of mass 323

17 Equivalent systems of forces 329
- 17.1 The meaning of equivalence 329
- 17.2 Couples 330
- 17.3 Replacement by a single force and couple 331
- 17.4 Coplanar systems 331
- 17.5 Resultants 336

18 Centres of gravity 346
- 18.1 Centre of parallel forces 346
- 18.2 Centre of gravity 347
- 18.3 Centre of gravity of a rigid body 347
- 18.4 Centres of gravity of some standard bodies 348
- 18.5 Centroids 355
- 18.6 List of standard results 355
- 18.7 Composite bodies 356

19 Rotation about a fixed axis 365
- 19.1 Equations of motion of a rigid body rotating about a fixed axis 365
- 19.2 Kinetic energy of a rotating body 366
- 19.3 Calculation of moments of inertia 367

CONTENTS

19.4 Parallel axis theorem 371
19.5 Perpendicular axis theorem for a lamina 374
19.6 Moments of inertia of standard bodies 375
19.7 Uniform angular acceleration 376
19.8 Angular simple harmonic motion (compound pendulum) 381
19.9 Reaction at the axis of a rotating body 387
19.10 Energy methods 390
19.11 Impulsive motion about a fixed axis 394

20 Motion in two dimensions 405
20.1 General equations of motion 405
20.2 Energy methods 411
20.3 Impulsive motion 415

21 Equilibrium 425
21.1 General conditions of equilibrium 425
21.2 Two-force problems 433
21.3 Three-force problems 434
21.4 Several bodies in contact 440
21.5 Jointed rods 442
21.6 Light frameworks 446
21.7 Alternative conditions for equilibrium 449
21.8 Miscellaneous examples 451

Answers 466

Index 494

* These sections may be omitted during the first reading of Part I.

CONTENTS

		Page
21.3.	THREE-FORCE PROBLEMS	418
21.4.	SEVERAL BODIES IN CONTACT	424
21.5.	JOINTED RODS	426
21.6.	LIGHT FRAMEWORKS	430
21.7.	ALTERNATIVE CONDITIONS FOR EQUILIBRIUM	433
21.8.	MISCELLANEOUS EXAMPLES	435
APPENDIX		450
SOLUTIONS		452
INDEX		477

PART I—PARTICLES

A particle is a body whose size is negligible but whose mass is not. In this part of the book large bodies, such as trains, cars etc., will be treated as if they were particles acted upon by the external forces. The justification of this will be given later in Part II (Section 16.2).

1
QUANTITIES AND UNITS

1.1. BASIC QUANTITIES

To answer problems about the motion of particles we must first be able to describe their motion. So a set of quantities is needed to measure how far the particle travels, how quickly it gets there, how difficult it is to stop, etc.

It has been found convenient to build up such a set of quantities from three basic ones: *length*, *mass* and *time*. These three are not defined in terms of other quantities, but are measured by comparing with specified standards.

Other quantities are derived from these three basic ones. For example, the average speed of a particle is defined as

$$\frac{\text{distance travelled}}{\text{time taken}},$$

and consists of a length divided by a time.

1.2. SYSTEMS OF UNITS

Once the units in which to measure length, mass and time have been decided, the units of derived quantities can be built up from them. Thus, if length is measured in metres and time in seconds, the units of average speed are metres/second. A set of units built up in this way is called a coherent system of units.

In this book the International System (SI) units are used. These form a coherent system in which the units of length, mass and time are the *metre* (m), *kilogramme* (kg) and *second* (s) respectively.

A metre is defined in terms of a wavelength associated with the Krypton-86 atom; a kilogramme is the mass of a standard bar kept in Sèvres, France; a second is a specified number of periods of a radiation associated with the Caesium-133 atom. (There are some plans for altering the definition of the standard metre. However, at the time of writing (1984) the above definition still stands.)

1.3. DERIVED QUANTITIES

A list of some of the more important derived quantities follows together with their SI units. Formal definitions and detailed consideration of these and other quantities are given in later chapters.

Quantity		SI unit
speed	rate of change of distance	m/s
acceleration	rate of change of speed	m/s^2
force	mass × acceleration	kg m/s^2 = newton (N)
work	force × distance	N m = joule (J)
power	rate of doing work	J/s = watt (W)

When two quantities have the same basic structure (in terms of mass, length and time), they will have the same units. Such quantities are said to have the same dimensions. For example, experiments show that in a spring

$$\text{modulus } (\lambda) = \text{tension} \times \frac{\text{natural length}}{\text{extension}}.$$

Hence the units in which λ is measured, in the International System, are newton m/m or newton and λ is said to have the dimensions of a force. (A quantity which is a ratio and has no units is called dimensionless.)

1.4. PRACTICAL UNITS

SI units are not always of convenient size for practical purposes, and other "practical" units of length, mass, speed etc. exist. A list of some of the more common practical units with their SI equivalents follows:

$$1 \text{ centimetre (cm)} = 0\cdot01 \text{ m}$$
$$1 \text{ kilometre (km)} = 1\,000 \text{ m}$$
$$1 \text{ gramme (g)} = 0\cdot001 \text{ kg}$$
$$1 \text{ tonne (t)} = 1\,000 \text{ kg}$$
$$1 \text{ hour (h)} = 3\,600 \text{ s}$$
$$1 \text{ km/h} = 5/18 \text{ m/s}$$
$$1 \text{ kilonewton (kN)} = 1\,000 \text{ N}$$
$$1 \text{ kilowatt (kW)} = 1\,000 \text{ W}.$$

Although information in a problem may be supplied in these practical units, it is usual, and often essential, to work throughout in one coherent system (in our case SI units). If this is not done, mistakes may occur since many formulae (e.g. $P = mf$, $H = Fv$, K.E. $= \frac{1}{2}mv^2$) depend upon the quantities they contain being expressed in coherent units.

EXERCISES

EXERCISES 1

1. The magnitude of the momentum of a body is measured by multiplying its mass by its speed. State the SI units of momentum.

2. The measurement of area in general is derived from the measurement of the area of a rectangle (length × breadth). State the SI units of area.

3. What are the SI units of volume and density?

4. The impulse of a constant force is the product of the force and the time for which it acts. State the SI units of impulse and show that impulse and momentum have the same dimensions.

5. In the formula $W = \tfrac{1}{2}I(v/r)^2$, W is work done, v a speed, r a length. Find the SI units in which I is measured.

6. When a truck is pulled along a level track, the resistances due to friction, air resistance, etc., total 70 newton per tonne. If the truck has a mass of 2 500 kg, calculate the total resistance in SI units.

7. Rewrite the following, expressing all quantities in SI units: "A train, of mass 300 t, ascends an incline of 1 in 70 at a constant speed of 18 km/h. If the resistances total 50 000 N, then the engine is working at 510 kW."

2
VECTOR QUANTITIES

2.1. SCALARS AND VECTORS

Definition—A *scalar* quantity has magnitude only and is not related to any definite direction in space, e.g. mass, temperature, time, work and electric charge. Scalars are completely specified by numbers which measure their magnitude in terms of some chosen unit. Operations with scalar quantities follow the usual laws of algebra.

Other quantities exist which have both magnitude and direction, e.g. velocity, acceleration force and magnetic field intensity. An example which does not have physical associations is the directed segment of a straight line $\overrightarrow{PQ}$. The length PQ is the magnitude of the vector. The direction is parallel to the line and in the sense from P to Q. We can represent all quantities which have both magnitude and direction by directed line segments.

Definition—A *vector* is a quantity which has both magnitude and direction and which can be compounded meaningfully* by the triangle or parallelogram rule described later in Section 2.3.

In addition to representing a vector by $\overrightarrow{PQ}$ we shall also use bold

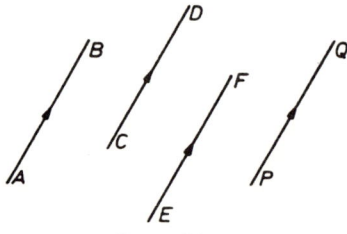

Figure 2.1

---

* Before a physical quantity is treated as a vector it should be proved theoretically, or shown experimentally, that it can be meaningfully compounded by the triangle rule. (Some directed quantities such as finite rotations do not give meaningful results when compounded in this way.) However, in this book we shall deal only with directed physical quantities which are vectors and their vector nature will be assumed without proof.

faced type such as A, a to indicate a vector. The *modulus* of a vector a is the positive number which is a measure of the length of the directed segment and will be denoted by $|a|$ or a.

Two special cases arise. If the modulus of a vector is zero (Q coincides with P), we refer to the zero vector 0 the direction of which is indeterminate. If the modulus of the vector is unity, it is referred to as a unit vector and will be denoted by $\hat{a}, \hat{b}, \ldots$, the circumflex indicating a unit vector.

It is important to realize that, in general (refer to Section 3.9), a vector is not fixed in space. In *Figure 2.1*, $\overrightarrow{AB}$, $\overrightarrow{CD}$, $\overrightarrow{EF}$ and $\overrightarrow{PQ}$ are all equal in length and parallel to one another and are four of the many possible representations of the same vector or, alternatively, they represent four equal vectors.

2.2. THE ANGLE BETWEEN TWO VECTORS

Let the two vectors, a and b, be represented by $\overrightarrow{OP}$ and $\overrightarrow{OQ}$.

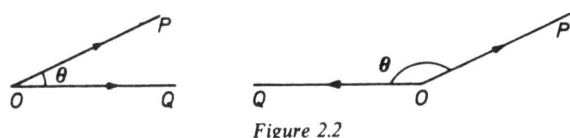

Figure 2.2

Definition—The angle between two vectors is θ, $0 \leqslant \theta \leqslant \pi$ (refer to *Figure 2.2*), the angle between OP and OQ where both vectors are drawn with the arrows pointing away from the point of intersection. When θ is equal to $\pi/2$, the vectors are said to be *perpendicular* and when θ is equal to 0 or π, they are said to be *parallel*.

Exercises 2a
1. Which of the following are vectors and which are scalars?
   (a) energy   (b) weight   (c) electric potential
   (d) work   (e) deceleration   (f) specific heat
   (g) volume   (h) temperature   (j) force
2. Draw, on the same diagram, directed line segments to represent the following horizontal vector quantities:
   (a) A force F_1 of 10 N in a direction 45° E. of N.
   (b) A force F_2 of 8 N in a direction 120° W. of N.
   (c) A displacement d of 10 m in a direction 150° E. of N.
   (d) A force F_3 of 10 N in a direction 135° W. of N.
   (e) A velocity v_1 of 10 m/s in a direction 45° E. of N.
   (f) An acceleration f of 8 m/s^2 in a direction due West.
   (g) A velocity v_2 of 10 m/s in a direction 135° E. of N.

VECTOR QUANTITIES

Are any of the *forces* equal and opposite? Can the same directed line segment represent two or more of these vectors?

3. In the previous question state the angle between the vectors:
(a) F_1 and d
(b) f and F_2
(c) F_2 and v_2.
(d) Are any two of the vectors perpendicular to each other?
(e) Are any two of the vectors parallel to each other?

4. Represent diagrammatically a horizontal acceleration f of 8 m/s² in a direction 20 degrees east of north. On the same diagram represent the following horizontal accelerations:
(a) 24 m/s² in the direction of f
(b) 2 m/s² in the direction of f
(c) 16 m/s² in a direction at right angles to f (two cases).

5. A man walks 10 km north and then $7\frac{1}{2}$ km east. Represent these two displacements graphically on the same diagram and find the resulting displacement: (a) graphically, (b) analytically. (Note this is an example of the triangular rule of vector addition.)

2.3. COMPOUNDING VECTORS

Vectors have both magnitude and direction and it is difficult to say from first-hand reasoning how they should be combined. However, consider the special case of displacements.

If from P, I walk 5 km due East this displacement can be represented completely by the directed segment $\overrightarrow{AB}$ (refer to *Figure 2.3*).

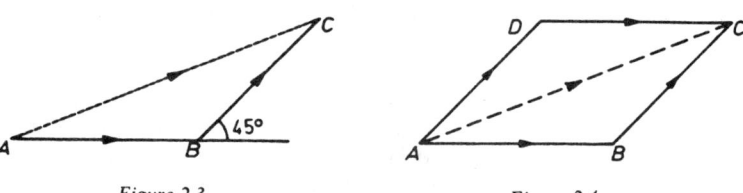

Figure 2.3   Figure 2.4

If I now walk a further distance of 3 km N.E., this further displacement can be represented by $\overrightarrow{BC}$. It would seem reasonable to say that the final result of my journey is the same as if I had walked a distance represented by the length of AC in the direction from A to C. $\overrightarrow{AC}$ is called the *resultant* of $\overrightarrow{AB}$ and $\overrightarrow{BC}$. It is found in practice that vector quantities such as force, acceleration, velocity, etc., can be compounded in the same way, representing them by directed

COMPOUNDING VECTORS

segments, and the resultants thus obtained are meaningful. This is known as the *triangle rule*.

Alternatively, we can represent our two displacements by the directed line segments $\overrightarrow{AB}$ and $\overrightarrow{AD}$ of *Figure 2.4*. Then we can construct $\overrightarrow{AC}$ representing our resultant displacement by completing the parallelogram ABCD. This is known as the *parallelogram rule*.

Example 1. In Figure 2.3, find the actual magnitude and direction of the displacement represented by $\overrightarrow{AC}$.

By the cosine rule:

$$AC^2 = AB^2 + BC^2 - 2AB \cdot BC \cos A\widehat{B}C.$$

Now $AB = 5$ units, $BC = 3$ units and $\angle ABC = 135°$.

∴ $\qquad AC^2 = 5^2 + 3^2 - 2 \cdot 5 \cdot 3 \cos 135°$

$\qquad\qquad = 34 + 30 \cos 45°.$

∴ $\qquad AC = 7\cdot43.$

By the sine rule:

$$\frac{BC}{\sin B\widehat{A}C} = \frac{AC}{\sin A\widehat{B}C}.$$

$$\frac{3}{\sin B\widehat{A}C} = \frac{7\cdot431}{\sin 135°}.$$

∴ $\qquad \sin B\widehat{A}C = 0\cdot2854$

$\qquad \angle BAC = 16° \ 35'.$

$\overrightarrow{AC}$ represents a displacement of 7·43 km in a direction 16° 35′ N. of E.

Example 2. A small aircraft is flying with a velocity of 100 km/h due north. It is also being blown by the wind which has a velocity of 40 km/h from the north west. Find the velocity of the aircraft over the ground.

The velocity of the aircraft over the ground is the result of compounding its own velocity with the wind's velocity. Representing

VECTOR QUANTITIES

these by $\overrightarrow{AB}$ and $\overrightarrow{BC}$, as shown in *Figure 2.5(b)*, then $\overrightarrow{AC}$ represents

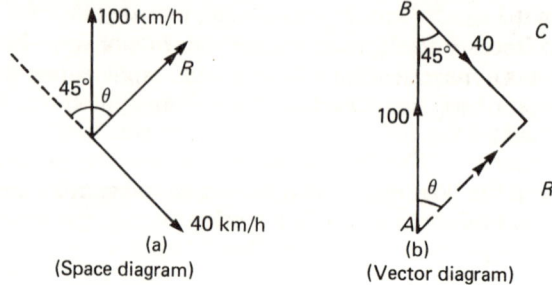

Figure 2.5

the velocity over the ground. Applying the cosine rule to the vector diagram,

$$R^2 = 100^2 + 40^2 - 2 \cdot 100 \cdot 40 \cos 45°.$$
$$\therefore \quad R = 77\cdot 10.$$

By the sine rule

$$\frac{40}{\sin \theta} = \frac{R}{\sin 45°}$$

$$\therefore \quad \sin \theta = \frac{40 \sin 45°}{77\cdot 10}$$

and, since θ is less than 180°, $\theta = 21° 31'$.

The velocity of the aircraft over the ground is 77·1 km/h in a direction N. 21° 31′ E.

Exercises 2b

1. Find the magnitude and direction of the resultant of two forces of magnitudes 5 N and 8 N, the angle between them being 120 degrees.

2. Find the magnitude and direction of the resultant of two forces of magnitudes 60 N and 120 N, the angle between them being 30 degrees.

3. Find the resultant of two vectors, of magnitudes 15 and 8 units, which are at right angles to one another.

4. A boat is sailing at a speed of 10 knots due north. It is also being carried by a strong current of 6 knots in a direction N. 126° 52′ E. Find the speed and direction of the boat as seen by an observer on the land.

10

COMPONENTS OF A VECTOR

5. A liner is travelling at 18 km/h and a passenger walks at 3 km/h across the ship in a direction inclined at 60 degrees to the forward motion of the ship. Find his actual velocity in space.

6. Two forces P and Q have magnitudes of 4 N and 5 N and their resultant has a magnitude of 6 N. Find the angle between P and Q.

7. A motor boat can travel at 20 km/h in still water. It crosses a river estuary 1 km wide in which there is a tide of 6 km/h running up the estuary. If the banks of the estuary are assumed to be parallel, and the boat moves at right angles to the bank throughout the journey, find the angle to the bank at which the boat must be steered. How long does it take the boat to cross?

8. Find the magnitudes of two forces such that if they act at right angles, their resultant is $\sqrt{13}$ N, but if they act at an angle of 60 degrees to each other, their resultant is $\sqrt{19}$ N.

9. Two vectors P and Q are at right angles to each other. The magnitude of P is 5 units and the magnitude of their resultant is 13 units. Find the magnitude of Q.

10. Find the vertical force and the force inclined at 30 degrees to the horizontal which together have a resultant of $6\sqrt{3}$ N horizontally.

2.4. COMPONENTS OF A VECTOR

We have discussed how to compound two vectors; now consider the reverse process of how to express a vector as the sum of two others, known as *components*. This can be done in many ways. If $\overrightarrow{AC}$ represents the vector a, then taking *any* point B and joining AB and BC gives two components $\overrightarrow{AB}$ and $\overrightarrow{BC}$. In most cases it is desirable to take the two components at right angles. If $\overrightarrow{AB}$ makes an angle θ with $\overrightarrow{AC}$, then the magnitudes of the component vectors represented by $\overrightarrow{AB}$ and $\overrightarrow{BC}$ are by trigonometry $a \cos \theta$ and $a \cos (90° - \theta)$ (*Figure 2.6*).

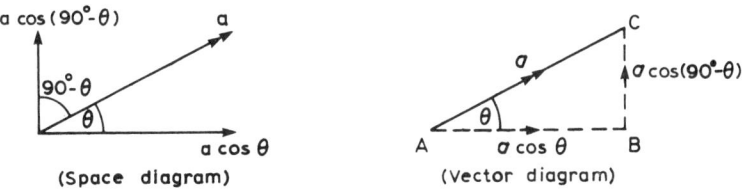

Figure 2.6

Example 1. A plate-layer's truck is being pulled along a railway line

VECTOR QUANTITIES

by means of a horizontal rope. The tension in the rope is 120 N and the rope makes an angle of 20 degrees with the railway line. Find the resolved parts of the force along and perpendicular to the line.

The two components (*Figure 2.7*) are 120 cos 20° N along the line and 120 cos (90° − 20°) = 120 cos 70° N perpendicular to the line.

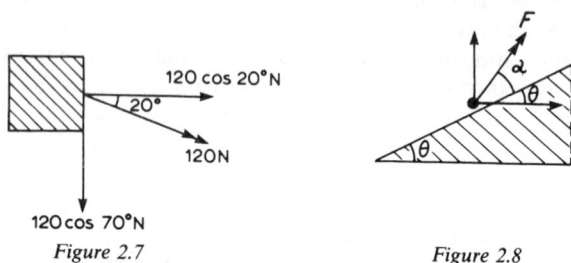

Figure 2.7   Figure 2.8

Example 2. A particle lies on a plane inclined at an angle θ to the horizontal. It is being pulled up the plane by a force of magnitude F, at an angle α to the line of greatest slope. Find the components of F vertically and horizontally.

The angle the force F makes with the horizontal is $\theta + \alpha$ (refer to *Figure 2.8*). Therefore the horizontal component of F is $F \cos(\theta + \alpha)$ and the vertical component of F is $F \cos\{90° - (\theta + \alpha)\}$.

Exercises 2c

1. Find the components of the following horizontal vectors in the directions north and east:
   (a) Force of 5 units in a direction due south.
   (b) Velocity of 30 units in a direction N. 120° E.
   (c) Acceleration of 12 units in a direction N. 45° W.
   (d) Force of 8 units in a direction N. 150° W.
   (e) Velocity of 25 units in a direction due west.
   (f) Acceleration of 8 units in a direction N. 30° E.

2. A stone leaves a man's hand with a velocity of 100 cm/s at an angle of 60 degrees to the upward vertical. Find the components of the velocity vertically and horizontally.

3. A particle on a plane inclined at an angle α to the horizontal, is acted on by a force mg vertically downwards, a force R perpendicular to the plane in an upwards direction and a force μR up the plane. Resolve all three forces horizontally and vertically upwards and give the sums of their components in the two directions.

4. A particle lying on a plane inclined at an angle α to the horizontal is acted on by a force F making an angle of $45° - \theta$ with the line of greatest slope. Resolve F vertically and horizontally (two cases).

5. A circular hoop stands on the ground with its plane vertical. A tangential force F is applied at a point P on its rim in an upward direction. If $\angle POQ = 90° + \theta$, where O is the centre of the hoop and Q its point of contact with the ground, find the horizontal and vertical components of F in the case when $(90° + \theta) > 90°$. Are they the same if $0 \leqslant (90° + \theta) \leqslant 90°$?

2.5. RESULTANT OF TWO OR MORE VECTORS

We have seen in Section 2.3 that two vectors can be compounded by the triangle rule (or parallelogram rule). Consider now the case of three vectors represented by $\overrightarrow{PQ}$, $\overrightarrow{QR}$ and $\overrightarrow{RS}$ (refer to *Figure 2.9*).

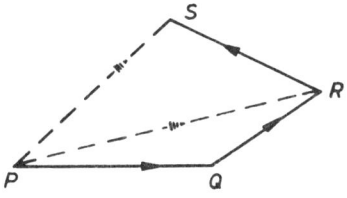

Figure 2.9

Combining $\overrightarrow{PQ}$ and $\overrightarrow{QR}$ gives $\overrightarrow{PR}$, and a further application of the triangle rule to PRS gives the combination of $\overrightarrow{PR}$ and $\overrightarrow{RS}$ as $\overrightarrow{PS}$. This can be extended to any number of directed segments and so we have the *polygon rule*, that if a number of vectors are represented by $\overrightarrow{PQ}$, $\overrightarrow{QR}$, $\overrightarrow{RS}$, ..., $\overrightarrow{YZ}$, taken in order, then their resultant is the vector represented by $\overrightarrow{PZ}$.

Example 1. A yacht is sailing with a velocity of 5 km/h due north. It is also being carried by a current of 2 km/h in a direction southwest. A man walks with a velocity of 3 km/h directly across the deck from starboard to port. Find the man's velocity relative to the shore.

The man is being carried by both the yacht and the current and he is also walking across the deck, thus the resultant of the three

VECTOR QUANTITIES

velocities is required (refer to *Figure 2.10*). An accurate scale drawing of the vector diagram gives $R = 5\cdot69$, $\theta = 50\cdot9$ degrees.

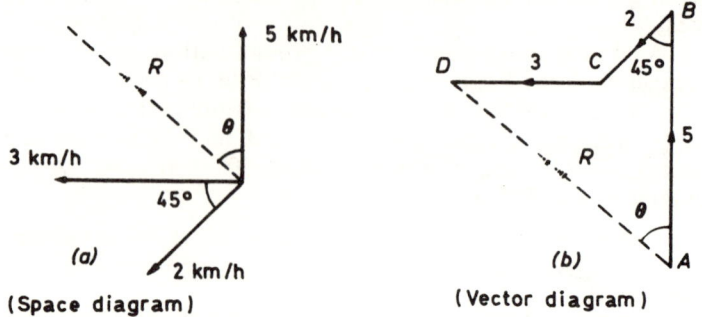

Figure 2.10

An alternative method of finding the resultant of two or more vectors is to find the components of the separate vectors and add these to find the resultant components which are then compounded. In 3-dimensional problems, components are required in three directions. However, to illustrate the method we shall confine ourselves to 2-dimensional problems.

Example 2. Consider again the problem posed in Example 1. Refer to *Figure 2.10(a)*. Resolving in the directions north and east we have:

	Components	
	North	East
Velocity of yacht	5	0
Velocity of current	$-2\cos 45°$	$-2\cos(90° - 45°)$
Velocity of man	0	-3
Totals	$5 - \dfrac{2}{\sqrt{2}}$	$-3 - \dfrac{2}{\sqrt{2}}$
i.e.	3·586	$-4\cdot414$

Thus $\qquad -R\sin\theta = -4\cdot414$

or $\qquad R\sin\theta = 4\cdot414$

and $\qquad R\cos\theta = 3\cdot586.$

EXERCISES

Squaring and adding
$$R^2 = 4\cdot414^2 + 3\cdot586^2$$
$$R = 5\cdot687.$$

Dividing
$$\tan\theta = \frac{4\cdot414}{3\cdot586}$$

with $\sin\theta$ and $\cos\theta$ positive.

∴ $\theta = 50°\ 54'.$

The resultant is 5·69 km/h in a direction N. 50° 54′ W. as before.

Exercises 2d

In the following Exercises 1–3, find the resultant vector by means of components and check your results by a scale drawing:

1. A velocity of 10 km/h in a direction due west.
A velocity of 6 km/h in a direction N. 30° E.
A velocity of 8 km/h in a direction N. 120° E.

2. A force of 80 N in a direction due south.
A force of 160 N in a direction N. 45° E.
A force of 30 N in a direction N. 150° W.

3. An acceleration of 32 m/s² vertically downwards. An acceleration of 40 m/s² at an angle of 5 degrees to the upward vertical. An acceleration of 20 m/s² vertically upwards.

4. Three forces of magnitudes 140, 100 and 160 N act on a particle. If the particle is in equilibrium, find the angle between the 100 and 140 N forces.

5. Three forces of magnitudes 2, 3 and 4 units act in one plane on a particle. The angle between any two of the forces is 120 degrees. Find their resultant.

6. Four horizontal forces have magnitudes and directions as follows: $3\sqrt{2}$ units, N. 45° W.; 8 units, N. 150° E.; 10 units, N. 120° W.; $5\sqrt{2}$ units, N. 135° E. Resolve the four forces in the directions north and east, and hence find the magnitude and direction of their resultant. Verify your result by a scale diagram.

EXERCISES 2

1. Two forces have magnitudes 12 and 15 and the angle between them is 60 degrees. Find the magnitude and direction of their resultant.

2. The angle between two vectors *P* and *Q* is 135 degrees, and

VECTOR QUANTITIES

the magnitude of their resultant is 7 units. If the magnitude of *P* is also 7 units, find the magnitude of *Q*.

3. A yacht is sailing due west at 6 km/h and is being carried by a current with a velocity of 3 km/h towards the north-west. A man walks at 3 km/h directly across the deck from port to starboard. Find the man's velocity relative to the shore.

4. An aircraft is flying with a velocity of 320 km/h in a direction N. 160° W. It is also being blown by the wind, which has a velocity of 35 km/h from the south. Find the velocity of the aircraft over the ground.

5. A cyclist starts from *A* and rides 6 km due west, 4 km due north and 8 km in a direction N. 135° W. Find his distance and bearing from *A*.

6. An aircraft is flying due north and is being carried by a cross wind towards the east. In one hour the aircraft flies 360 km in a direction N. 10° E. over the ground. Find the aircraft's speed and the wind's speed.

7. Forces of magnitude 4, 2, and 2 N act on a particle and are parallel respectively to the sides *AB*, *AC* and *BC* of an equilateral triangle in the sense indicated by the letters. Find the magnitude and direction of their resultant.

8. In Question 7, if the forces had been acting *along* the sides of of the triangle *ABC*, would the magnitude of the resultant have been altered?

9. Two forces *P* and *Q* have magnitudes 25 N and 7 N. If the magnitude of their resultant is 24 N, find the angle between *P* and *Q*.

10. A bomb is moving vertically downwards with a speed of 400 m/s. At a height of 20 m above the ground it explodes and all the particles are thereby given an additional speed of 200 m/s in all directions. Find the diameter of the smallest circle within which the particles will strike the ground (neglect the effect of gravity).

11. The angle between two forces *P* and *Q* is 112° 35′, and the magnitude of their resultant is 12 units. If the magnitude of *P* is 13 units, find the magnitude of *Q*.

12. Six forces each of magnitude 10 units are parallel to, and in the same sense, as the six sides of a regular hexagon *ABCDEF*. Find the resolved parts of the six forces in two perpendicular directions and verify that their resultant is zero.

13. Show that the resultant of the following five horizontal forces is zero: 1 unit, N. 30° W.; 3 units, due west; 4 units, N. 30° E.; 4 units, due east; 5 units, N. 150° W.

14. A passenger in a train travelling at 60 km/h notices that the rain, which is falling vertically, makes streaks on the window.

EXERCISES

If the streaks are inclined at $\tan^{-1} 3$ to the vertical, what is the speed of a raindrop (assumed constant)?

15. A particle has two velocities whose magnitudes are equal. When the magnitude of one of the velocities is halved, the angle which their resultant makes with the other velocity is also halved. Show that the angle between the velocities is 120 degrees.

16. The magnitude of the resultant of two forces P and Q is $Q\sqrt{3}$ and it makes an angle of 30 degrees with the direction of P. Prove that the magnitude of P is either equal to, or double the magnitude of Q.

17. A particle at the corner P of a cube is acted on by three forces each of magnitude F, acting along the three diagonals of the three adjacent faces of the cube which pass through P. Find the resultant force.

18. Forces P and Q act along lines OA and OB respectively, and their resultant is a force of magnitude P; if the force P along OA is replaced by a force $2P$ along OA, the resultant of $2P$ and Q is also a force of magnitude P. Find

(a) the magnitude of Q in terms of P,
(b) the angle between OA and OB,
(c) the angles which the two resultants make with OA.

(Oxford)

3
VECTOR ALGEBRA

3.1. INTRODUCTION

Consider the following problem. A rectangular piece of land of length L m and width W m is to have a strip added to it, parallel to the shorter edge, to increase its area to A m^2. Find a formula for determining the width of the strip.

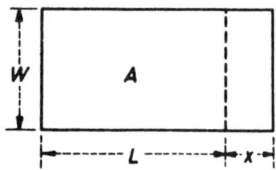

Figure 3.1

Let the width of the strip be x m (refer to *Figure 3.1*). Then since the area of a rectangle is the product of its length and breadth

$$W \times (L + x) = A. \qquad \ldots \text{(i)}$$

This equation is a 'shorthand' way of expressing the relationships between the various dimensions involved: W, L, x and A being symbols standing for the magnitudes of the width, length etc. (which are real numbers); and $\times$, $+$ standing for the operations of multiplying and adding real numbers.

However, because of our knowledge of the way real numbers behave, we can go further than simply writing down a shorthand statement of the information given. We can convert equation (i) into the form:

$$x = \frac{1}{W}(A - WL). \qquad \ldots \text{(ii)}$$

It is instructive to examine step by step one way of converting equation (i) to equation (ii):

INTRODUCTION

$$W(L + x) = A$$
$$\therefore \quad WL + Wx = A \qquad \{a(b + c) = ab + ac\}$$
$$\therefore \quad (WL + Wx) - WL = A - WL$$
$$\therefore \quad (Wx + WL) - WL = A - WL \qquad \{a + b = b + a\}$$
$$\therefore \quad Wx + (WL - WL) = A - WL \quad \{(a + b) + c = a + (b + c)\}$$
$$\therefore \quad Wx = A - WL$$
$$\therefore \quad \frac{1}{W}(Wx) = \frac{1}{W}(A - WL)$$
$$\therefore \quad \left(\frac{1}{W}W\right)x = \frac{1}{W}(A - WL) \qquad \{a(bc) = (ab)c\}$$
$$x = \frac{1}{W}(A - WL).$$

The properties of real numbers that we have assumed are indicated in brackets on the right. Naturally equation manipulation is not usually broken down in this way and these real number properties are neither stated nor thought about, but it is on them that our manipulation depends.

The main properties upon which real number algebra depends are:

I Commutative $a + b = b + a,$ $ab = ba,$
II Associative $(a + b) + c = a + (b + c),$ $(ab)c = a(bc),$
III Distributive $a(b + c) = ab + ac.$

The commutative laws enable us to alter the order of symbols when required; the associative laws to omit brackets; the distributive law to expand brackets and to take out factors.

Suppose now that the letters a, b, c etc. stood, not for real numbers but for some other entities (say, for example, for different chemical compounds). Suppose too, that $+, \times$ stood for some operations on these entities—perhaps mixing together ($+$) and heating together ($\times$) for our chemical compounds. Then we could use the letters and symbols to write shorthand statements about our entities, e.g. $a \times b = c$ could be a short way of saying chemical a when heated with chemical b gives chemical c.

If, in addition, our operations were such that most if not all of laws I, II, III were true, then the shorthand statements could be manipulated in a manner similar to that of real number algebra.

For our chemical compounds, however, it would seem that while

VECTOR ALGEBRA

the commutative laws might apply, the associative and distributive laws would not. So it is unlikely that any useful algebra could be developed here. Many such algebras, however, have been created. For example, operations called $+$, $\times$ can be defined for complex numbers and for certain classes of polynomials which obey all the laws, and for patterns of numbers called matrices which obey all except the commutative law of multiplication. Their algebras are very similar to the algebra of real numbers.

For vectors too, operations can be defined which satisfy laws like these, and the development of a vector algebra is the subject matter of the remainder of this chapter. Use will be made of conventional algebraic signs but care will be taken to verify that the laws of algebra hold for combinations of vectors.

3.2. ADDITION AND SUBTRACTION OF VECTORS

In Section 2.3 we described a method of compounding two or more vectors by the triangle or parallelogram rule. We shall use the $+$ sign from algebra and define the addition of vectors as follows:

Definition—If a is represented by the directed segment $\overrightarrow{PQ}$ and b by the directed segment $\overrightarrow{QR}$, then $a + b$ is defined as the vector which is represented by $\overrightarrow{PR}$ (refer to *Figure 3.2*).

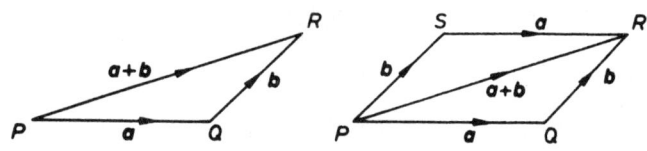

Figure 3.2

Complete the parallelogram $PQRS$. Then the opposite sides being equal and parallel, we have

$$*\overrightarrow{PQ} = \overrightarrow{SR} \quad \text{and} \quad \overrightarrow{PS} = \overrightarrow{QR}$$

and since both $\overrightarrow{PS} + \overrightarrow{SR}$ and $\overrightarrow{PQ} + \overrightarrow{QR}$ are equal to $\overrightarrow{PR}$, it follows that:

$$b + a = a + b.$$

That is, vectors satisfy the commutative law of addition.

* The equals sign is being used to indicate that PQ is both equal and parallel to $\overrightarrow{SR}$. The arrows indicate the sense in which $PQ, SR, \ldots$ are described. It follows that $\overrightarrow{PQ}$ and $\overrightarrow{SR}$ can be regarded as representations of the same vector.

ADDITION AND SUBTRACTION OF VECTORS

Now consider the addition of another vector *c* represented by $\overrightarrow{RT}$ in *Figure 3.3* (not necessarily in the same plane).
Now

$$\overrightarrow{PR} = (a + b) \quad \text{and} \quad \overrightarrow{QT} = (b + c)$$

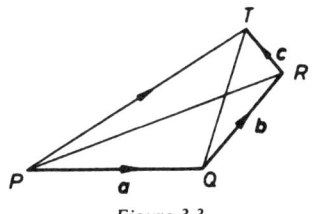

Figure 3.3

Applying the law of addition again

$$(a + b) + c = \overrightarrow{PR} + \overrightarrow{RT} = \overrightarrow{PT} = \overrightarrow{PQ} + \overrightarrow{QT} = a + (b + c).$$

The argument can be extended to any number of vectors. Thus vectors satisfy the associative law of addition and we can write $a + b + c + \cdots$ without brackets being needed to indicate the order of summation.

Definition—If the vector *b* is represented by $\overrightarrow{QR}$ we define $-b$ to be the vector represented by $\overrightarrow{RQ}$, hence,

$$b + (-b) = 0.$$

We now interpet $a - b$ as the sum of the vectors *a* and $-b$

i.e. $$a - b = a + (-b).$$

Referring to *Figure 3.4*

$$\overrightarrow{SQ} = \overrightarrow{SP} + \overrightarrow{PQ} = \overrightarrow{RQ} + \overrightarrow{PQ}$$
$$= -b + a$$
$$= a - b$$

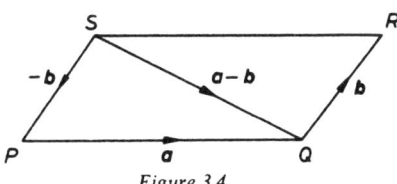

Figure 3.4

Note that the opposite diagonal of the parallelogram $\overrightarrow{PR} = a + b$.

21

VECTOR ALGEBRA

Example 1. Show that $|a + b| \leq |a| + |b|$. In what circumstances does the equality hold?

Referring to *Figure 3.4*, $|a|$ is the number of units in the length of PQ, $|b|$ is the number of units in the length of QR, $|a + b|$ is the number of units in the length of PR.

Since PQR is a triangle,

$$PR \leq PQ + QR$$

or

$$|a + b| \leq |a| + |b|.$$

The equality holds when $PR = PQ + QR$, that is, when P, Q, R (in that order) are collinear, i.e. a and b are parallel and in the same sense.

Exercises 3a

1. Sketch directed line segments to represent vectors a 3 units north, b 2 units north-east and c 4 units west. Now sketch directed line segments to represent the following vectors: (a) $\hat{c}$ (b) $-c$ (c) $a + b$ (d) $a + c$ (e) $a - c$ (f) $b + b$ (g) $a + b - c$ (h) $a + b + c$ (j) $a - b + c$ (k) $a - b - c$.

2. For the three vectors a, b, c defined in Question 1, find the values of the following quantities: (a) $|b|$ (b) $|a + c|$ (c) $|-c|$ (d) $|a + b|$ (e) $|c + b|$ (f) $|c - a|$.

3. Show graphically that for any vectors a and b, $-(b - a) = a - b$. Is it true that $|b - a|$ is equal to $|a - b|$?

4. Show that $|a - b| \geq ||a| - |b||$. In what circumstances does the equality hold?

5. Show that $|a + b + c| \leq |a| + |b| + |c|$, where a, b and c are any three vectors.

6. If a and b are two vectors such that $|a| = |b| = |a + b|$, find the angle between a and b.

7. If a and b are two vectors such that $|a| = |b| = |a - b|$, find the angle between a and b.

8. Given that for any two vectors a and b, $|a + b| = |a - b|$ and also that the two vectors $a - b$ and $a + b$ are at right angles, show that $|a| = |b|$ and that a and b are at right angles.

9. $ABCDEF$ is a regular hexagon. If $\overrightarrow{AB}$, $\overrightarrow{BC}$ represent vectors a, b respectively, find the vectors represented by $\overrightarrow{DE}$, $\overrightarrow{FE}$ and $\overrightarrow{DF}$.

10. A man walks from A to B: his displacement from a point O at A, $\overrightarrow{OA} = a$ and his displacement from O at B $\overrightarrow{OB} = b$. What is his displacement from A when at B?

MULTIPLICATION OF A VECTOR BY A SCALAR

3.3. MULTIPLICATION OF A VECTOR BY A SCALAR

From the law of addition of vectors it follows that $a + a$ is a vector in the same direction as a and whose magnitude is twice the magnitude of a. We can denote it by $2a$. By continued addition, $a + a + a + \cdots$ to n terms, is a vector in the direction of a whose magnitude is equal to n times the magnitude of a. It can be denoted by na. We generalize this for any n, fractional or negative as follows:

Definition—The product of a scalar λ and a vector a (written λa) is a vector, in the direction of a if λ is positive and in the opposite direction to a if λ is negative, of magnitude $|\lambda| \cdot |a|$.

Example. F is a force of 3 units in a direction N. 50° E., *what is meant by* $-4F$?

$-4F$ is a force of magnitude $|-4| \cdot |3| = 12$ units. The direction of $-4F$ is opposite to the direction of F, that is, N. 130° W. Therefore, $-4F$ is a force of 12 units in a direction N. 130° W.

It follows from the definition that if μ is also a scalar $\mu(\lambda a) = (\mu\lambda)a = \lambda(\mu a)$ and the associative law of multiplication holds. Also from definition $(\lambda + \mu)a = \lambda a + \mu a$.

We can also show that $\lambda(a + b) = \lambda a + \lambda b$ as follows: consider Figure 3.5.

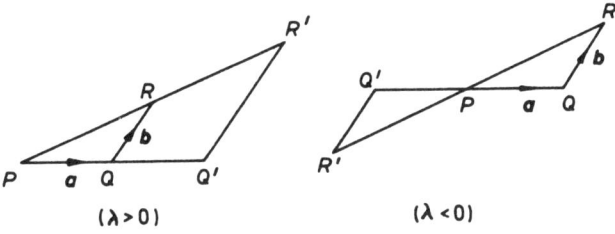

Figure 3.5

$\overrightarrow{PQ} = a$ and $\overrightarrow{QR} = b$, thus $\overrightarrow{PR} = a + b$. Extend PQ to Q' and PR to R' so that

$$PQ' = \lambda PQ \quad \text{and} \quad PR' = \lambda PR.$$

Hence $\quad \overrightarrow{PQ'} = \lambda a \quad$ and $\quad \overrightarrow{PR'} = \lambda(a + b) \quad \ldots$ (i)

For any λ (positive or negative) $Q'R'$ is parallel to QR and by similar triangles $Q'R' = \lambda QR$.

$\therefore \qquad\qquad\qquad \overrightarrow{Q'R'} = \lambda b \qquad\qquad \ldots$ (ii)

VECTOR ALGEBRA

Now $\overrightarrow{PR'} = \overrightarrow{PQ'} + \overrightarrow{Q'R'}$. Hence from (i) and (ii),
$$\lambda(a + b) = \lambda a + \lambda b,$$
and the distributive law of algebra holds for multiplication of vectors by a scalar.

Division of a vector by a scalar m is defined as multiplication by $1/m$.

Notes: (i) If $\hat{a}$ is a unit vector in the direction of a, $\hat{a} = a/|a|$.
(ii) If b is parallel to a, $b/|b| = a/|a|$.

Theorem

If two vectors are represented by $\lambda\overrightarrow{OP}$ and $\mu\overrightarrow{OQ}$, then their resultant is represented by $(\lambda + \mu)\overrightarrow{OR}$, R being the point which divides PQ in the ratio $\mu:\lambda$.

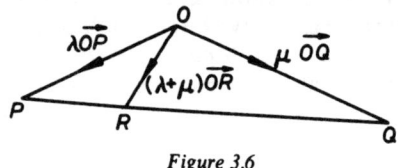

Figure 3.6

Referring to *Figure 3.6*,
$$\overrightarrow{OP} = \overrightarrow{OR} + \overrightarrow{RP} \quad \text{(by the triangle law).}$$
$$\therefore \quad \lambda\overrightarrow{OP} = \lambda\overrightarrow{OR} + \lambda\overrightarrow{RP}.$$

Similarly,
$$\mu\overrightarrow{OQ} = \mu\overrightarrow{OR} + \mu\overrightarrow{RQ}.$$

By addition
$$\lambda\overrightarrow{OP} + \mu\overrightarrow{OQ} = (\lambda + \mu)\overrightarrow{OR} + \lambda\overrightarrow{RP} + \mu\overrightarrow{RQ}. \quad \ldots\text{(i)}$$

But $\lambda\overrightarrow{RP}$ and $\mu\overrightarrow{RQ}$ are collinear and in opposite senses. Also since
$$\frac{PR}{RQ} = \frac{\mu}{\lambda} \quad \text{(given),} \quad \lambda RP = \mu RQ.$$

$\therefore \quad \lambda\overrightarrow{RP}$ and $\mu\overrightarrow{RQ}$ are equal and opposite.

$\therefore \quad \lambda\overrightarrow{RP} + \mu\overrightarrow{RQ} = 0 \quad \ldots\text{(ii)}$

RESOLUTION OF A VECTOR

From (i) and (ii),
$$\lambda\overrightarrow{OP} + \mu\overrightarrow{OQ} = (\lambda + \mu)\overrightarrow{OR}.$$

Note: If $\lambda = \mu = 1$, R is the mid-point of PQ and
$$2\overrightarrow{OR} = (\overrightarrow{OP} + \overrightarrow{OQ}).$$

Exercises 3b

1. If $a = \lambda b$, state the relationship between a and b and express λ in terms of $|a|$ and $|b|$.

2. If $ma + nb = 0$ and a, b are not in the same direction, state the values of m, n.

3. Sketch directed line segments to represent vectors a, 5 units north; b, 12 units east; and c, 8 units south-west. Now sketch line segments to represent the following vectors: (i) $-2b$ (ii) $-\frac{1}{2}c$ (iii) $2(c - a)$ (iv) $-\frac{1}{4}(a + b)$.

4. V is a velocity of 5 m/s in a direction N. 123° W. What is meant by $-6V$?

5. For any two vectors a and b, draw a scale diagram to verify that $3(a - b) = 3a - 3b$.

6. Simplify the following expressions (a) $a + 2a$ (b) $(a - 3a) + a$ (c) $3(a + 2b) - 6(\frac{1}{2}a + b)$.

7. Solve the following equations for x in terms of a, b and c (a) $a + (x + b) = 2b$ (b) $3(2a - 3x) + b = c - b$.

8. ABC is a triangle. A particle at A is acted on by forces $2\overrightarrow{AB}$, $\overrightarrow{BA}$ and $\overrightarrow{AC}$ N. Find the resultant force on the particle.

9. ABC is a triangle and G is the intersection of its medians. Resolve a vector $6\overrightarrow{AG}$ into components parallel to $\overrightarrow{AB}$ and $\overrightarrow{AC}$.

10. The sides $\overrightarrow{AB}, \overrightarrow{BC}$ of a regular hexagon $ABCDEF$ represent the vectors p and q, respectively. Find the vectors which are represented by the remaining sides.

3.4. RESOLUTION OF A VECTOR

We first define collinear and coplanar vectors.

Definition—Two or more vectors are said to be *coplanar* if they are all parallel to the same plane.

Definition—Two or more vectors are said to be *collinear* if they are all parallel to the same line.

Given a vector r and any three non-coplanar vectors, a, b and c, then r can be expressed as the sum of three vectors parallel to a, b and c.

Let $\hat{a}, \hat{b}, \hat{c}$, be unit vectors parallel to a, b, c, respectively. With any

VECTOR ALGEBRA

point O as origin construct a parallelepiped whose co-terminous edges OA, OB, OC are parallel to a, b, c, respectively (refer to Figure 3.7).

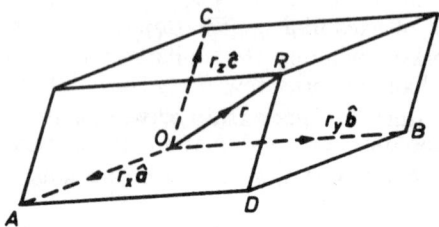

Figure 3.7

Then there exist real numbers r_x, r_y, r_z such that $\overrightarrow{OA} = r_x\hat{a}$, $\overrightarrow{OB} = r_y\hat{b}, \overrightarrow{OC} = r_z\hat{c}$. [$r_x$ is positive or negative according as $\overrightarrow{OA}$ is in the same or opposite direction to $\hat{a}$, similarly for r_y and r_z].

Thus $\qquad r = \overrightarrow{OR} = \overrightarrow{OA} + \overrightarrow{AD} + \overrightarrow{DR}.$

Opposite edges of a parallelepiped are equal and parallel thus:

$$\overrightarrow{AD} = \overrightarrow{OB} \quad \text{and} \quad \overrightarrow{DR} = \overrightarrow{OC}.$$

$\therefore \qquad r = \overrightarrow{OA} + \overrightarrow{OB} + \overrightarrow{OC}$

$$r = r_x\hat{a} + r_y\hat{b} + r_z\hat{c}. \qquad \ldots\ldots \text{(i)}$$

$r_x\hat{a}, r_y\hat{b}, r_z\hat{c}$ are known as the *component vectors* and r_x, r_y, r_z as the *components* of r in the three given directions. Only one such parallelepiped can be constructed and the resolution is unique. Thus equal vectors have equal components, conversely, if all three components of two vectors are equal, the vectors are equal. Also, from equation (i), the sum of several vectors is expressible in the form:

$$\sum r = \sum (r_x a + r_y b + r_z c)$$

$$= \left(\sum r_x\right) a + \left(\sum r_y\right) b + \left(\sum r_z\right) c,$$

and since the direction of a is arbitrary we have that:

The component of the resultant of a number of vectors, in any direction is equal to the sum of the components of the individual vectors in that direction.

RECTANGULAR RESOLUTION OF A VECTOR

3.5. RECTANGULAR RESOLUTION OF A VECTOR

The most important case is when $r = r_x\hat{a} + r_y\hat{b} + r_z\hat{c}$ and the three directions defined by $\hat{a}, \hat{b}$ and $\hat{c}$ are mutually at right angles and form a rectangular frame of reference. The lines are labelled to form a right-handed set of axes, that is, rotation from Ox to Oy takes a right-handed corkscrew along Oz, similarly, a rotation from Oy to Oz takes a right-handed corkscrew along Ox, similarly for Oy (refer to *Figure 3.8*).

In this case, the unit vectors $\hat{a}, \hat{b}$ and $\hat{c}$ are known as $i, j,$ and k [without circumflexes] and

$$r = r_x i + r_y j + r_z k.$$

r_x, r_y and r_z are now referred to as the *rectangular components*, or simply, *components* if there is no fear of a misunderstanding.

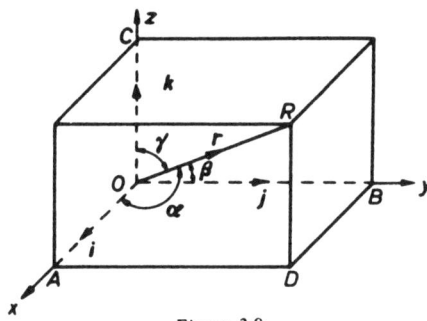

Figure 3.8

If α is the angle OR makes with OA,

$$OA = OR \cos \alpha$$

or $\qquad r_x = r \cos \alpha \quad$ (where $r = OR$).

Similarly, $\qquad r_y = r \cos \beta$

$\qquad r_z = r \cos \gamma$

Now $\qquad OR^2 = OB^2 + BR^2 \quad$ (refer to *Figure 3.8*)

$\qquad\qquad = OB^2 + BD^2 + DR^2$

$\qquad\qquad = OB^2 + OA^2 + OC^2$

or $\qquad r^2 = r_x^2 + r_y^2 + r_z^2.$

From which we see that the modulus of a vector r with rectangular

VECTOR ALGEBRA

components r_x, r_y, r_z is $|r| = \sqrt{(r_x^2 + r_y^2 + r_z^2)}$. (Note the positive square root.)

It follows from the associative and distributive laws that

(I) $\qquad a + b = (a_x + b_x)i + (a_y + b_y)j + (a_z + b_z)k$

(II) $\qquad ma = ma_x i + ma_y j + ma_z k \quad$ (m a scalar).

Example 1. If $a = 3i + 4j - 12k, b = i + 12k, c = i - j + k$. Find $|a|, |b|, |c|, |a + b|$ and $|a + b + c|$.

From our equation $|r| = \sqrt{(r_x^2 + r_y^2 + r_z^2)}$,

$$|a| = \sqrt{[3^2 + 4^2 + (-12)^2]} = 13$$
$$|b| = \sqrt{(1^2 + 0^2 + 12^2)} = \sqrt{145}$$
$$|c| = \sqrt{(1^2 + (-1)^2 + 1^2)} = \sqrt{3}$$
$$|a + b| = |4i + 4j| = \sqrt{(4^2 + 4^2 + 0^2)} = 4\sqrt{2}$$
$$|a + b + c| = |5i + 3j + k| = \sqrt{(5^2 + 3^2 + 1^2)} = \sqrt{35}$$

Vectors and the vector algebra we are developing are eminently suitable for work in 3-dimensions (and can be generalized to n-dimensions). However, many of the examples used in this book will be 2-dimensional applications, the r_z component being suppressed thus:

$$r = r_x i + r_y j$$
and $\qquad |r| = \sqrt{(r_x^2 + r_y^2)}.$

Example 2. P and Q are the points (1, 2) and (3, 7), respectively; express $\overrightarrow{PQ}$ in terms of unit vectors i and j which are parallel to Ox and Oy respectively.

Deduce the general result when P and Q are the points (x_1, y_1) and (x_2, y_2).

Refer to *Figure 3.9* where PR and RQ are drawn parallel to Ox and Oy respectively.

$$\overrightarrow{PQ} = \overrightarrow{PR} + \overrightarrow{RQ}.$$

The length of PR is $3 - 1 = 2$ units and it is parallel to and in the same sense as Ox, therefore, $\overrightarrow{PR} = 2i$.

Similarly, $\overrightarrow{RQ} = (7 - 2)j = 5j$. Therefore

$$\overrightarrow{PQ} = 2i + 5j$$

RECTANGULAR RESOLUTION OF A VECTOR

In general, $\vec{PQ} = (x_2 - x_1)\mathbf{i} + (y_2 - y_1)\mathbf{j}$.

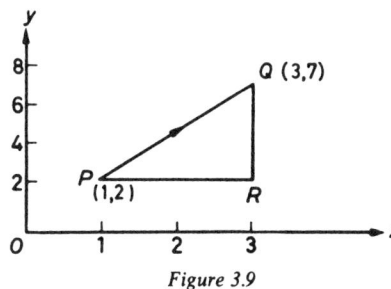

Figure 3.9

Exercises 3c

1. Express the following vectors in terms of $\mathbf{i}$ and $\mathbf{j}$, unit vectors, due east and due north, respectively.

(a) $10\sqrt{2}$ units in a direction north-east.
(b) $10\sqrt{2}$ units in a direction south-west.
(c) 5 units in a direction N. 60° W.
(d) $8\sqrt{2}$ units in a direction N. 135° E.
(e) 20 units in a direction N. 120° W.

2. Given that $r_1 = 2\mathbf{i} + \mathbf{j}, r_2 = \mathbf{i} + \mathbf{j}$, find (a) $r_1 + r_2$; (b) $2r_1 - r_2$; (c) $(4r_1 + 2r_2)/3$.

3. If $a = \mathbf{i} + \mathbf{j} - 2\mathbf{k}, b = \mathbf{i} + \mathbf{k}, c = 2\mathbf{i} - \mathbf{j} + 3\mathbf{k}$, find (a) $a + b - 3c$; (b) $|a + b + c|$; (c) $a - 2b + c$; (d) $|2a + b + 2c|$.

4. If $r = 2\mathbf{i} - \mathbf{j} + 2\mathbf{k}$, find $|r|$ and the cosines of the angles between r and $\mathbf{i}, \mathbf{j}$ and $\mathbf{k}$, respectively (these are known as the direction cosines of r). Write down, in terms of $\mathbf{i}, \mathbf{j}$ and $\mathbf{k}$, a unit vector in the direction of r.

5. If $\mathbf{i}, \mathbf{j}$ and $\mathbf{k}$ are unit vectors parallel to rectangular axes Ox, Oy and Oz, respectively, find a vector of magnitude 6 units equally inclined to the three axes.

6. Given four points $A(2, 3); B(5, 4); C(6, -3)$ and $D(-5, 2)$, express $\vec{AB}, \vec{BC}, \vec{CD}$ and $\vec{DA}$ in terms of $\mathbf{i}$ and $\mathbf{j}$. Hence verify that $\vec{AB} + \vec{BC} + \vec{CD} = \vec{AD}$.

7. Given the points $A(0, 4), B(4, 10), C(7, 8)$, find $\vec{AB}, \vec{BC}, \vec{CA}$ in terms of $\mathbf{i}$ and $\mathbf{j}$. Hence find $|\vec{AB}|, |\vec{BC}|$ and $|\vec{CA}|$ and prove that ABC is a right-angle triangle.

8. Find the resultant of the following displacements: 8 m due north, 6 m north-west, 7 m due west and 10 m N. 150° W.; both analytically (by expressing each displacement in terms of $\mathbf{i}$ and $\mathbf{j}$) and graphically.

VECTOR ALGEBRA

9. Ox, Oy, Oz are a right-handed set of mutually perpendicular axes. $\overrightarrow{OP}$ has a length of 8 units and makes an angle of ϕ with its projection $\overrightarrow{OQ}$ on the xOy plane. $\overrightarrow{OQ}$ makes an angle θ with $\overrightarrow{Ox}$. Express a velocity of 8 m/s in a direction parallel to $\overrightarrow{OP}$ in terms of unit vectors i, j, k parallel to $Ox, Oy,$ and Oz, respectively.

10. Given three points in 3-dimensional space $A(2, 3, 4)$; $B(3, 1, 7)$; $C(4, 3, 8)$, express $\overrightarrow{AB}, \overrightarrow{BC}, \overrightarrow{CA}$ in terms of i, j, k. Hence find $|\overrightarrow{AB}|$, $|\overrightarrow{BC}|$ and $|\overrightarrow{CA}|$ and prove that the triangle ABC is right angled.

3.6. DIFFERENTIATION AND INTEGRATION OF A VECTOR

A vector can change in direction and/or magnitude. The changes can be arbitrary or be a function of one or more variables. We shall deal only with cases of one variable. First let us consider three typical examples of variable vectors.

(a) If a particle is moving in a circle of radius r with constant angular speed ω rad/s and v represents its velocity at any instant, then $|v| = r\omega$ and is constant. However, the direction of v is always changing, thus v is a variable vector.

(b) A particle P is moving along a straight line with velocity v whose magnitude at any time t is kt. In this case the direction of v is constant but its magnitude is always changing therefore v is a variable vector.

(c) A planet P moves in an elliptical orbit around the sun with varying speed. In this case the velocity v is changing both in magnitude and in direction.

We now consider the general case of the differentiation of a vector r, whose magnitude and direction both depend on the value of a scalar variable t.

Suppose r is a continuous and single valued function of t, say $r(t)$. For each value of t there exists only one value of r and as t varies so does r. Consider two values $t, t + \delta t$; to these correspond $r(t), r(t + \delta t)$ which we can call r and $r + \delta r$.

Then we define
$$\frac{dr}{dt} = \lim_{\delta t \to 0} \frac{\delta r}{\delta t}$$

In the case of a point P moving along a curve we can give a pictorial representation of dr/dt (Figure 3.10).

DIFFERENTIATION AND INTEGRATION OF A VECTOR

Let O be a fixed point and $\overrightarrow{OP} = r(t)$ which varies as t varies and P moves along the curve. (Note that r is called the *position vector* of P with respect to O, an expression we shall use later.) If t changes to $t + \delta t$, r changes to $r + \delta r$ and P moves to the point Q. Then $\delta r/\delta t$ is a vector parallel to δr, i.e. to $\overrightarrow{PQ}$.

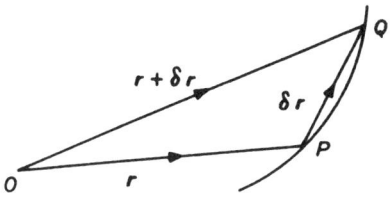

Figure 3.10

As $\delta t \to 0$, $Q \to P$, and the chord PQ tends to coincide with the tangent at P. Therefore $dr/dt = \lim_{\delta t \to 0} \dfrac{\delta r}{\delta t}$ is a vector in the direction of the tangent to the curve traced out by P. dr/dt is, in general, a function of t and thus has a derivative which we denote by d^2r/dt^2 and call the second derivative of r. Similarly, the derivative of this d^3r/dt^3 is the third derivative of r.

Two cases of special importance are:

(a) If r depends on the parameter s which measures distance along the curve from some fixed point, then

$$\frac{dr}{ds} = \lim_{\delta s \to 0} \frac{\delta r}{\delta s}$$

and the modulus of this is

$$\lim_{\delta s \to 0} \frac{\text{chord } PQ}{\text{arc } PQ} = 1.$$

So that *in this case* dr/ds is the *unit* tangent to the curve at P.

(b) When t denotes time and r is the position vector of P relative to some fixed origin O, then δr is the displacement of the point P during an interval of time δt, and $\lim_{\delta t \to 0} \delta r/\delta t = dr/dt$ is called the velocity v of P whose direction is along the curve. The acceleration f of the particle is dv/dt. For further details see Chapters 4 and 5.

VECTOR ALGEBRA

In general, the normal differentiation formulae apply, that is, if vectors *A* and *B* and scalar ϕ are differentiable functions of t.

(a) $\dfrac{d}{dt}(A + B) = \dfrac{dA}{dt} + \dfrac{dB}{dt}.$

(b) $\dfrac{d}{dt}(\phi A) = \dfrac{d\phi}{dt}A + \phi\dfrac{dA}{dt}.$

(c) If *C* is a constant vector (constant in *both* magnitude and direction), $dC/dt = 0$.

Example 1. A particle P moves along a curve whose parametric equations are $x = k(1 + \cos t)$; $y = k(t + \sin t)$ where t is time. Find the magnitudes of its velocity and acceleration at any time t.

The position vector $\overrightarrow{OP}$ of the particle is given by

$$r = xi + yj$$
$$= k(1 + \cos t)i + k(t + \sin t)j.$$

Velocity $\qquad v = dr/dt = -k \sin t\, i + k(1 + \cos t)j \qquad \ldots\ldots(i)$

$\therefore \qquad |v| = \sqrt{(-k \sin t)^2 + k^2(1 + \cos t)^2}$
$\qquad\qquad = \sqrt{(k^2 \sin^2 t + k^2 + 2k^2 \cos t + k^2 \cos^2 t)}$
$\qquad\qquad = k\sqrt{(2 + 2 \cos t)}$
$\qquad\qquad = k\sqrt{4 \cos^2 (t/2)}$
$\qquad\qquad = 2k \cos (t/2).$

Acceleration $f = dv/dt = -k \cos t\, i - k \sin t\, j$ [from (i)].

$\therefore \qquad |f| = \sqrt{(-k \cos t)^2 + (-k \sin t)^2}$
$\qquad\qquad = k.$

The *indefinite integral*, with respect to a scalar variable *t*, of a vector *g* is defined as follows:

$$G = \int g\, dt,$$

where *G* is such that $dG/dt = g$.

DIFFERENTIATION AND INTEGRATION OF A VECTOR

The integral G is indefinite because any constant vector C can be added to it.

Example 2. *Find the indefinite integral with respect to θ of the vector* $g = 3\cos\theta a + 4\sin\theta b$, *where a and b are constant vectors.*

$$\int g\, d\theta = \int (3\cos\theta a + 4\sin\theta b)\, d\theta$$
$$= 3\sin\theta a - 4\cos\theta b + C.$$

The *definite integral* of a vector g with respect to a scalar variable t is defined as follows:

Let $g(t)$ be a vector function of the scalar variable t and suppose $g(t)$ is finite and continuous in the range a to b. Let the range a to b be sub-divided into n sub-ranges $\delta t_1, \delta t_2, \ldots, \delta t_n$ and let t_1 be some value in the first sub-range, t_2 be some value in the second sub-range, ..., t_n be some value in the nth sub-range, and let

$$S = g(t_1)\delta t_1 + g(t_2)\delta t_2 + \ldots + g(t_n)\delta t_n$$
$$= \sum_{a}^{b} g(t)\, \delta t.$$

Let $n \to \infty$ so that the width of each sub-range tends to zero. Then the $\lim_{n\to\infty} S$ is known as the definite integral of g from a to b and is written

$$\int_a^b g(t)\, dt.$$

As for real functions* it can be shown that

$$\int_a^b g(t)\, dt = G(b) - G(a)$$

where G is the indefinite integral of g.

Example 3. *Find the definite integral with respect to t of the vector* $g(t) = t^2 i - (3t + 1)j + 3k$ *over the range $t = 0$ to $t = 3$.*

$$\int_0^3 g(t)\, dt = \int_0^3 [t^2 i - (3t+1)j + 3k]\, dt$$
$$= \left[\frac{t^3}{3}i - \left(3\frac{t^2}{2} + t\right)j + 3tk\right]_0^3$$
$$= [9i - 16\tfrac{1}{2}j + 9k] - [0]$$
$$= 9i - 16\tfrac{1}{2}j + 9k.$$

* See B. D. Bunday and H. Mulholland, *Pure Mathematics for Advanced Level,* 2nd edn, Butterworths (1983).

VECTOR ALGEBRA

In addition, we can define the following definite integral associated with vectors:

$$\int_{r_1}^{r_2} \phi \, d\mathbf{r} = \lim_{\delta r \to 0} \sum_{r_1}^{r_2} \phi \, d\mathbf{r} \quad (\phi \text{ a scalar variable}).$$

3.7. DIFFERENTIATION OF A UNIT VECTOR

If $\hat{a}$ is a unit vector which varies in direction, to find the magnitude and direction of $d\hat{a}/dt$ where t is a scalar variable. Let $\hat{a}$ make an angle θ with an initial line Ox. Referring to *Figure 3.11*, when t

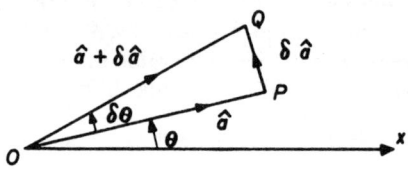

Figure 3.11

increases by δt, P moves to Q where $\angle POQ = \delta\theta$. Since $\hat{a}$ is a unit vector and only varies in direction, $OP = OQ(=1)$

so that

$$\angle OQP = \angle OPQ.$$

As $\delta t \to 0$, $Q \to P$, $\delta\theta \to 0$ so that $\angle OQP = \angle OPQ \to \pi/2$, and $\delta\hat{a}$ [$\overrightarrow{PQ}$ in *Figure 3.11*] is perpendicular to $\hat{a}$. Division by a scalar does not alter the line of action of a vector.

$\therefore \quad \dfrac{d\hat{a}}{dt}$ is perpendicular to $\hat{a}$.

If $Q \to P$, then chord $PQ \to$ arc PQ, i.e. $|\delta\hat{a}| \to 1 \times \delta\theta$, and

$$\left|\frac{d\hat{a}}{dt}\right| = \lim_{Q \to P} \left|\frac{\delta\hat{a}}{\delta t}\right| = \lim_{Q \to P} \frac{\delta\theta}{\delta t}$$

$$= \frac{d\theta}{dt}$$

Therefore $d\hat{a}/dt$ is a vector perpendicular to $\hat{a}$ in the direction of θ increasing and equal in magnitude to $d\theta/dt$.

SCALAR (OR DOT) PRODUCT

Exercises 3d

1. Given that a is a constant vector, find the derivatives of the following expressions with respect to t: (a) $t^2 a$ (b) a/t (c) $5a$ (d) $a/7$ (e) $3t^2 a + 5a$.

2. If $A = 3t^2 i + (1 - t)j$, $B = ti - (t^2 + t^3)j$ and $C = 3i + 5tj$, find $A + B + C$, dA/dt, dB/dt and dC/dt and hence verify that $d(A + B + C)/dt = dA/dt + dB/dt + dC/dt$.

3. Find a value of r which satisfies the differential equation $d^2 r/dt^2 = a$ (a constant vector).

4. Given that $d^2 r/dt^2 = at + b$ and that $r = dr/dt = 0$ when $t = 0$, find r.

5. If $r = a\cos^3\theta\, i + a\sin^3\theta\, j$, find $|dr/d\theta|$.

6. A particle moves along a curve whose parametric equations are $x = 2\sin 5t$, $y = 3\cos 5t$, where t is time. Find, in terms of i and j (unit vectors parallel to Ox and Oy respectively),

   (a) its velocity and acceleration at any time t,

   (b) the magnitude of its velocity and acceleration at time $t = \pi/2$.

7. Find the definite integral with respect to θ of the vector $g(\theta) = \cos 2\theta\, i - \sin 2\theta\, j$ over the range $\theta = 0$ to $\theta = \pi/4$.

8. Verify that $g(\theta) = \cos\theta\, i - \sin\theta\, j$ is a unit vector and find $dg/d\theta$. By means of a diagram verify that g is perpendicular to $dg/d\theta$.

9. A particle moves along a curve with velocity $v = -2ti + 2j$, where t is time. Its position vector with respect to the origin O is r. If at time $t = 0$, $r = i$, show that at time t, $|r| = 1 + t^2$.

As remarked earlier, the ways in which we combine vectors depend on the ways in which they combine in practice. Two kinds of product arise one of which is a scalar and the other a vector.

3.8. SCALAR (OR DOT) PRODUCT

Definition—The *scalar product* of two vectors a and b is the *scalar* quantity $ab\cos\theta$, where a and b are the moduli of a and b and θ is the angle between them. The product is written as $a \cdot b$ and is sometimes referred to as the dot product.

$$a \cdot b = ab\cos\theta.$$

One use of this definition is seen when we consider the work done by a force F whose point of application moves a distance r in a direction making an angle θ with the line of action of F.

VECTOR ALGEBRA

Work done = (magnitude of force in the direction of motion) × (distance moved)

$$= F \cos \theta r$$

$\therefore$ Work done $= Fr \cos \theta$

Using a vector notation because both force and displacement are vectors

$$\boldsymbol{F} \cdot \boldsymbol{r} = Fr \cos \theta = \text{work done.}$$

From the definition:

$$\boldsymbol{b} \cdot \boldsymbol{a} = ba \cos \theta = ab \cos \theta = \boldsymbol{a} \cdot \boldsymbol{b},$$

that is, scalar multiplication is commutative.

We note two special cases:
(a) If $\boldsymbol{a} \cdot \boldsymbol{b} = 0$ then either $\boldsymbol{a} = 0$, $\boldsymbol{b} = 0$ or $\cos \theta = 0$ [$\theta = 90°$]. Conversely, if $\boldsymbol{a}$ and $\boldsymbol{b}$ are perpendicular, then $\cos \theta = 0$ and $\boldsymbol{a} \cdot \boldsymbol{b} = 0$.
(b) If the vectors are collinear (refer to Section 3.4), $\cos \theta = \pm 1$ and $\boldsymbol{a} \cdot \boldsymbol{b} = \pm ab$ according as they are in the same or opposite sense.

In the case of the unit vectors $\boldsymbol{i}, \boldsymbol{j}, \boldsymbol{k}$, since they are mutually perpendicular,

$$\boldsymbol{i} \cdot \boldsymbol{j} = \boldsymbol{j} \cdot \boldsymbol{k} = \boldsymbol{k} \cdot \boldsymbol{i} = 0$$
$$\boldsymbol{i} \cdot \boldsymbol{i} = \boldsymbol{j} \cdot \boldsymbol{j} = \boldsymbol{k} \cdot \boldsymbol{k} = 1 \qquad \ldots\text{(i)}$$

For any vector $\boldsymbol{a}$, $\boldsymbol{a} \cdot \boldsymbol{a} = |\boldsymbol{a}|^2$ which is often written $\boldsymbol{a}^2$ or a^2.

Another interpretation of the scalar product is

$$\boldsymbol{a} \cdot \boldsymbol{b} = ab \cos \theta = a \times (b \cos \theta)$$
$$= a \times (\text{projected length of } \boldsymbol{b} \text{ on } \boldsymbol{a})$$
$$= a \times (\text{component of } \boldsymbol{b} \text{ in the direction } \boldsymbol{a}).$$

Now the component of a sum of vectors in any direction is equal to the sum of the components of the individual vectors in that direction. Hence

$$\boldsymbol{a} \cdot (\boldsymbol{b} + \boldsymbol{c}) = a \times (\text{component of } \boldsymbol{b} + \boldsymbol{c} \text{ in direction of } \boldsymbol{a})$$
$$= a \times (\text{component of } \boldsymbol{b}) + a \times (\text{component of } \boldsymbol{c})$$

$\therefore$ $\boldsymbol{a} \cdot (\boldsymbol{b} + \boldsymbol{c}) = \boldsymbol{a} \cdot \boldsymbol{b} + \boldsymbol{a} \cdot \boldsymbol{c}.$

SCALAR (OR DOT) PRODUCT

Also, it is evident from the definitions of $a \cdot b$ and λa that

$$\lambda(a \cdot b) = \lambda ab \cos \theta = (\lambda a) \cdot b = a \cdot (\lambda b).$$

Thus scalar multiplication is distributive with respect to addition and commutative with respect to multiplication by a scalar, so that brackets can be removed and inserted as in ordinary algebra.

Let $\quad a = a_x i + a_y j + a_z k \quad$ and $\quad b = b_x i + b_y j + b_z k$

Then

$$\begin{aligned}
a \cdot b &= (a_x i + a_y j + a_z k) \cdot (b_x i + b_y j + b_z k) \\
&= a_x b_x i \cdot i + a_x b_y i \cdot j + a_x b_z i \cdot k \\
&\quad + a_y b_x j \cdot i + a_y b_y j \cdot j + a_y b_z j \cdot k \\
&\quad + a_z b_x k \cdot i + a_z b_y k \cdot j + a_z b_z k \cdot k
\end{aligned}$$

Therefore, using the set of equations (i)

$$a \cdot b = a_x b_x + a_y b_y + a_z b_z \qquad \ldots \text{(ii)}$$

Example 1. *A force $F = -2i + 3j$ units has its point of application moved from the point $A(1, 3)$ to the point $B(5, 7)$, find the work done.*

By definition of the scalar product

$$\text{work done} = F \cdot \overrightarrow{AB}$$

$$\overrightarrow{AB} = (5 - 1)i + (7 - 3)j = 4i + 4j$$

∴ $\quad$ work done $= (-2i + 3j) \cdot (4i + 4j)$

$$= (-2) \times 4 + 3 \times 4 \text{ (by equation ii)}$$

$$= 4 \text{ units.}$$

The positive sign indicates that the work is done *by* the force.

Example 2. *Find the projection of the vector $a = 2i - 8j + k$ in the direction of the vector $b = 3i - 4j - 12k$.*

If θ is the angle between a and b, then $a \cos \theta$ is the required projection. By definition $a \cdot b = ab \cos \theta$.

∴ $\qquad\qquad\qquad a \cos \theta = a \cdot b/|b|$

∴ the projection of a on b is $a \cdot \hat{b}$ where $\hat{b}$ is the unit vector in the direction of b. Now

VECTOR ALGEBRA

$\therefore \quad |b| = \sqrt{3^2 + (-4)^2 + (-12)^2} = 13$

$\therefore \quad \hat{b} = (3i - 4j - 12k)/13$

$\therefore$ the projection of a on $b = (2i - 8j + k) \cdot (3i - 4j - 12k)/13$

$$= \{(2) \cdot (3) + (-8) \cdot (-4) + 1 \cdot (-12)\}/13 \text{ by (ii)}$$

$$= 2.$$

The normal differentiation formula for a product applies, that is, if A and B are differentiable functions of t,

$$\frac{d}{dt}(A \cdot B) = \frac{dA}{dt} \cdot B + A \cdot \frac{dB}{dt} \qquad \ldots \text{(iii)}$$

Example 3. Given the two vectors $A = t^2 i + 2j - tk$ and $B = 4t^2 i + tj + (t-1)k$, find dA/dt, dB/dt, $d(A \cdot B)/dt$.

$$\frac{dA}{dt} = 2ti - k, \qquad \frac{dB}{dt} = 8ti + j + k$$

Hence using relation (iii) above

$$\frac{d(A \cdot B)}{dt} = (2ti - k) \cdot \{4t^2 i + tj + (t-1)k\}$$
$$+ (t^2 i + 2j - tk) \cdot (8ti + j + k)$$
$$= \{(2t)(4t^2) + 0 + (-1)(t-1)\}$$
$$+ \{(t^2)(8t) + (2)(1) + (-t)(1)\} \text{ from (ii)}$$
$$= 16t^3 - 2t + 3.$$

We shall also have occasion to use the following integral:

$$\int_{r_1}^{r_2} g \cdot dr = \lim_{\delta r \to 0} \sum g \cdot \delta r \quad (g \text{ a vector variable})$$

but the method of evaluation will not be pursued here.

SCALAR (OR DOT) PRODUCT

Exercises 3e

1. Given that $r_1 = 2i + j$ and $r_2 = i - 3j$, find (a) $r_1 . r_2$ (b) $r_1 . (r_2 - r_1)$ (c) $(10r_1 + r_2) . r_2$.

2. If $a = i + j - 2k$, $b = i + k$ and $c = 2i - j + 3k$, evaluate (a) $a . b$ (b) $2a . (b + 3c)$ (c) $(5a - 37b + 15c) . c$.

3. Given that $\overrightarrow{AB} = i - 2j + 3k$ and $\overrightarrow{AC} = 3i - 4k$, find the length of the projection of AB on AC and of the projection of AC on AB.

4. Show that the vectors $a \cos \theta i + b \sin \theta j$ and $b \sin \theta i - a \cos \theta j$ are perpendicular.

5. Three points A, B, C are such that $\overrightarrow{AB} = 2i + j$, $\overrightarrow{BC} = i + 3j$, find the angle ABC.

6. Three points A, B, C are such that $\overrightarrow{AB} = 2i - 2j + k$ and $\overrightarrow{BC} = -4i - 8j + k$. Find the angle ABC.

7. If $a = i + uj + u^2 k$ and $b = \sin ui + \cos uj$, calculate $d(a . b)/du$.

8. Given $A = 3t^2 i + (1 - t)j$ and $B = ti - (t^2 + t^3)j$, find dA/dt, dB/dt and $d(A . B)/dt$.

9. Given $A = ti + 3t^2 j$, $B = 5ti - 2t^3 j$, $C = (1 + t)i + (1 - t)j$ and $\phi = A . B$, find ϕ, $d\phi/dt$, dC/dt and verify that

$$\frac{d(\phi C)}{dt} = \frac{d\phi}{dt} C + \phi \frac{dC}{dt}.$$

10. Expand the following and simplify where possible (a) $a . (a + b + c)$ (b) $(a + b) . (a - b)$ (c) $(a + b)^2$.

11. If $a . b = a . c$ and $a \neq 0$, what can be said about the relation between b and c?

12. Prove, using a vector method, that the perpendicular bisectors of the sides of a triangle meet in a point.

13. Given the three vectors $a = 2i + 3j + 5k$, $b = 3i - j + 2k$ and $c = i + j + k$, find $(a . b)c - (a . c)b$ in terms of i, j, k.

14. a and b are any two nonzero vectors: (a) if $|a| = |b|$, show that $(a + b)$ and $(a - b)$ are at right angles. (b) if a and b are perpendicular show that $|a - b| = |a + b|$.

15. Show that $d(v^2)/dt = 2v . dv/dt$. Hence show that if a is a vector of constant magnitude da/dt is perpendicular to a.

3.9. VECTOR (OR CROSS) PRODUCT*

Definition—The *vector product* of two vectors a and b, written $a \times b$, is the vector $ab \sin \theta \hat{n}$, where $\hat{n}$ is a unit vector perpendicular to the

---
* This section may be omitted during the first reading of Part I.

VECTOR ALGEBRA

plane of *a* and *b*, such that *a*, *b*, *n̂* (in that order) form a right-handed set (*Figure 3.12*)

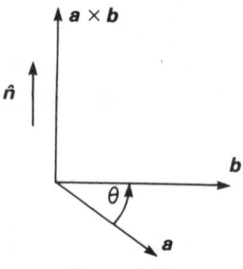

Figure 3.12

From the definition, $a \times b = -b \times a$, that is, the vector product is not commutative.

$|a \times b| = ab \sin \theta$ from which it follows that, if $a \times b = 0$, either $|a| = 0, |b| = 0$ or $\sin \theta = 0$. If $\sin \theta = 0, \theta = 0$ or π and the vectors *a* and *b* are collinear and have the same or opposite sense according as $\theta = 0$ or π respectively. Conversely, if *a* and *b* are collinear, $a \times b = 0$. In particular, $a \times a = 0$.

In the special case of the unit vectors *i*, *j*, *k* since they form a right-handed system and are mutually at right angles

$$i \times j = k, \quad j \times k = i, \quad k \times i = j$$
$$j \times i = -k, \ k \times j = -i, \ i \times k = -j \qquad \ldots (i)$$
$$i \times i = j \times j = k \times k = 0$$

From the definition of $a \times b$ it follows that

$$\lambda(a \times b) = \lambda ab \sin \theta \hat{n} = (\lambda a) \times b = a \times (\lambda b).$$

†Also it can be shown that

$$a \times (b + c) = a \times b + a \times c.$$

Thus the distributive law holds for vector products and the vector products of the sums of two sets of vectors can be expanded, as in ordinary algebra, provided the order of the factors is maintained.

---

† The reader is referred to any standard introductory textbook on vectors.

VECTOR (OR CROSS) PRODUCT

Example. Simplify $(a + b) \times (a + b + c)$.

$$\begin{aligned}(a + b) \times (a + b + c) &= (a + b) \times \{(a + b) + c\} \\ &= (a + b) \times (a + b) + (a + b) \times c\end{aligned}$$

(since $u \times u = 0$) $\qquad\qquad = 0 + (a + b) \times c$

$$(a + b) \times (a + b + c) = (a + b) \times c$$

Using the set of equations (i),

$$\begin{aligned}a \times b &= 0 + a_x b_y k - a_x b_z j \\ &\quad - a_y b_x k + 0 + a_y b_z i \\ &\quad + a_z b_x j - a_z b_y i + 0 \\ &= (a_y b_z - a_z b_y)i - (a_x b_z - a_z b_x)j + (a_x b_y - a_y b_x)k,\end{aligned}$$

which may be written in determinant form as

$$a \times b = \begin{vmatrix} i & j & k \\ a_x & a_y & a_z \\ b_x & b_y & b_z \end{vmatrix}.$$

Note that by the laws governing the manipulation of determinants we have,

$$a \times b = - \begin{vmatrix} i & j & k \\ b_x & b_y & b_z \\ a_x & a_y & a_z \end{vmatrix} = -b \times a.$$

The normal differentiation formula for a product applies, that is, if A and B are differentiable functions of t,

$$\frac{d}{dt}(A \times B) = \frac{dA}{dt} \times B + A \times \frac{dB}{dt}$$

but the *order* of the terms in *each* product must be preserved.

VECTOR ALGEBRA

3.10. MOMENTS*

When the line of action of a vector is fixed as well as its magnitude and direction, it is called a *localized* vector. Examples of localized vectors are a force acting along a given line or the momentum of a particle.

Definition—The *moment* of a localized vector v about a point O is the vector quantity

$$r \times v$$

where r is the position vector (refer to Section 3.6) relative to O of any point P on the line of action of v (*Figure 3.13*).

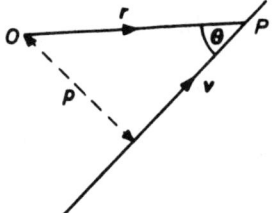

Figure 3.13

Thus the moment M (say) is a vector perpendicular to the plane containing v and O, such that r, v and M form a right-handed set. Its magnitude is $vr \sin \theta$ or vp, where p is the perpendicular distance from O to the line of action of v. Hence the moment of v about O is unaffected by the position of P on its line of action.

Example 1. The line of action of a force F acts through a point P whose position vector is $i - 2j + k$. If $F = 2i - 3j + 4k$, find the moment of F about the point Q whose position vector is $2i + j + k$.

The position vector of P relative to $Q = \overrightarrow{QP}$

$$= (i - 2j + k) - (2i + j + k)$$
$$= -i - 3j$$

* This section may be omitted during the first reading of Part I.

MOMENTS

∴   the moment of F about $Q = \overrightarrow{QP} \times F$

$$= (-i - 3j) \times (2i - 3j + 4k)$$

$$= \begin{vmatrix} i & j & k \\ -1 & -3 & 0 \\ 2 & -3 & 4 \end{vmatrix}$$

$$= -12i + 4j + 9k.$$

As we shall see later in Chapter 16, the moment of a force and the moment of the momentum of a particle are useful quantities in determining the rotational motion of a rigid body.

Definition—The moment of a localized vector v about an *axis* (or line) l is the component in the direction of l of the moment of v about any point on l, see *Figure 3.14a*; the moment of v about l is $\hat{l}.(r \times v)$, where $\hat{l}$ is a unit vector parallel to l.

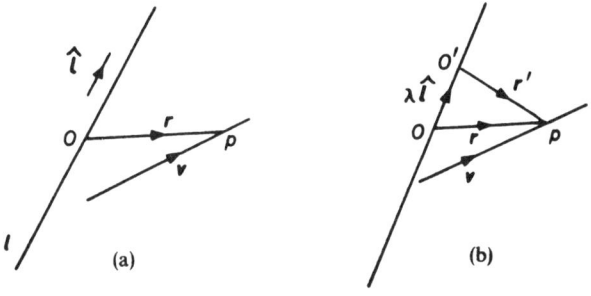

Figure 3.14

The moment of v about l is independent of the choice of the point O. For if O' is any other point on l (*Figure 3.14b*), then

$$\overrightarrow{O'P} = \overrightarrow{O'O} + \overrightarrow{OP}$$

i.e. $r' = r - \lambda\hat{l}$ (where λ is some scalar)

∴ $\hat{l}.(r' \times v) = \hat{l}.(r \times v) - \hat{l}.(\lambda\hat{l} \times v).$

But $\hat{l}.(\lambda\hat{l} \times v) = 0$ being the scalar product of two perpendicular vectors,

∴ $\hat{l}.(r' \times v) = \hat{l}.(r \times v).$

VECTOR ALGEBRA

Example 2. Using the data of Example 1 above, find the moment of F about a line l through Q parallel to the vector $3i - 4j$.

A unit vector in direction of l, $\hat{l} = \frac{3}{5}i - \frac{4}{5}j$

∴ the moment of F about $l = \hat{l} \cdot (\overrightarrow{QP} \times F)$

$$= (\tfrac{3}{5}i - \tfrac{4}{5}j) \cdot (-12i + 4j + 9k)$$
$$= -\tfrac{36}{5} - \tfrac{16}{5} + 0$$
$$= -\tfrac{52}{5}.$$

The moment of F about l is $10\tfrac{2}{5}$ units.

In this book we shall be particularly concerned in finding the moments of vectors about lines to which they are perpendicular. See *Figure 3.15*, where v and $\hat{l}$ are at right angles. Let O be

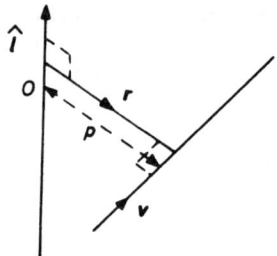

Figure 3.15

the point on l nearest to the line of action of v. Then r is perpendicular to both $\hat{l}$ and to v and hence $r \times v$ is parallel to l.

∴ $\hat{l} \cdot (r \times v) = |r \times v||\hat{l}| \cos 0°$
$$= |r \times v|$$
$$= pv$$

i.e. the moment of v about the *perpendicular* axis l is the magnitude of v multiplied by its perpendicular distance from l.

Example 3. A table has a square top ABCD of side 4 m. If a vertical force of 10 N acts downwards at D, find its moment (a) about the line AB, (b) about AC.

Referring to *Figure 3.16*,

the moment of the force about AB

$\qquad = 10 \times$ its perpendicular distance from AB

$\qquad = 10 \times AD$

$\qquad = 40$ N m,

the moment of the force about $AC = 10 \times OD$

$\qquad\qquad\qquad\qquad\qquad = 10 \times 2\sqrt{2}$

$\qquad\qquad\qquad\qquad\qquad = 20\sqrt{2}$ N m.

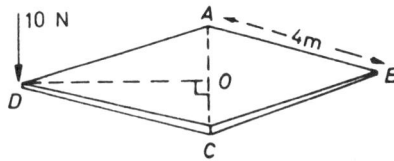

Figure 3.16

Exercises 3f*

1. Expand and simplify the following expressions: (a) $a \times (a + b)$ (b) $(a + b) \times a$ (c) $(a + b) \times (a + b)$ (d) $(a + b) \times (a - b)$.

2. Given that $r_1 = 2i + j$, $r_2 = i - j$, find (a) $r_1 \times r_2$ (b) $r_1 \times (r_1 \times r_2)$.

3. a, b and c are, respectively, vectors of 4 units north, 3 units west, and 8 units N. 150° E., describe carefully the following (a) $a \times b$ (b) $a \times c$ (c) $b \times c$.

4. If a and b are not zero and $a \times x = b \times x$, what is the relationship between them?

5. Given that $a = 5i - j + 2k$, $b = i - 2j + 3k$,
   (a) find $a \times b$ and verify that $a \cdot (a \times b) = 0$ and $b \cdot (a \times b) = 0$.
   (b) If, in addition, $c = i + j + k$, find $b \times c$ and hence $a \times (b \times c)$.

6. Using the data of Question 5, find $(a \cdot c)b - (a \cdot b)c$ and verify that it equals $a \times (b \times c)$.

7. The momentum of a particle is defined as mv where m is its mass, v its velocity. A particle of mass 2 kg is moving with velocity $2i + j - k$ m/s. Find the moment of the momentum of the particle about O when the particle is at a point P such that $\overrightarrow{OP} = i - j - 2k$ m.

8. ABC is a triangle in which $\angle ACB = 30$ degrees and $AC = 4$ m. Find the perpendicular distance of A from BC. If a force of 6 N acts along BC, find its moment about A.

* These exercises may be omitted during the first reading of Part I.

VECTOR ALGEBRA

9. The line of action of a vector $v = 2a + b$ passes through a point whose position vector is $3a - b$. Find the moment of v about the origin and about the point $a + b$.

10. If $a = 6i - 3j + 2k$, find $\hat{a}$. The line of action of a vector $v = -i + 2j - 3k$ passes through the point with position vector $-i - 4j + k$. Find the moment of v about an axis through the origin parallel to a.

11. $ABCD$, $A'B'C'D'$ are the opposite faces of a rectangular block in which $AB = 2a$, $BC = 3a$ and $CC' = 4a$. A force of magnitude P acts along the edge AD. Find the moment of this force about (a) BC (b) BB' (c) $A'B'$ and (d) AB.

3.11. POSITION VECTORS AND GEOMETRICAL APPLICATIONS*

Given a point O as origin, then a point P is uniquely specified by the vector $\overrightarrow{OP}$, known as the *position vector* of P with respect to O. We shall use $a, b, \ldots$ for the position vectors of the points $A, B, \ldots$.

Example 1. The point R divides the straight line joining the points A and B in the ratio $m:n$. If A and B have position vectors a and b with respect to an origin O, find the position vector of R with respect to O.

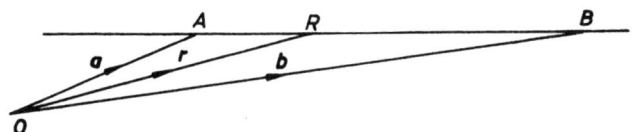

Figure 3.17

Referring to *Figure 3.17*, since ARB is a straight line, $\overrightarrow{AR}$ and $\overrightarrow{RB}$ are collinear. Also

$$\frac{AR}{RB} = m/n$$

or $\qquad nAR = mRB$

∴ $\qquad n\overrightarrow{AR} = m\overrightarrow{RB}$

or $\qquad n(r - a) = m(b - r)$

Whence $\qquad r = \dfrac{na + mb}{m + n} \qquad (m + n \neq 0)$.

---
* This section may be omitted during the first reading of Part I.

POSITION VECTORS AND GEOMETRICAL APPLICATIONS

The result is true whether the ratio m/n is positive or negative. In the case of m/n negative, the point R is outside the segment AB.

Note that if R is the mid-point of AB, $m = n$ and $r = (a + b)/2$.

Example 2. Show that the medians AD, BE, CF, of a triangle ABC have a common point of intersection, which divides each median in the ratio $2:1$.

Let the position vectors of A, B, C be a, b and c, respectively. Since D, E, F are mid-points of the sides BC, CA, AB, respectively, their position vectors are $D, \frac{1}{2}(b + c)$; $E, \frac{1}{2}(c + a)$; $F, \frac{1}{2}(a + b)$ (*Figure 3.18*).

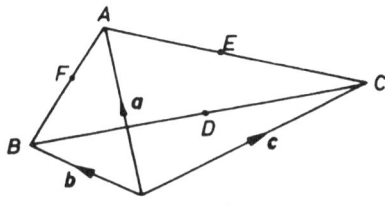

Figure 3.18

Let G be the point which divides AD in the ratio $2:1$. Then, by the result of Example 1, the position vector of G is given by

$$g = \frac{1 \times a + 2 \times \frac{1}{2}(b \times c)}{1 + 2}$$

$$= \tfrac{1}{3}(a + b + c),$$

and the symmetry of this result shows that G lies on BE and CF and divides them in the ratio $2:1$.

The position of a point $A(x, y, z)$ can be specified by the position vector $\overrightarrow{OA} = a = xi + yj + zk$ where i, j, k are unit vectors parallel to Ox, Oy and Oz respectively.

Exercises 3g*

1. The position vectors of the points A and B are respectively $a = 2i + 3j$ and $b = i + 5j$. Find the position vector of the point R which divides AB in the ratio $3:-2$.

2. P, Q have position vectors a, b respectively. Find the position vector of R, the point which divides PQ internally in the ratio $3:2$.

* These exercises may be omitted during the first reading of Part I.

VECTOR ALGEBRA

Show that R, $S(2a)$, $T\{(9b - 14a)/5\}$ are collinear and find the ratio in which R divides ST. Write down an expression for the length ST.

3. $ABCD$ is a parallelogram and a, b, c are the position vectors of A, B, C. What is the position vector of D?

4. A, B, C, D are any four points in three-dimensional space and P, Q, R, S are the mid-points of AB, BC, CD, DA respectively. Show that $PQRS$ is a parallelogram. (Use a vector method.)

5. Given that AD, BE, CF are the medians of a triangle ABC show that (a) $\overrightarrow{FE} = \frac{1}{2}\overrightarrow{BC}$ and (b) $\overrightarrow{AD} + \overrightarrow{BE} + \overrightarrow{CF} = 0$.

6. If a, b are the position vectors of the points A, B respectively, find the position vector of a point C in AB produced, such that $AC = 4BC$ and the position vector of a point D in BA produced, such that $BD = 3BA$.

7. A, B, C, D, E, F are any six points in three-dimensional space. P is the point of intersection of the medians of the triangle ABC, Q is the point of intersection of the medians of the triangle ABD, similarly, R of triangle DEF, and S of triangle CEF. Show that P, Q, R, S are the vertices of a parallelogram (refer to *Example 2*, Section 3.11).

8. a, b, c are the position vectors of the three points A, B, C. A' is the mid-point of BC and G divides AA' internally in the ratio $2:1$. Find the position vector of G. If d is the position vector of another point D, non-coplanar with the points A, B and C, and H divides DG internally in the ratio $3:1$, find the position vector of H.

3.12. THE VECTOR EQUATION OF A STRAIGHT LINE*

A straight line can be uniquely specified in a number of ways, for example, by (a) its direction and the position of a point on it, and (b) the position of two points on it.

The equation of the straight line is obtained by expressing the position vector r of a general point P on the locus in terms of the given conditions. To find the vector equation of a straight line through a given point A (a) and parallel to a given direction b.

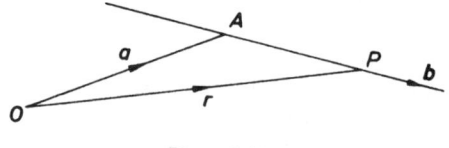

Figure 3.19

* This section may be omitted during a first reading of Part I.

THE VECTOR EQUATION OF A STRAIGHT LINE

Let $P(r)$ be any point on the line, then $\overrightarrow{AP}$ is parallel to b (*Figure 3.19*) and is therefore equal to tb, where t varies according to the position of P. Now

$$r = \overrightarrow{OP} = \overrightarrow{OA} + \overrightarrow{AP}$$

∴ $$r = a + tb$$

which is the required equation.

For the particular case of a straight line through the origin, $a = 0$ and we have

$$r = tb.$$

(**Note:** t is not equal to AP unless b is a unit vector.)

Example 1. *To find the vector equation of a straight line through two given points, $A(a)$ and $B(b)$.*

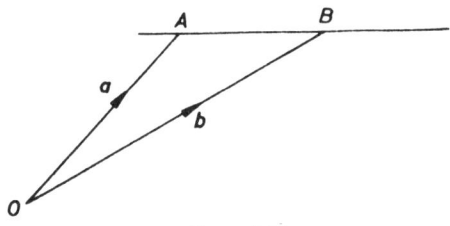

Figure 3.20

We note that $AB = b - a$ (*Figure 3.20*), so that we have a straight line through $A(a)$ parallel to $b - a$. Its equation is therefore:

$$r = a + t(b - a)$$

or $$r = (1 - t)a + tb$$

Example 2. *The vector equations of two coplanar lines are $r = a + tb$ and $r = (2a + b) + s(a - b)$. (a and b being given vectors.) Find the position vector of their point of intersection.*

Suppose the lines meet at $P(r_p)$. Since P lies on both lines,

$$r_p = a + tb \quad \text{and} \quad r_p = (2a + b) + s(a - b)$$

∴ $$a + tb = (2a + b) + s(a - b).$$

For this to be true the coefficients of a and b must be equal.

VECTOR ALGEBRA

$$\therefore \quad 1 = 2 + s$$
and
$$t = 1 - s.$$

From these two equations $s = -1$ and $t = 2$, so that, from $r_p = a + tb$, the position vector of P is $a + 2b$.

Check. From
$$r_p = (2a + b) + s(a - b),$$
$$r_p = (2a + b) - (a - b)$$
$$= a + 2b$$

as before.

*Exercises 3h**

1. Three points A, B, C, have position vectors $(2a + b)$; $(a + 3b)$; $(4a - 3b)$ respectively. Write down the vector equation of the line AB and hence verify that A, B, C are collinear.

2. Find the position vector of the point of intersection of the two lines $r = 3a + t(4a + 3b)$ and $r = (a + 3b) + s(2a - 3b)$.

3. The equation of a straight line through the origin is $r = tb$. If (x, y, z) are the coordinates of the point $P(r)$ and $b = (b_1 i + b_2 j + b_3 k)$; i, j, k being unit vectors parallel to Ox, Oy, Oz, respectively, deduce that,

$$\frac{x}{b_1} = \frac{y}{b_2} = \frac{z}{b_3} = t.$$

4. P, Q, R are the points with position vectors $i + 2j + k$, $2j + 2k$, $3i - j + k$ respectively. Find vector equations for the following lines: (a) through the origin parallel to PQ; (b) through R parallel to PQ; (c) through Q and R.

5. Show that the lines $r = 2i + 2j - 4k + t(i + 3j - 3k)$ and $r = i + j + k + s(i + 2j - 4k)$ intersect and find the position vector of their point of intersection.

6. If the lines $r = i + 2j + k + t(i - 2j + k)$ and $r = 2i + s(i + pj + 2k)$ intersect, find p and the position vector of their point of intersection.

7. Find the angle between the lines $r = (1 - 3t)i + (1 - 4t)j$ and $r = (1 - 5s)i + 12sj$.

8. If in Question 3, the equation of the line had been $r = a + tb$ and the point $A(a)$ had coordinates (a_1, a_2, a_3), deduce a similar set of results to those given in Question 3.

9. The three lines $r = (3a + b) + ta$; $r = (a + 4b) + s(a - 3b)$;

* These exercises may be omitted during a first reading of Part I.

EXERCISES

$r = (-a + 4b) + u(3b - 2a)$ lie in the same plane. Find the position vectors of their points of intersection.

EXERCISES 3

1. The sides $\overrightarrow{AB}, \overrightarrow{AC}$ of the triangle ABC are represented by p and q respectively. D, E and F are the mid-points of BC, CA, AB respectively. Express $\overrightarrow{DE}, \overrightarrow{EF}, \overrightarrow{FD}$ in terms of p and q.

2. $ABCD$ is a quadrilateral. L is the mid-point of AB, N is the mid-point of CD, and M is the mid-point of LN. Prove that $\overrightarrow{MA} + \overrightarrow{MB} + \overrightarrow{MC} + \overrightarrow{MD} = 0$.

3. Three points A, B and C have position vectors $a = -i + 5j$, $b = 2i + 4j$, $c = 2j$ with respect to a given origin. Show that ABC is an isosceles triangle.

4. Write down the vector algebra of the following statements:

(a) The vector a is parallel to the vector b in the same sense and of twice its magnitude.

(b) The vector a is equal and opposite to the vector b.

(c) The vector a is parallel to the vector b.

(d) The resultant of F_1 and F_2 is parallel to a.

(e) x and y are perpendicular.

(f) The resultant of P and Q is equal and opposite to the resultant of R and S.

(g) x, y, z (in that order) form a right-handed set and are mutually perpendicular.

(h) a, b, c are mutually perpendicular.

5. Depict on the same diagram the forces F; $-5F$; $F/3$; $-1\cdot5F$ and a force of magnitude $2F$ at right angles to F (two cases). Is it possible to depict $1/F$?

6. The resultant of two forces $2P$ and $2Q$ is equal and opposite to the resultant of Q, another force R and a force equal and opposite to P. Express P in terms of Q and R.

7. For what values of λ are the vectors $3\lambda i - \lambda j - 3k$ and $2\lambda i + 7j + k$ perpendicular?

8. Find a unit vector in the direction of $i - 2j$ and hence find the component of the vector $3i + 2j$ in that direction.

9. Find the work done when the point of application of the force $3i + 2j$ moves in a straight line from the point $(2, -1)$ to the point $(6, 4)$.

10. $ABCDEF$ is a regular hexagon. Show that the resultant of the five forces represented by $\overrightarrow{AB}, \overrightarrow{AC}, \overrightarrow{AD}, \overrightarrow{AE}$ and $\overrightarrow{AF}$ is the force represented by $6\overrightarrow{AO}$ where O is the centre of the hexagon.

VECTOR ALGEBRA

11. Form the vector product of both sides of the equation $a + b = c$ with a. Hence prove the sine rule for a triangle. Use scalar products to obtain the cosine rule.

12. Given that $u = i - 2j + 3k$ and $v = 4i - 3j - k$. Find (a) unit vectors along u, v and $u + v$ (b) $u \cdot v$ (c) the component of u in the direction of v (d) the angle between u and v.

13. Given the three points A, B, C with position vectors $(3a + 2b)$; $(-a + 6b), (-2a + 7b)$ respectively, write down the vector equation of the line AC. Hence show that A, B, C are collinear and find the ratio $AB:BC$.

14. Using vector methods, find the angle between two diagonals of a cube.

15. A particle moves so that its displacement r at time t is $r = (4\cos t + 3\sin t + 2)i + (3\cos t - 4\sin t - 1)j$. If v and f are its velocity and acceleration, respectively, show that $|v| = 5 = |f|$.

16. Given two vectors $a = 2i + 3j$ and $b = 4i - 3j$, find $|a|$ and $|b|$. Hence express $a \cdot b$ in terms of $\cos \theta$ where θ is the angle between a and b. Also evaluate $a \cdot b$ using components and hence find θ.

17. Two vectors a and b are such that their resultant is perpendicular to b. If the resultant of a and $2b$ is perpendicular to a, find a relation between $|a|$ and $|b|$.

18. $ABCD$ is a quadrilateral and X and Y are the mid-points of its diagonals AC and BD respectively, prove that $\overrightarrow{AB} + \overrightarrow{AD} + \overrightarrow{CB} + \overrightarrow{CD} = 4\overrightarrow{XY}$.

19. A particle moves along a curve whose parametric equations are $x = 10 + 7\cos t$, $y = 8 + 7\sin t$, where t is time. Find the magnitudes of its velocity and acceleration at any time t.

20. (a) The position vectors of the points A, B and C are a, b and c respectively, referred to the point O as origin. Given that $3a + b = 4c$, prove that the points A, B and C are collinear and find the ratio $AB:AC$.

(b) Three forces $7i + 5j$, $2i + 3j$ and λi act at the origin O, where i and j are unit vectors parallel to the x-axis and the y-axis respectively. The unit of force is the Newton. If the magnitude of the resultant of the three forces is 17 N, calculate the two possible values of λ. Show that the two possible directions of the line of action of the resultant are equally inclined to Oy. (AEB)

21. A particle moves so that its position vector r at time t is $(a\cos^3 t)i + (a\sin^3 t)j$ (i.e. along a curve called the Astroid). Find its velocity v at time t and show that $|v|$ has a maximum value $3a/2$ when $t = \pi/4$ or $3\pi/4$.

22. $\hat{a}$ and $\hat{b}$ are unit vectors in the x-y plane, they make angles α, β respectively, with Ox. Express $\hat{a}$ and $\hat{b}$ in terms of component

52

EXERCISES

vectors parallel to Ox and Oy and hence prove the formulae:
$$\cos(\alpha - \beta) = \cos\alpha\cos\beta + \sin\alpha\sin\beta$$
$$\cos(\alpha + \beta) = \cos\alpha\cos\beta - \sin\alpha\sin\beta$$

23. The line of action of $P = i - 2j$ passes through the point whose position vector is $-j + k$, where i, j and k are unit vectors parallel to rectangular axes Ox, Oy, Oz. Find (a) the moment of P about O; (b) the moment of P about the point $i + k$; (c) the moment of P about the y-axis.

24. $ABCDEF$ is a regular hexagon of side a. Forces of magnitude 1, 2, 3, 4 and 2 units act along AB, BC, DC, DE and DB respectively. Find the magnitude of the sum of their moments about an axis through the centre of the hexagon perpendicular to its plane.

25. A vector of magnitude 12 units acts along the line $r = i - j + t(2i + 2j + k)$. Find the moment of this vector about the point $i + j + k$ and about the line $r = i + j + k + s(3i - 4j)$.

26. A table has a triangular top which is an equilateral triangle ABC in which $AB = 2a$. D is the mid-point of BC. Three forces of magnitude $P, 2P, 3P$ act vertically downwards at A, B and C respectively. Find the sum of their moments (a) about AB (b) about AD.

27. A force $F_1 = 4i - j + 2k$ acts at a point whose position vector is $4i - 2j + 2k$. Write down the equation of its line of action. A second force $F_2 = i + j + 2k$ acts at a point P on the line $r = 5i + 3j + k + \lambda(4i - 10j - k)$. If F_1 and F_2 intersect, find P and the point of intersection.

28. The three lines $r = 6i + 4j + 4k + s(ai + bj + 3k)$, $r = -10i + 2j - 3k + t(2j - ck)$ and $r = -8i + 10j - 10k + u(-4i + 4j + 4k)$ are concurrent. Find the values of a, b and c.

29. Given that $F = \sin\theta i + \cos\theta j$ and $dr = [(1 + \cos\theta)i - \sin\theta j] d\theta$, evaluate $\int_0^{\pi/2} F \cdot dr$.

30. (a) In a parallelogram $ABCD$, X is the mid-point of AB and the line DX cuts the diagonal AC at P. Writing $\overrightarrow{AB} = a$, $\overrightarrow{AD} = b$, $\overrightarrow{AP} = \lambda\overrightarrow{AC}$ and $\overrightarrow{DP} = \mu\overrightarrow{DX}$, express $\overrightarrow{AP}$ (i) in terms of λ, a and b (ii) in terms of μ, a and b. Deduce that P is a point of trisection of both AC and DX.

(b) Define the scalar product $a \cdot b$ and the vector product $a \times b$ of two vectors a and b.

The points P, Q, R have coordinates $(1, 1, 1)$, $(1, 3, 2)$, $(2, 1, 3)$ respectively, referred to rectangular axes $Oxyz$. Calculate the products $\overrightarrow{PQ} \cdot \overrightarrow{PR}$, $\overrightarrow{PQ} \times \overrightarrow{PR}$ and deduce the values of the cosine of the angle QPR and the area of the triangle PQR. (JMB)

VECTOR ALGEBRA

31. (a) If p is a unit vector of varying direction, prove that $p \cdot dp/dt$ is zero. Hence, show that the component of d^2p/dt^2 in the direction of p is $-v^2$, where v is the vector dp/dt.

(b) By considering the scalar product of the vectors $a_1i + a_2j + a_3k$ and $b_1i + b_2j + b_3k$, show that for any real numbers $a_1, a_2, a_3, b_1, b_2, b_3$ $(a_1^2 + a_2^2 + a_3^2)(b_1^2 + b_2^2 + b_3^2) \geq (a_1b_1 + a_2b_2 + a_3b_3)^2$.
(JMB)

32. Four points P, Q, R, S in a plane through the origin O have position vectors $\overrightarrow{OP}, \overrightarrow{OQ}, \overrightarrow{OR}, \overrightarrow{OS}$, given by $2i + 3j$, $3i + 2j$, $4i + 6j$, $9i + 6j$, respectively, where i and j are given non-parallel vectors, Express the vectors $\overrightarrow{PR}$ and $\overrightarrow{QS}$ in terms of i and j.

Show that the position vectors $\overrightarrow{OA}$ and $\overrightarrow{OB}$ of the points A and B on PQ and RS respectively, and such that $PA/PQ = a$ and $RB/RS = b$, are $(2 + a)i + (3 - a)j$ and $(4 + 5b)i + 6j$ respectively. Hence determine the position vector with respect to O of the point of intersection of the lines PQ and RS.
(JMB)

33. Given the two lines L_1, L_2 whose equations are respectively, $r = (\hat{a} + 2\hat{c}) + t(\hat{b} - 3\hat{c})$ and $r = (-2\hat{a} - 2\hat{b} - \hat{c}) + s(\hat{a} + \hat{b})$, show that they intersect at P and find the position vector of P. What is the condition that L_1, L_2 are at right angles? If this condition is satisfied, write down the equation of a third line L_3 through P such that L_1, L_2, L_3 form a right-handed set of mutually perpendicular axes. [$\hat{a}, \hat{b}, \hat{c}$ are constant unit vectors.]

34. Differentiate the following expressions, given that r is a vector function of t and r is its modulus, a, b are constant vectors and a, b their moduli.

(a) $\dfrac{1}{2}a\left(\dfrac{dr}{dt}\right)^2$ (b) $(a \cdot b)r$ (c) $(a \cdot r)b$

(d) $r^2 + \dfrac{1}{r^2}$ (e) $r \times \dfrac{dr}{dt}$ (f) $r \cdot \dfrac{dr}{dt}$.

35. Show that
$$2\frac{dr}{dt} \cdot \frac{d^2r}{dt^2} = d\left(\frac{dr}{dt}\right)^2 \bigg/ dt.$$

Hence, given that $d^2r/dt^2 = -n^2r$, prove that $(dr/dt)^2 = c - n^2r^2$, where c is an arbitrary constant.

36. Forces $\lambda\overrightarrow{OA}$ and $\mu\overrightarrow{OB}$ act along the lines $\underline{OA}$ and OB respectively. Show that the resultant is a force $(\lambda + \mu)\overrightarrow{OC}$ where C lies on AB and $AC:CB = \mu:\lambda$.

EXERCISES

Forces $3\overrightarrow{AB}$, $2\overrightarrow{AC}$ and $\overrightarrow{CB}$ act along the sides AB, AC and CB respectively of a triangle ABC. Their resultant meets BC in P and AC in Q and its magnitude is kPQ. Find $BP:PC$, $AQ:QC$ and k.
(London)

37. Show that the three vectors $i + j + k$, $2i - 3j + k$ and $4i + j - 5k$ are mutually perpendicular.

38. O is any point in the plane of a square $ABCD$ whose diagonals intersect at E. Four forces are represented completely by $3\overrightarrow{OA}$, $2\overrightarrow{OB}$, $3\overrightarrow{OC}$ and $2\overrightarrow{OD}$. Show that their resultant passes through E and find its magnitude in terms of OE. (London, part)

39. O, A, B, C are four points in a plane. The position vectors of A, B, C with respect to O as origin are a, b, c respectively. Prove that scalar quantities λ, μ, r, not all zero, exist to satisfy the identity

$$\lambda a + \mu b + rc = 0.$$

40. Three points A, B and C have position vectors $i + j + k$, $i + 2k$ and $3i + 2j + 3k$ respectively, relative to a fixed origin O. A particle P starts from B at time $t = 0$, and moves along BC towards C with constant speed 1 unit per second. Find the position vector of P after t seconds. (a) relative to O and (b) relative to A.

If the angle $PAB = \theta$, find an expression for $\cos \theta$ in terms of t.
(London)

41. Show that the vectors u, v and w, where

$$u = i + j + k,$$
$$v = i - \tfrac{1}{2}j - \tfrac{1}{2}k,$$
$$w = j - k,$$

are mutually perpendicular and find the unit vector in the direction of the vector u.

Find constants α, β and γ such that

$$i = \alpha u + \beta v + \gamma w.$$

If P, Q and R have position vectors $5i - j - k$, u and v respectively with respect to the origin O, show that O, P, Q and R are coplanar. Find the cosine of the angle POQ.

If the vector c lies in the xy plane, find c such that $c \times w = u$. Hence find the value of $w \cdot (c \times w)$. (AEB)

42. Two forces $2i + 3j - k$ and $i - 2j + k$ act at points whose position vectors are $3i + 6j + k$ and $3i - j + 4k$ respectively. Show that the lines of action of these forces intersect and find the position vector of the point of intersection. Find also

(i) the magnitude of the resultant and a vector equation of its line of action,
(ii) the vector equation of the plane containing the two forces,
(iii) the moment vector of the resultant about the origin.

(AEB)

43. (i) In the quadrilateral $ABCD$, X and Y are the mid-points of the diagonals AC and BD respectively. Show that

(a) $\overrightarrow{BA} + \overrightarrow{BC} = 2\overrightarrow{BX}$,

(b) $\overrightarrow{BA} + \overrightarrow{BC} + \overrightarrow{DA} + \overrightarrow{DC} = 4\overrightarrow{YX}$.

(ii) The point P lies on the circle through the vertices of a rectangle $QRST$. The point X on the diagonal QS is such that $\overrightarrow{QX} = 2\overrightarrow{XS}$. Express $\overrightarrow{PX}$, $\overrightarrow{QX}$ and $(\overrightarrow{RX} + \overrightarrow{TX})$ in terms of $\overrightarrow{FQ}$ and $\overrightarrow{PS}$.

(London)

44. Find a vector equation of the line L through the point with position vector $i - 2j + k$ and in the direction $3i + 4k$. Show that the line L meets the plane π, whose equation is $r.(i - j + 2k) = -6$, at the point A with position vector $-2i - 2j - 3k$.

The point B is at a distance 10 units from A in the direction $3i + 4k$. A force $F = 3i + 2j + 4k$ acts on a particle of unit mass and moves it from A to B.

Find
(i) the work done by the force,
(ii) the position vector of the point in the plane, π, which is nearest to the point B.

(AEB)

4

SPEED AND VELOCITY

4.1. AVERAGE SPEED

Definition

$$\text{Average speed} = \frac{\text{distance travelled}}{\text{time taken}}.$$

We note that no reference is made to the direction of the motion and thus average speed has magnitude only and is a scalar (refer to Section 2.1). For example, if a train travels 200 km in 4 h, its average speed is 50 km/h. It will be noted that the average speed gives no indication of the speed at any instant, which could have any value up to the top speed of the train.

Exercises 4a

The following is a portion of a train time-table giving the departure times of four different trains from four consecutive stations:

Station	Trains			
	1	2	3	4
A	08·05	09·00	10·21	11·00
B	08·15			11·10
C	08·28		10·41	11·25
D	08·38			11·35
E (arrive)	08·49	09·30	10·54	11·44

The distances between successive stations are A to B, 5 km, B to C, 6 km, C to D, 5 km and D to E, 4 km. Use this information to answer Questions 1 to 4. (Time spent at a station is ignored.)

1. What are the average speeds of each of the four trains in travelling from A to E.

2. Find the average speeds of the first train between the successive stations A to B, B to C, C to D and D to E.

3. In Question 2, does the average of the four answers give the

SPEED AND VELOCITY

same result as in the first part of Question 1? If not, explain why they are different.

4. What is the speed of the fastest train between A and C?

5. A car is travelling round a semi-circular bend of radius 1 000 m. If the time taken is 2 min, find the average speed of the car around the bend in kilometres per hour.

4.2. SPEED

The average speed gives no indication of the motion of a particle at any instant during the interval considered. To find the *instantaneous speed* at any instant, we consider the average speed over any small interval of time, δt, which includes that instant. If, as we make δt smaller and smaller, the average speed approaches a definite value, v, we call v the speed at that instant. This can be stated more precisely using the notation of limits and the differential calculus.

Definition—If during a time interval, t to $t + \delta t$, a particle moves a distance, δs, then its speed v, at time t, is given by

$$v = \lim_{\delta t \to 0} \frac{\delta s}{\delta t} = \frac{ds}{dt}.$$

SI unit of speed ... m/s.

If the relation between the distance travelled and the time taken is known from theoretical considerations, then the distance–time curve can be drawn. For example, for a falling body $s = 16t^2$ and the curve is shown in *Figure 4.1*.

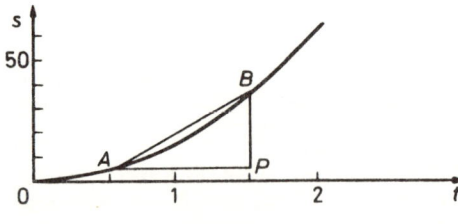

Figure 4.1

The average speed between any two points is given by BP/AP, and its speed v, at any time t, is ds/dt which is the slope of the curve at time t, so v can be found by differentiation. Here $s = 16t^2$,

SPEED

$$\therefore \quad v = \frac{ds}{dt} = 32t.$$

(When the equation connecting s and t is not known, but a set of pairs of values of s and t are given, then a smooth curve can be drawn. An approximate value of v can be found by drawing a tangent to the curve and finding its slope.)

Example 1. *The equation of motion of a particle subject to a force proportional to its speed is* $s = 10(1 - e^{-5t})$ *where* s *m is the distance moved in* t *seconds. Find its speed,* v, *at the start of the motion and after* $\frac{1}{5}$ *s.*

$$s = 10(1 - e^{-5t})$$

$$v = \frac{ds}{dt} = 50e^{-5t}$$

Initially $\quad t = 0 \quad$ and $\quad v = 50e^0 = 50$ m/s.

At time $\quad t = \frac{1}{5}, \quad\quad v = 50e^{-1}$

$$= \frac{50}{e} = 18.4 \text{ m/s}.$$

In the special case when the speed is constant,

$$\frac{ds}{dt} = v \; (v \text{ constant}).$$

Therefore by integration $s = vt + c$ and the distance–time curve is a straight line whose slope is v. If, initially, when $t = 0$, $s = 0$, then it follows that $c = 0$ and we have that

$$s = vt.$$

Example 2. B *is 10 km from* A *and a motorway runs directly from* A *to* B. *Three cars* P, Q, R, *are travelling with constant speeds of 45 km/h, 70 km/h and 60 km/h respectively,* P *and* Q *from* A *to* B *and* R *from* B *to* A. P *and* Q *pass* A *at 16·00 and 16·02 h,* R *passes* B *at 16·05 h. Find, graphically, when* R *meets* P *and* Q *and when* Q *overtakes* P.

Since the speeds are constant, $s = vt$, that is, $t = s/v$. Since P covers the 10 km from A to B at 45 km/h.

$\therefore \quad\quad P$ takes $\frac{10}{45}$ h $= \frac{10}{45} \times 60$ min $= 13\frac{1}{3}$ min

SPEED AND VELOCITY

∴   P arrives at B at $16 \cdot 13\frac{1}{3}$ h.

Similarly   Q takes $\frac{10}{70} \times 60 = 8\frac{4}{7}$ min

∴   Q arrives at B at $16 \cdot 10\frac{4}{7}$ h

and   R takes $\frac{10}{60} \times 60 = 10$ min

∴   R arrives at A at $16 \cdot 15$ h.

Tabulating these results we have,

	P	Q	R
Time at A	16·00	16·02	16·15 h
Time at B	$16 \cdot 13\frac{1}{3}$	$16 \cdot 10\frac{4}{7}$	16·05 h

The answers are: R meets P at $16 \cdot 08\frac{4}{7}$ h; R meets Q at 16·08 h and Q overtakes P at $16 \cdot 05\frac{2}{3}$ h. (Refer to *Figure 4.2*.)

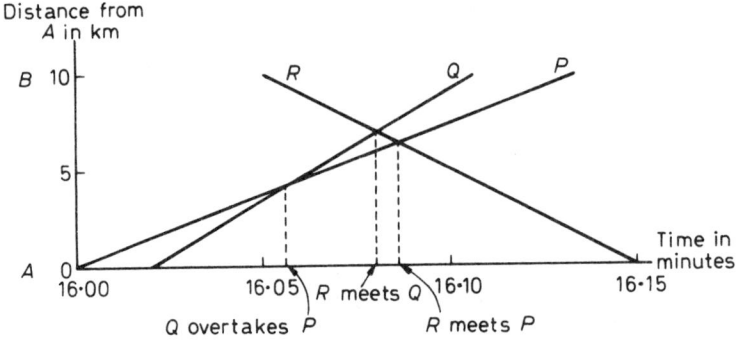

Figure 4.2

Exercises 4b

1. Draw a distance-time graph for each of the following:
   (a) A body moves with a constant speed of 6 m/s for 10 s.
   (b) A body moves from A to B with a constant speed of 8 m/s, the distance A to B being 384 m.
   (c) A body moves for 10 s at a constant speed of 18 m/s, stops for 6 s, and continues at a constant speed of 18 m/s for a further 63 m.
   (d) The distance s metres moved by a body in t seconds is given by $s = t^2 + 6t$ [$t = 0$ to $t = 20$].

2. In questions 1(c) and 1(d), find the average speed of the body for the first 18 s in each case.

3. The distances from A to B and B to C are respectively, 20 km

SPEED

and 15 km. A cyclist takes 1 h 40 min to travel from A to B, where he stops for 20 min. He then cycles from B to C. Assuming that he travels with the same constant speed while cycling, draw the distance–time graph of the journey. Find his speed while cycling and his average speed for the whole journey.

4. The distance s metres travelled by a particle in t seconds is given by the formula $s = 16t^2$. Draw a distance–time graph for $t = 0$ to $t = 6$, and from the graph find:
   (a) The average speed in the first 3 seconds.
   (b) The average speed in the second 3 seconds.
   (c) The average speed for the whole 6 seconds.
Find also the instantaneous speed at $t = 2\frac{1}{2}$ and $t = 5$.

5. If a cyclist goes from A to B at 12 km/h and returns at 15 km/h, what is his average speed for the whole journey?

6. The distance s metres travelled by a particle in t seconds is given by $s = 10 \sin(4t - \pi/2)$. Find its speed when $t = \pi/12$ and $t = \pi/3$. Also find its distance from its starting point when it first comes to rest.

7. The distance travelled by a body in t seconds is given by $s = 5t + t^2$ cm. Find its average speeds for: the first second, the first $\frac{1}{10}$ second and the first $\frac{1}{100}$ second. Deduce the average speed for the first $\frac{1}{1000}$ second and hence deduce the instantaneous speed at $t = 0$.

8. A particle A leaves a point O and travels in a straight line at a constant speed of 10 m/s. Simultaneously, another particle B leaves O and travels along the same line in the same direction, its distance from O is given by $s = 2t^2$ [s in metres, t in seconds]. On the same diagram draw the distance–time curves for the two particles from $t = 0$ to $t = 6$ and find when and where B overtakes A.
If A stops after 3 seconds, find when and where B overtakes A.

9. A cyclist travels at a constant speed of 12 km/h from A to B and, 35 min after he leaves A, a second cyclist sets out from A, also at a constant speed, and overtakes him 35 km from A. Both cyclists continue without varying their speeds until they reach B. If the second cyclist takes a further 3 h 56 min to reach B, find, by a graphical method, how long he has to wait at B for the first cyclist to arrive.

10. A train leaves Preston at 12·00 h and reaches London at 15·00 h. During the journey it meets another train travelling to Preston which left London at 12·40 h. If the first train had travelled $\frac{13}{20}$ ths of the journey when they meet, find the time at which the second train arrives at Preston. Assume that the trains travel at constant speeds.

SPEED AND VELOCITY

4.3. VELOCITY

Definition—The velocity of a point is its rate of change of displacement.

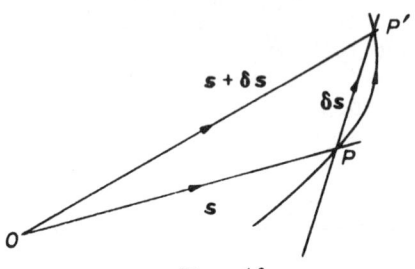

Figure 4.3

Referring to *Figure 4.3*, if at times t and $t + \delta t$, a point P has position vectors $\overrightarrow{OP}(= s)$ and $\overrightarrow{OP'}(= s + \delta s)$ with respect to an origin O, then

$$\text{velocity} = \lim_{\delta t \to 0}\left(\frac{\overrightarrow{PP'}}{\delta t}\right)$$

i.e.
$$v = \frac{ds}{dt}$$

In the limit as $P' \to P$, $P'P$ becomes a tangent at P to the path of P. Therefore the direction of v is tangential to the path of P.

Also
$$|v| = \lim_{P' \to P}\frac{\text{chord } PP'}{\delta t} = \lim_{P' \to P}\frac{\text{length of path } PP'}{\delta t}$$

$$= \lim_{P' \to P}\frac{\delta s}{\delta t}$$

$$= \text{speed of } P$$

Thus the magnitude of the velocity v is the speed.

For uniform, or constant velocity, a particle must be moving with a constant speed in a constant direction.

Note that both magnitude *and* direction must be constant for a

VELOCITY

velocity to be uniform. A particle describing a circle uniformly has constant speed but not a constant velocity because its direction is constantly changing. In the case of motion in a straight line, the velocity is constant if the speed is constant.

Example. A particle moves so that after t seconds its displacement s is given by $s = (3t^2 + 1)i + (t^4 - 5t)j$ m, where i and j are unit vectors due east and north, respectively. Find its velocity v after 3 seconds in magnitude and direction.

$$s = (3t^2 + 1)i + (t^4 - 5t)j$$

∴ $$v = \frac{ds}{dt} = 6ti + (4t^3 - 5)j \text{ m/s}.$$

When $t = 3$,

$$v = 18i + 103j \text{ m/s}.$$

and $$|v| = \sqrt{(18^2 + 103^2)} = 104\cdot 6 \text{ m/s}.$$

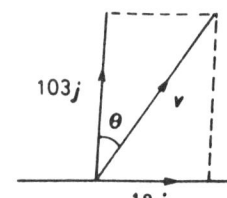

Figure 4.4

Referring to *Figure 4.4*,

$$\tan \theta = 18/103 = 0\cdot1748$$

∴ $$\theta = 9° 55'.$$

The velocity is 105 m/s in a direction north 9° 55' east.

Exercises 4c

In all the following exercises, *i* and *j* and *k* denote unit vectors due east, due north and vertical respectively, unless otherwise stated.

1. A particle moves so that after *t* seconds its displacement is $2t^2 i - tj$ cm. Find its velocity after 2 seconds in magnitude and direction.

2. A particle moves so that its displacement after *t* seconds is $\sin 2t i + \cos 2t j$ m. Show that the magnitudes of both its displacement and its velocity are constant. What path does it follow?

3. A particle moves so that its displacement after *t* seconds is

$4t^3a$ m, where $a = 3i - j$. Find the magnitude and direction of its velocity after 1 second. What is its position at that time and what shape is its path?

4. A particle P has a velocity of $3i + 2j$ m/s, where i and j are unit vectors parallel to Ox and Oy respectively. If P is initially at the point $(3, -4)$ m, find its position after 3 seconds.

5. If, in Question 4, a second particle Q has a velocity of $2i - j$ m/s, and was initially at the point $(1, 0)$, find its position after 3 s. Hence find how far apart P and Q are after 3 seconds.

6. A particle has a constant velocity v. In one second it moves in a straight line from the point $P(5, 3)$ to the point $Q(9, 2)$. Express its velocity in terms of i and j (unit vectors parallel to Ox and Oy respectively). Hence, find its position (a) 3 seconds (b) 7 seconds and (c) t seconds after passing P.

7. Find the magnitude of the resultant of the three velocities, $a = (5i + 2j - k)$ m/s, $b = (2i - 3j + 4k)$ m/s and $c = (-4i + 5j + 9k)$ m/s.

8. A particle moves so that its displacement after t seconds is $\cos t \cos 2t i + \sin t \cos 2t j + \sin 2t k$ metres. Find the magnitudes of both its displacement and its velocity after t seconds.

9. A particle moves so that its displacement after time t is $at^2 i + 2at j$. Show that at any time t the direction of its velocity is north $\tan^{-1}(t)$ east. What is the cartesian equation of its path?

10. The position of a particle at time t is $\cos 2t \cos 3t i + \cos 2t \sin 3t j + \sin 2t k$. Show that the magnitude of the displacement is constant and find the magnitude of its velocity.

4.4. RELATIVE VELOCITY

Consider two aeroplanes Q and P flying with different velocities. In a simplified case they can be flying in the same straight line with P ahead of Q and flying faster. To an observer in Q, P will appear to be flying away from him with an apparent speed equal to the rate at which the distance between them increases.

In general, both the magnitudes and directions of the velocities of Q and P will be different. An observer in Q looking at P will estimate P's apparent velocity, called the velocity of P relative to Q, by the rate at which the displacement between them alters in magnitude and direction.

Definition—The velocity of P relative to Q (v_{PQ}) is the rate of change of the displacement $\overrightarrow{QP}$

i.e. $$v_{PQ} = d(\overrightarrow{QP})/dt$$

RELATIVE VELOCITY

Now, if at time t the displacement of P from some fixed point is s_P and that of Q, s_Q, then we have

$$\overrightarrow{QP} = s_P - s_Q$$
$$\therefore \quad v_{PQ} = d(s_P - s_Q)/dt$$
$$= \frac{ds_P}{dt} - \frac{ds_Q}{dt}$$
$$\therefore \quad v_{PQ} = v_P - v_Q.$$

In problems, the relative velocity can be found by the usual methods of finding resultants (refer to Chapter 2), that is, from vector diagrams or by resolving. Also, writing the velocities in terms of unit vectors i and j may simplify presentation. The methods are illustrated in the examples which follow.

When the relative velocity is given we can use the derived relation:

$$v_P = v_{PQ} + v_Q.$$

Example 1. A ship A is steaming due north at 16 km/h and a ship B is steaming due west at 12 km/h. Find the relative velocity of A with respect to B.

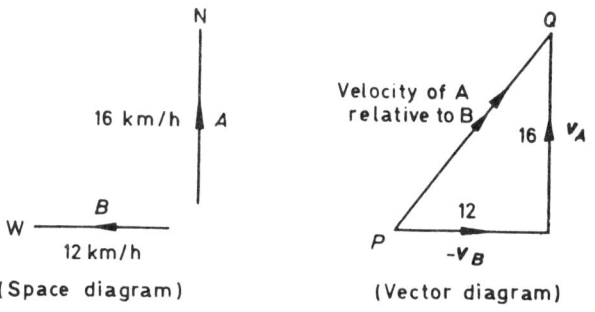

Figure 4.5

Referring to the vector diagram in *Figure 4.5*,

$$\overrightarrow{PQ} = v_A - v_B$$

represents the required relative velocity, and

$$PQ^2 = 12^2 + 16^2 \quad \therefore \quad PQ = 20$$
$$\tan N\widehat{P}Q = \tfrac{12}{16} \quad \therefore \quad \angle NPQ = 36° \, 52'.$$

65

The relative velocity of A with respect to B is 20 km/h, in a direction N. 36° 52′E.

Example 2. *A ship is steaming due west at 18 km/h. To an observer in the ship a hovercraft appears to be moving in a direction north west at 12 km/h Find the velocity of the hovercraft.*

Using the derived relation (see page 64)

$$v_P = v_{PQ} + v_Q$$

where v_{PQ} is the velocity of P relative to Q. We have that v_{PQ} is 12 km/h in a direction north-west, and v_Q is 18 km/h due west. The resultant of these two is v_P.

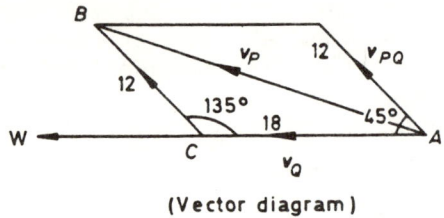

(Vector diagram)

Figure 4.6

Referring to *Figure 4.6*

$$|v_P|^2 = AB^2 = AC^2 + BC^2 - 2AC \cdot BC \cos 135°$$
$$|v_P|^2 = 18^2 + 12^2 + 2 \cdot 18 \cdot 12 \cdot 0.7071$$
$$|v_P| = 27.81 \text{ km/h.}$$

Also
$$\frac{BC}{\sin B\hat{A}C} = \frac{BA}{\sin 135°}$$

∴
$$\sin B\hat{A}C = \frac{12 \times 0.7071}{27.81}$$
$$= 0.3051$$

∴ $\angle BAC = 17° 46′$.

v_P is 27·81 km/h in a direction N. 72° 14′ W.

It will be remembered (refer to Section 3.5) that a vector may be expressed in terms of components, and that in solving problems it is particularly useful if the components are at right angles.

Example 3. *A steamer is travelling north-west at 10 km/h. What is the*

RELATIVE VELOCITY

apparent velocity of the steamer as seen by a man in a boat which is travelling N. 60° E. at 5 km/h?

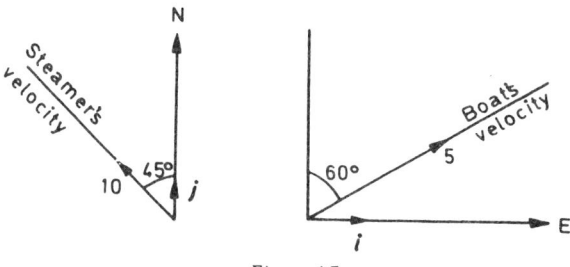

Figure 4.7

Referring to *Figure 4.7* we see that the components of the velocity of the steamer are 10 cos 45° north and −10 sin 45° east. The velocity of the steamer is therefore:

$$v_S = -5\sqrt{2}i + 5\sqrt{2}j.$$

Similarly, the components of the velocity of the boat are 5 cos 60° north and 5 cos 30° east. The velocity of the boat is therefore:

$$v_B = \tfrac{5}{2}\sqrt{3}i + \tfrac{5}{2}j.$$

The apparent velocity of the steamer from the boat is

$$v_{SB} = v_S - v_B$$
$$= (-5\sqrt{2}i + 5\sqrt{2}j) - (\tfrac{5}{2}\sqrt{3}i + \tfrac{5}{2}j)$$
$$= (-5\sqrt{2} - \tfrac{5}{2}\sqrt{3})i + (5\sqrt{2} - \tfrac{5}{2})j$$
$$= -11\cdot40i + 4\cdot57j.$$

∴ Apparent speed $= \sqrt{[(4\cdot57)^2 + (-11\cdot40)^2]} = 12\cdot29$ km/h.

If θ is the angle v_{SB} makes with j,

$$\tan\theta = \frac{-11\cdot40}{4\cdot57}.$$

∴ $\theta = -68° 9'$.

Apparent speed is 12·3 km/h in a direction N. 68° 9′ W.

Example 4. A man travelling north at 14 km/h finds that the wind appears to blow from the west. On doubling his speed it appears to come from the north-west. Find the velocity of the wind.

67

SPEED AND VELOCITY

i and j will be taken as unit vectors due east and due north respectively.

Any velocity coplanar with i and j can be written $xi + yj$ (refer to Section 3.4).

Let the wind's velocity be
$$xi + yj \qquad \ldots \text{(i)}$$

The original velocity of the man is $14j$, therefore the velocity of the wind relative to the man is
$$xi + yj - 14j$$
i.e. $\qquad xi + (y - 14)j$

but this is from the west and is therefore of the form ki (k positive) the coefficient of j being zero.

$\therefore \qquad\qquad y - 14 = 0$

$\therefore \qquad\qquad y = 14 \quad \text{and} \quad x > 0 \qquad\qquad \text{(ii)}$

When he doubles his speed, the man's velocity becomes $28j$ and the velocity of the wind relative to the man is
$$xi + yj - 28j \qquad\qquad \text{(iii)}$$
or $\qquad\qquad xi + (y - 28)j$

but this is from the north-west and is equally inclined to $-i$ and j, therefore it is of the form $+pi - pj$. That is, the coefficients of i and j are equal in value but opposite in sign.

$\therefore \qquad\qquad +x = -y + 28 \qquad\qquad \text{(iv)}$

From (ii) and (iv) $x = 14$ and $y = 14$

$\therefore \qquad$ from (i) the wind's velocity is $14i + 14j$

or $14\sqrt{2}$ km/h from the south-west.

Check: Substituting in equation (iii) we have $14i - 14j$ which is of the required form $pi - pj$; and *NOT* $-pi + pj$.]

Because we are only interested in the change of relative position between two points, P and Q, we do not alter the relative velocity if we impress on both P and Q equal velocities. Thus, an alternative way of regarding the relative velocity of P with respect to Q, is to impress on both a velocity of $-q$, Q is thus reduced to rest and P has has the velocity $p - q$, which is the relative velocity of P with respect to Q.

RELATIVE VELOCITY

Similarly we can impress a velocity of $-p$ on both, P is thus reduced to rest and Q has the velocity $q - p$, which is the relative velocity of Q with respect to P.

Example 5. A ship B is steaming on a straight course south-east at a uniform speed of 15 km/h. *Another ship A, is at a distance of* 10 km *due north of B and steams at a speed of* 12 km/h. *Find the course that A must steer in order to get as close to B as possible, and their minimum distance apart.*

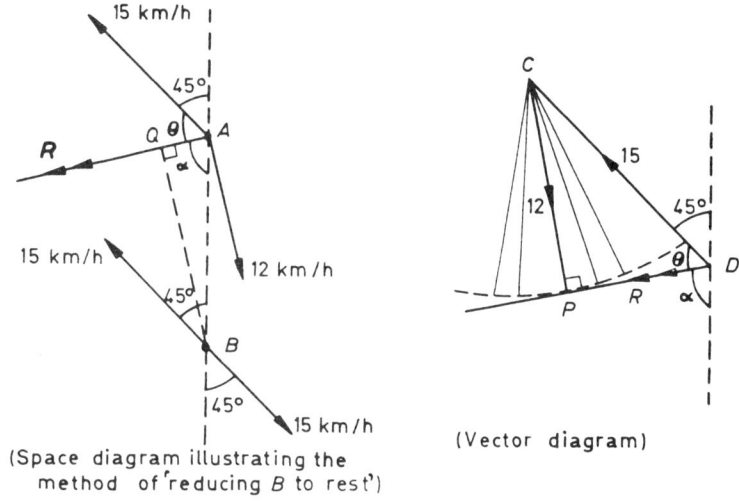

Figure 4.8

Let A and B be the initial positions of the ships. *Figure 4.8*, shows the position after B has been brought to rest by impressing a velocity of 15 km/h in a direction north-west onto both ships.

B is at rest and A has a fixed velocity 15 km/h north-west represented by $\overrightarrow{DC}$ (refer to the vector diagram in *Figure 4.8*). To $\overrightarrow{DC}$ we can add a vector $\overrightarrow{CP}$ representing 12 km/h. The length of $\overrightarrow{CP}$ is fixed at 12 units, but its direction, which is the direction in which A steers, can vary. Various positions of $\overrightarrow{CP}$ are indicated and P lies on a circle centre C radius 12 units.

$\overrightarrow{DP}$ is the resultant of $\overrightarrow{DC}$ and $\overrightarrow{CP}$ and represents the relative velocity R of A with respect to B. In order for A to approach as near to B as possible, it must sail as close as possible to the direction AB, that is, angle α must be a minimum. Since $\overrightarrow{DC}$ is fixed, we must

SPEED AND VELOCITY

therefore make $\angle CDP$ a maximum. This is so when $\overrightarrow{DP}$ is tangential to the locus of P and $\angle DPC = 90$ degrees. When $\triangle DPC$ is right angled at P,

$$\sin \angle CDP = \tfrac{12}{15} = 0\cdot 8.$$

$$\therefore \quad \theta = \angle CDP = 53° 8'.$$

$\therefore$ $\overrightarrow{CP}$ makes an angle of $45° - 36° 52' = 8° 8'$ with the north–south line and the required course is S. $8° 8'$ E.

To find the minimum distance apart, from B, draw BQ perpendicular to R. Then

$$BQ = AB \sin \alpha$$

and $\quad \alpha = 180° - 45° - \theta = 180° - 45° - 53° 8' = 81° 52'.$

$$\therefore \quad BQ = 10 \sin 81° 52'$$
$$= 9\cdot 90 \text{ km}.$$

In the following example distances are measured in kilometres, speeds in kilometres per hour and i and j are unit vectors due east and due north respectively.

*Example 6. *Two ships are observed from a coastguard station at 10·00 h and 11·00 h respectively. They have the following displacement (s) and velocity (v) vectors*

$$s_1 = i + 3j \quad \text{and} \quad v_1 = i + 2j \quad \text{at 10·00 h.}$$
$$s_2 = i + 2j \quad \text{and} \quad v_2 = 5i + 6j \quad \text{at 11·00 h.}$$

If they continue with the same velocities, find the smallest distance between the two ships in the subsequent motion and at what time this occurs. Also if at 11·00 h the first ship had changed its velocity to $\tfrac{11}{3}i + 2j$ show that the ships would have collided and find the time of collision.

At 11·00 h the displacement of the first ship will be

$$s'_1 = (i + 3j) + 1 \cdot (i + 2j)$$
$$= 2i + 5j.$$

Hence at 11·00 h,

relative displacement $s_{21} = s_2 - s'_1 = -i - 3j,$

relative velocity $\quad v_{21} = v_2 - v_1 = 4i + 4j,$

* This example may be omitted on the first reading of Part I.

giving the situation in *Figure 4.9* which shows displacement and velocity relative to the first ship.

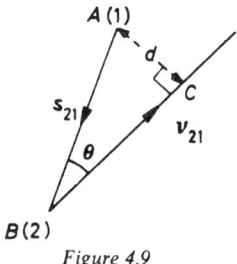

Figure 4.9

Shortest distance $= AC = AB \sin \theta$

$$= |s_{21}| \sin \theta$$

$$= \frac{|v_{21}||s_{21}| \sin \theta}{|v_{21}|}$$

$$= \frac{|v_{21} \times s_{21}|}{|v_{21}|} \text{ (refer to Section 3.9)}$$

$$= \frac{1}{4\sqrt{2}} \begin{vmatrix} i & j & k \\ -1 & -3 & 0 \\ 4 & 4 & 0 \end{vmatrix}$$

$$= \frac{1}{4\sqrt{2}} |8k|$$

$$= \sqrt{2} \text{ km.}$$

Relative distance sailed $BC = AB \cos \theta$

$$= |s_{21}| \cos \theta$$

$$= \frac{|v_{21}||s_{21}| \cos \theta}{|v_{21}|}$$

$$= \frac{|v_{21} \cdot s_{21}|}{|v_{21}|} \text{ (refer to Section 3.8)}$$

$$= \frac{|-4 - 12 - 0|}{4\sqrt{2}}$$

$$= \frac{16}{4\sqrt{2}} \text{ km.}$$

SPEED AND VELOCITY

Relative speed is $4\sqrt{2}$ km/h

$$\therefore \quad \text{time taken is } \frac{16/4\sqrt{2}}{4\sqrt{2}} = \tfrac{1}{2} \text{ h.}$$

The shortest distance between the ships is $\sqrt{2}$ km and this occurs at 11·30 h.

If we put $v_1 = \tfrac{11}{3}i + 2j$, then

$$s_{21} = -i - 3j \quad \text{and} \quad v_{21} = \tfrac{4}{3}i + 4j,$$

which are parallel but in opposite directions. Hence the ships will collide.

If they collide after t hours, then

$$i + 3j = t(\tfrac{4}{3}i + 4j)$$
$$t = \tfrac{3}{4} \text{ h.}$$

They will collide at 11·45 h.

Exercises 4d

1. Three ships are travelling with velocities v_1, v_2, v_3 respectively, where $v_1 = 2i - 3j, v_2 = 5i + 2j; v_3 = -i + j$. Find the magnitudes of the relative velocities of A with respect to B, B with respect to C and C with respect to A.

2. One man, A, is travelling due south at 20 km/h, another man, B, is travelling due east at 14 km/h. Find the magnitude and direction of the relative velocity of A w.r.t. B.

3. A man on a hovercraft, which is moving with velocity $v_1 = 5i + 6j$, finds that three ships appear to be moving with velocities $v_2 = 2i - 4j$, $v_3 = -13i + j$, and $v_4 = 7i - 8j$. Find the actual velocities of the ships.

4. Two roads meet at right angles at P. One man, A, is walking at 3 km/h along one road towards P and observes another man, B, trotting at 6 km/h along the other road towards P. Find the relative velocity of A with respect to B.

5. Rain is falling vertically. A man in a train travelling at 60 km/h notices that the rain makes lines on the window inclined at 10 degrees to the horizontal. What is the speed of the rain?

6. The racing pennant of a yacht, which is moving north west at 5 km/h, shows the wind apparently coming from the west, and its apparent speed is 10 km/h. What is the true velocity of the wind?

7. A river of width, d, has a current of constant velocity, λu. A man rows a boat with constant speed, $u (= |u|)$, relative to the water. He

crosses by the shortest path. Find his time for the journey if $\lambda < 1$.

8. To an observer on shore a ship appears to be travelling N. 75° 58′ W. at 12·37 km/h. The ship is being carried by a current which is flowing with a velocity of 4·24 km/h in a direction north-east. Find the velocity of the ship relative to the water.

9. A man travelling west at 15 km/h finds that the wind appears to blow from the direction S. $\tan^{-1}\frac{1}{4}$ E. If he reduces his velocity to 10 km/h towards the west, the wind appears to blow from a direction S. $\tan^{-1}\frac{1}{3}$ E. Find the velocity of the wind.

10. With distances measured in kilometres and speeds in kilometre per hour, two ships are observed, at the same time, to have the following displacement (*s*) and velocity (*v*) vectors:

$$s_1 = 2i + 3j \quad \text{and} \quad v_1 = -i - j$$
$$s_2 = 4i + 7j \quad \text{and} \quad v_2 = -3i - j$$

If they continue to sail with the same velocities, find the time at which they will be closest and what this closest distance is in kilometres.

4.5. ANGULAR SPEED

Definition—If a point P is moving in a plane and O is a fixed point and OA a fixed line in the plane, then the *angular speed* of P about O is defined as the rate at which the $\angle POA$ increases (refer to *Figure 4.10*).

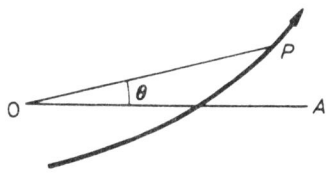

Figure 4.10

The instantaneous angular speed is obtained by considering any small interval of time, δt, which includes that instant. If $\delta\theta$ is the corresponding angle turned through by OP then

$$\text{angular speed } \omega = \lim_{\delta t \to 0} \frac{\delta\theta}{\delta t} = \frac{d\theta}{dt}.$$

(The angular speed is said to be uniform when equal angles are turned through in equal times, i.e. ω is constant. In this case only, $\theta = \omega t$). The SI unit of angular speed is rad/s.

SPEED AND VELOCITY

THEOREM

When a point P moves around the circumference of a circle of radius r, then its angular speed ω at any instant is given by $\omega = v/r$, where v is its speed at that instant.

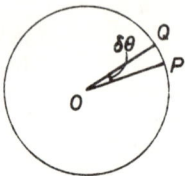

Figure 4.11

Let PQ be an arc of the circle length, δs, where P is the position of the point at the instant under consideration (refer to *Figure 4.11*). Let $\angle POQ = \delta\theta$ and let P move from P to Q in time δt. Since

$$\text{arc } PQ = r \times \angle POQ \quad \text{(in radians)},$$

$$\delta s = r\, \delta\theta$$

$$\therefore \quad \frac{\delta s}{\delta t} = r \frac{\delta\theta}{\delta t}$$

and in the limit as $Q \to P$ this gives

$$v = r\frac{d\theta}{dt} = r\dot\theta$$

or $$v = r \cdot \omega$$

$$\therefore \quad \omega = \frac{v}{r}.$$

Example 1. A wheel centre O, radius r is rolling uniformly, without sliding, along a straight line. Find the velocity of any point on its perimeter.

Suppose that, after time t, P comes into contact with point P' on the ground. Since there is no slipping, each point of the arc AP touches the ground in succession.

$$\therefore \quad \text{arc } AP = AP',$$

because both these distances are described in the same time.

ANGULAR SPEED

The speed of P relative to centre O = speed of P' along the ground = speed of centre O (because O is always directly above the point P'). Hence, if v is the speed of the centre of the wheel, any point P on its

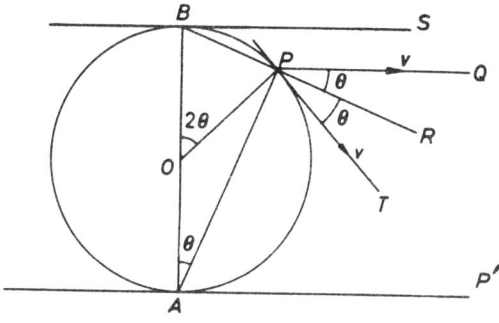

Figure 4.12

circumference has two velocities (refer to *Figure 4.12*), v horizontally, due to the forward motion of the wheel, v tangentially, due to the rotation of the wheel.

Let $\angle PAB = \theta$

∴ $\angle POB = 2\theta$ (angle at centre equals twice the angle at the circumference)

∴ $\angle QPT = 2\theta$ [PT perpendicular to OP and PQ perpendicular to AB].

Using horizontal and vertical components, P has velocities $v\mathbf{i}$ and $v \cos 2\theta \mathbf{i} - v \sin 2\theta \mathbf{j}$.

∴
$$v_P = v(1 + \cos 2\theta)\mathbf{i} - v \sin 2\theta \mathbf{j}$$
$$= v2\cos^2 \theta \mathbf{i} - v2 \sin \theta \cos \theta \mathbf{j}$$
$$= 2v \cos \theta [\cos \theta \mathbf{i} - \sin \theta \mathbf{j}]$$
$$= 2v \cos \theta [\cos(-\theta)\mathbf{i} + \sin(-\theta)\mathbf{j}].$$

∴ v_P has a magnitude $2v \cos \theta$ in a direction of $-\theta$ with PQ.

Now $\angle QPR = \angle SBP$ (PQ ∥ to BS)
$\qquad = \angle BAP$ (angle between chord and tangent)
$\qquad = \theta$.

Therefore, the direction of v_P is along PR, the continuation of BP,

SPEED AND VELOCITY

which is perpendicular to PA. The angular speed ω about A is

$$\frac{2v\cos\theta}{PA} = \frac{2v\cos\theta}{2r\cos\theta} = \frac{v}{r}.$$

Since the wheel is rigid, all points in AP have the same angular speed, ω, about A (the point in contact with the ground) as the wheel about its centre.
  Speed of A is $v - v = 0$.
  Speed of B, the highest point of the wheel, is $v + v = 2v$.

Example 2. Two model cars A and B are moving with speeds of 1·5 m/s and 2 m/s respectively, around two tracks made up of two pairs of parallel straights connected by two pairs of concentric semi-circular ends of radii 0·5 m and 0·6 m. B enters on its outer semi-circular track slightly behind A which has traversed 0·2 m of its inner semi-circular track. Will B overtake A before reaching the next straight part of the track?
  A's speed is 1·5 m/s on a circular path of radius 0·5 m.

A travels at $1\cdot 5/0\cdot 5 = 3$ rad/s.

A has travelled a distance of 0·2 m.

A is $0\cdot 2/0\cdot 5 = 0\cdot 4$ radians ahead.

B's speed is 2 m/s on a circular path of radius 0·6 m.

B travels at $2/0\cdot 6 = 3\frac{1}{3}$ rad/s.

∴ B overtakes A at a rate of $3\frac{1}{3} - 3 = \frac{1}{3}$ rad/s.

If still travelling on the semi-circular bend, B would overtake A after $0\cdot 4/\frac{1}{3} = 1\cdot 2$ s. In this time A would have gone $1\cdot 2 \times 3 = 3.6$ rad, which is greater than π. Therefore B will not overtake A before reaching the next straight part of the track.

4.6. ANGULAR VELOCITY AS A VECTOR QUANTITY*

Consider a rigid body rotating about an axis ON (*Figure 4.13*). Then its motion is uniquely specified by a vector $\boldsymbol{\omega}$ whose magnitude is the angular speed ω, and whose direction is parallel to ON (in the sense in which a right-handed screw would advance with the rotation). This vector $\boldsymbol{\omega}$ is called the angular velocity of the body;

i.e.   angular velocity $\boldsymbol{\omega} = \omega\hat{\boldsymbol{n}}$

where $\hat{\boldsymbol{n}}$ is a unit vector in the direction described.

* This section may be omitted on the first reading of Part I.

ANGULAR VELOCITY AS A VECTOR QUANTITY

Let O be a fixed point on the axis ON and P any point in the body with position vector r with respect to O (refer to *Figure 4.13*). If PA is perpendicular to the axis of rotation meeting it at A, then the point

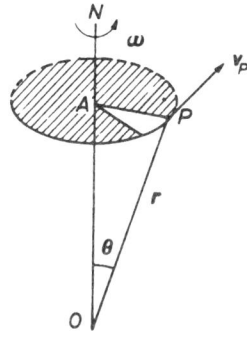

Figure 4.13

P is moving in a circular path of radius AP, and centre A, whose plane is perpendicular to ON. Its velocity v_P is given by:

$$|v_P| = \omega\, AP$$
$$= \omega\, OP \sin \theta$$
$$= \omega r \sin \theta$$
$$= |\omega \times r| \quad \text{(refer to Section 3.9)}$$

and its direction is perpendicular to the plane OPA, that is, the plane of r and ω.

$$\therefore \qquad v_P = \omega \times r$$

by definition of the cross product the sense of $\omega \times r$ is the same as that of the velocity v_P.

Exercises 4e

1. Find the angular speed in rad/s of any point on a disc if
   (a) it is making 200 rev/min,
   (b) it is making 225 rev/min,
   (c) a point on its circumference has a constant speed of 5 m/s and its radius is 2 m.

2. Find the speed of a point moving in a circle of radius 3 m with angular speed: (a) 4 rad/s (b) 300 rev/min (c) 120°/min.

3. A flywheel is rotating at 3 000 rev/min about its centre. Find

SPEED AND VELOCITY

its angular speed in radians per second. Also find the speed of a point on the flywheel 4 cm from its centre.

4. Assuming the Earth to be a sphere of radius 6 400 km and that it rotates on its axis once every 24 h, find its angular speed, and the speed (in metres per second) of a point on its equator.

5. An engine is travelling at 25 m/s and one of its wheels is 1·2 m in diameter. Find the speed of each of the two points of the wheel which are 0·3 m above the ground.

6. Find the time between 4·0 and 5·0 o'clock at which the minute and hour hands of a clock are concurrent.

7. A disc is set spinning and turns through $\theta°$ in t seconds, where $\theta = 120t - 6t^2$. Calculate the rate, in revolutions per minute, at which the disc is rotating when $t = 6$. (London)

8. At time t, the position vector $\overrightarrow{OP}$ is $2\mathbf{i} + t\mathbf{j}$. Write down an expression for $\tan \theta$, where θ is the angle which $\overrightarrow{OP}$ makes with the vector $\mathbf{i}$, and deduce that the angular speed of $\overrightarrow{OP}$ at time t is

$$\frac{2}{4 + t^2}.$$
(London)

EXERCISES 4*

The following is a portion of a train time-table giving the arrival and departure times from three consecutive stations A, B, C. The distances A to B and B to C are respectively 6 km and 9 km. Use the information given to answer Questions 1–3.

	OUT				IN		
		Trains				Trains	
Station	1	2	3	Station	4	5	6
A dep.	12·05	12·32	13·00	C dep.	12·10	12·40	13·12
B arr.	12·16	—	13·10	B arr.	12·25	—	13·24
B dep.	12·20	—	13·13	B dep.	12·30	—	13·30
C arr.	12·35	12·52	13·28	A arr.	12·40	12·55	13·39

1. What are the average speeds of each of the six trains between A and C.

2. On the same diagram draw the space-time graphs for all the six trains. Assuming that their speeds between stations are constant,

* Exercises marked thus † have been metricized, *see* Preface.

EXERCISES

find all the times between 12·00 h and 13·30 h when the trains are passing one another.

3. Check the answers to Question 2 by calculation.

4. A cyclist leaves A at 12·00 h and travels from A to B at a uniform speed of 15 km/h. After cycling 43 km he meets another cyclist, also travelling at a uniform speed, who left B at 12·07 h. If the second cyclist arrives at A at 16·54 h, find graphically, the distance from A to B and the speed of the second cyclist.

5. A and B are two places 90 km apart. A cyclist leaves A at 06·00 h and cycles at a uniform speed of 15 km/h to B. At 06·10 h, a second cyclist leaves B for A and also cycles at a uniform speed. After 30 km he has a 20 min rest and then continues his journey at the same uniform speed as before. He meets the first cyclist 45 km from A, find the time at which he arrives at A.

6. A particle leaves a point O and travels along a straight line, its distance s metres from O after t seconds being given by $9s = t^2$. Simultaneously, another particle leaves O and travels along the same line in the same direction, its distance from O being given by $3s^2 = 8t$. On the same diagram draw the distance-time graphs for the two particles from $t = 0$ to $t = 8$ and find the position and time that A overtakes B. If, after 3 seconds, B stops, when will A overtake it?

In the following questions, where necessary, i and j are to be taken as horizontal unit vectors due east and due north respectively.

7. A ship A is sailing due north at 20 km/h, and a second ship B is sailing N. 45° E. at $10\sqrt{2}$ km/h. Find the velocity of A relative to B.

8. P has a velocity of $3i - 2j$, the relative velocities of Q and R with respect to P are $2i + 4j$ and $i - 5j$, respectively. Find the velocity of Q relative to R. Is it necessary to know P's velocity in order to answer the question?

9. A hovercraft is 8 km due west of a ship, which is sailing with a speed of 16 km/h due north. If the speed of the hovercraft is 34 km/h, what is the shortest time in which it can intercept the ship?

10. A man travelling south-east at $2\sqrt{2}$ km/h notices that the wind appears to come from the north-east. On trebling his speed, the wind appears to come from the east. Find the velocity of the wind.

11. A ship is sailing north at 20 km/h. A second ship B is 10 km from A on a bearing from A of S. 45° W. If B can sail at 16 km/h, what course must B steer in order to get as close to A as possible? How near can they be?

12. An aeroplane and a helicopter are maintaining the same

height above the ground. The aeroplane is moving due north at 80 m/s and the helicopter is moving in a direction N. 60° W. at 60 m/s. The helicopter is 32 m due east of the plane and they both continue with the same velocities. Find the shortest distance between them in the subsequent motion and the time at which this occurs.

13. A ship A is moving due east at 18 km/h and another ship B is moving in a direction N. 120° E. at 10 km/h. B is 12 km due north of A and both ships continue with the same velocities. Find the shortest distance between the ships and the time at which this occurs.

14. A ship is steaming due north at 12 knots. To an observer in this ship a second ship appears to be moving in a direction north-east at 8 knots. Graphically, or otherwise, find the actual magnitude and direction of the velocity of the other ship.   (London, part)

15. Two straight roads cross at right angles at O. A cyclist is travelling at 5 m/s due north. When passing through O, he notices a motorcar on the other road 200 m from O. The motorcar is travelling at 12 m/s due east towards O. Find the velocity of the car relative to the cyclist and the least distance between the cyclist and the car.

16. A ship, A, is sailing with a speed of 5 km/h in a direction N. $\tan^{-1} 2$ E. Another ship, B, is sailing with a speed of $5\sqrt{2}$ km/h in a direction N. $\tan^{-1}(\frac{1}{3})$ W. The displacement of B from A is $\sqrt{17}$ km in a direction N. $\tan^{-1} 4$ E. Express in i, j form the velocities of the two ships and the displacement of B from A. Hence find, to the nearest minute, when the two ships will be closest together if their velocities remain constant.

17. A boat is sailing S. 60° E. at 10 knots, and the wind appears to come from the south. If the boat sails S. 30° E. at the same speed, the wind appears to come from S. 15° W. Find the speed and direction of the wind.

18. An aeroplane is to fly in a straight line from a point A to a point B due north of A, and back to A. If the aeroplane flies at 250 km/h relative to the air, and there is a wind blowing from the south-west at 50 km/h, show in a diagram the courses which must be steered on the outward and return journeys. Calculate the ratio of the times taken from A to B and from B to A.   (London)†

19. With distances measured in kilometres and speeds in kilometres per hour, two ships are observed at the same time to have the following displacement (s) and velocity (v) vectors:

$$s_1 = 3i - 2j \quad \text{and} \quad v_1 = 10i + 10j$$
$$s_2 = 4i + 6j \quad \text{and} \quad v_2 = i + 3j.$$

EXERCISES

If the velocities remain constant, find the time at which the two ships will be nearest together. Find also their distance apart at that time.

20. If, in Question 19, the first ship's velocity had been $2i + 11j$ instead of $10i + 10j$, show that the ships would have collided. Find also the time at which this would have occurred and the displacement vector of the point of collision.

21. With distances measured in kilometres and speeds in kilometres per hour, three ships are observed from a coastguard station at half-hour intervals. They have the following distance (s) and velocity (v) vectors.

$s_1 = 2i + 6j$ and $v_1 = 5i + 4j$ at 12 noon

$s_2 = 6i + 9j$ and $v_2 = 4i + 3j$ at 12.30 p.m.

$s_3 = 11i + 6j$ and $v_3 = 2i + 7j$ at 1 p.m.

Prove that, if the ships continue with the same velocities, two of them will collide and find the time of collision. If, at that instant, the third ship changes course and then proceeds directly to the scene of collision at its original speed, find at what time it will arrive. (London)†

22. A and B are two points a km apart with B due east of A. An aircraft flies from A to B and back to A. A wind is blowing at v km/h from the south-west. The ground speeds of the aircraft during the outward and return journeys are u_1 km/h and u_2 km/h, respectively. If the aircraft flies throughout at a speed u km/h relative to the air ($u > v$), show that $u_1 - u_2 = v\sqrt{2}$ and $u_1 u_2 = u^2 - v^2$.

Hence find, in terms of u and v, the time taken for the whole journey. (London)†

23. A steam engine is moving at 20 km/h due north and the wind is blowing at 10 km/h from the south-east. Assuming that, as the smoke leaves the engine, it immediately takes the velocity of the wind, find the angle the smoke trail makes with the engine.

24. A pilot keeps his aeroplane headed due S. at a constant speed of 200 km/h relative to the air, and after 15 min finds he has travelled 40 km in a direction S. 15° E. Show clearly in a diagram the velocity of the wind and find its magnitude and direction by measurement or calculation. Find also the direction in which he must now steer in order to bring his aeroplane due S. of his starting-point in the shortest possible time, with the same air-speed of 200 km/h.

(London)†

25. An aircraft A, which can fly at a maximum speed of 750 km/h,

sets out to intercept a second aircraft B, which is 100 km away in a direction 60° W. of S. and is flying due east at 500 km/h.

(a) Find the course (in degrees E. or W. of S.) which A should set if it is to fly in a straight line at maximum speed, and the time required to reach B.

(b) Find the least speed V_1 km/h at which A can fly to intercept B.

(c) Show that if A flies at a speed greater than V_1 km/h but less than a certain speed V_2 km/h (which should be found explicitly), it has a choice of two courses.

(A solution by drawing and measurement will not obtain full marks.) (WJEC)†

26. A and B are two points at sea with B 1 km due east of A. Three boats start at the same instant, the first from A sailing due north at a constant speed 8 km/h, the second from B sailing due east at a constant speed 5 km/h and the third from B sailing due north at such a speed that the three boats are always in one straight line. Show that t minutes later the third boat has a speed

$$8\left[1 - \frac{144}{(12+t)^2}\right] \text{ km/h.}$$

Find the velocity of the third boat relative to each of the other boats 4 min from the start. (London)†

27. A ship A is moving due N. at u km/h, and another ship, B, is moving at v km/h in a direction $\theta°$ W. of N. Initially, B is a km due E. of A. If B passes to the south of A, show that the shortest distance between the ships in the subsequent motion is $a(u - v\cos\theta)/V$, where $V^2 = u^2 + v^2 - 2uv\cos\theta$, and occurs after a time $av\sin\theta/V^2$. (London)†

28. A battleship, whose top speed is 24 km/h, is detailed to intercept a convoy steaming due east at 16 km/h. The convoy is 29 km away in a direction S. 30° E., and it maintains its course. Find the shortest time in which the battleship can reach the convoy.

29. Two points, A_1 and A_2, moving in a plane have coordinates (x_1, y_1) and (x_2, y_2) respectively, referred to axes fixed in the plane. State the components (parallel to the axes) of (a) the displacement $\overrightarrow{A_1A_2}$ and (b) the velocity of A_2 relative to A_1.

At a certain instant a ship, A_1, is sailing due east with speed u_1, a second ship, A_2, north-east of A_1, is sailing due north with speed u_2, a third ship, A_3, north-east of A_2, is sailing due south with speed u_3, also $A_1A_2 = A_2A_3 = d$. Assuming that the velocities of the three ships remain constant, find the easterly and northerly components of the velocities of A_2 and A_3 relative to A_1. Deduce expressions for the components of the displacements $\overrightarrow{A_1A_2}$ and $\overrightarrow{A_1A_3}$ after time t. Hence, or otherwise, show that the three ships are again in a straight line after a time.

$$\frac{u_1 + 2u_2 + u_3}{u_1(u_2 + u_3)} \cdot \frac{d}{\sqrt{2}}.$$ (JMB)

30. An aeroplane flies round a horizontal square of side a with constant air-speed v. There is a horizontal wind of constant speed kv ($k < 1$) making a constant angle α with one side of the square. Prove that the time taken for the aeroplane to complete the course is

$$2a[\sqrt{(1 - k^2 \sin^2 \alpha)} + \sqrt{1 - k^2 \cos^2 \alpha}]/v(1 - k^2) \quad \text{(Oxford)}$$

31. A particle P describes a circle centre O with constant speed while a second particle Q moves with constant velocity along a diameter AB of this circle. Both particles start from A at the same moment with P moving faster than Q. Initially the speed of Q relative to P is $\sqrt{80}$ m/s and when the angle AOP is 30 degrees it is $\sqrt{48}$ m/s. Find the speeds of P and Q.

When P has traced out an arc AP subtending an angle θ at O show that the relative velocity makes an angle ϕ with BA where $\tan \phi = 2 \cos \theta/(1 - 2 \sin \theta)$. (London)†

32. Two model cars, A and B, are moving around two concentric circular tracks, centre O, radii r and $2r$, in the same sense. The constant angular velocities of OA and OB are $4k$ and k respectively. If initially, OAB is a straight line, show that after time $\frac{4}{15}k$ the angular velocity of AB is zero. Find the angle OBA at this instant.

33. Two ships A and B move with constant velocities and have the following velocity and position vectors at time $t = 0$:

	Velocity Vector	Position Vector
Ship A	$-i + 3j$	$5i + j$
Ship B	$2i + 5j$	$3i - 3j$

Show that any point on the line $r = 4i - j + \lambda(2i - j)$, where λ is a scalar, is equidistant from the ships A and B at time $t = 0$. Find the position vector of the point which is equidistant from both ships at time $t = 0$ and from the point with position vector $i - j$.

Calculate the value of t when the distance between A and B is least. (AEB)

34. A motor launch whose speed in still water is V km h^{-1} has to travel directly from a harbour X to a buoy lying 5 km from X

SPEED AND VELOCITY

and in a direction N 30° E from X. There is a steady current of 6 km h^{-1} flowing *from* the direction N $\theta°$ W. The journey from X to Y under these conditions takes exactly one hour. Prove that

$$\sqrt{3} \cos \theta - \sin \theta = (V^2 - 61)/30.$$

If the return journey under the same conditions takes only half an hour, calculate the value of V and the direction of the current. (C)

5

ACCELERATION

5.1. INTRODUCTION

As we have seen in Chapter 4, velocity is a vector quantity having both magnitude and direction. If either of these change, the velocity changes.

Definition—The acceleration of a point is its rate of change of velocity, i.e., if a point is moving with a velocity v (represented by $\overrightarrow{OP}$, in *Figure 5.1*), at some time t and with velocity, $v + \delta v$ (repre-

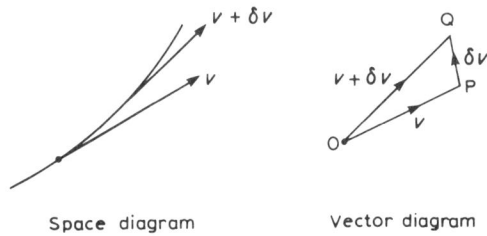

Space diagram   Vector diagram

Figure 5.1

sented by $\overrightarrow{OQ}$), at time, $t + \delta t$, then its acceleration, f, is given by

$$f = \lim_{\delta t \to 0} \frac{\delta v}{\delta t} = \frac{dv}{dt}$$

SI unit of acceleration ... m/s².

In *Figure 5.1* the point is instantaneously moving in the direction $\overrightarrow{OP}$ and we note that δv is *not* in this direction. Also the magnitude of δv is *not* equal to the change in the scalar speed v of the point. So f is not necessarily along the path of P and $|f|$ is not in general equal to dv/dt.

ACCELERATION

5.2. MOTION IN A STRAIGHT LINE

However, in the special case of a point moving in a straight line, the direction of motion remains constant and we consider only changes in the magnitude of the velocity. Hence in this case

$$|f| = f = \frac{dv}{dt}.$$

Note two alternative forms for f:

(i) $f = \dfrac{d^2s}{dt^2}$, since $v = \dfrac{ds}{dt}$.

(ii) $f = v\dfrac{dv}{ds}$, since $\dfrac{dv}{dt} = \dfrac{ds}{dt} \cdot \dfrac{dv}{ds}$.

We shall make use of all three forms in subsequent sections.

5.3. UNIFORM ACCELERATION

When a point is moving in a straight line with *constant* acceleration we have

$$\frac{dv}{dt} = f \text{ (a constant)}.$$

This gives

$$v = ft + C.$$

Now suppose that the initial speed is u so that $v = u$ when $t = 0$. Then we have $u = 0 + C$ and our equation becomes

$$v = u + ft.$$

Then this can be written as $ds/dt = u + ft$, and integrating again,

$$s = ut + \tfrac{1}{2}ft^2 + D.$$

Measuring s from the starting point makes $s = 0$ when $t = 0$ so that $0 = 0 + 0 + D$.

Hence

$$s = ut + \tfrac{1}{2}ft^2.$$

UNIFORM ACCELERATION

It is to be noted that this equation gives the displacement in t seconds from a fixed point (the position of the particle at $t = 0$). This is not the distance travelled in the tth second which is obtained by the displacement in t seconds − displacement in $(t-1)$ seconds. (Refer to Example 3.)

Also since acceleration can be written $v\dfrac{dv}{ds}$.

$$v\frac{dv}{ds} = f \quad \text{(constant)}$$

Integrating with respect to s

$$\tfrac{1}{2}v^2 = fs + E$$

and since when $\quad s = 0, \quad v = u$

$$\tfrac{1}{2}u^2 = 0 + E$$

∴ $\qquad \tfrac{1}{2}v^2 = fs + \tfrac{1}{2}u^2$

or $\qquad v^2 = u^2 + 2fs.$

Summarizing we have

$$\boxed{\begin{array}{c} v = u + ft \\ s = ut + \tfrac{1}{2}ft^2 \\ v^2 = u^2 + 2fs \end{array}}$$

Example 1. *A train, moving in a straight line with a speed of 72 km/h, is brought to rest in 2 min with a uniform retardation. Find how far the train travels before being brought to rest.*

$$u = 72 \text{ km/h} = 72 \times \tfrac{5}{18} \text{ m/s}$$
$$= 20 \text{ m/s}$$
$$t = 2 \text{ min} = 120 \text{ s}.$$

The train is brought to rest, therefore $v = 0$ and we are required to find s.

The last two of the three equations we have summarized deal with s, but in both cases f is included and this is not known. Therefore we use the first equation to find f.

i.e. $\qquad v = u + ft$

∴ $\qquad 0 = 20 + f\,120$

∴ $\qquad f = -\tfrac{1}{6} \text{ m/s}^2.$

ACCELERATION

Now substituting in the second equation,

$$s = ut + \tfrac{1}{2}ft^2$$
$$s = 20 \cdot 120 + \tfrac{1}{2}(-\tfrac{1}{6})120^2$$
$$s = 1\,200 \text{ m}$$
$$s = 1.2 \text{ km.}$$

Alternatively, using the third equation,

$$v^2 = u^2 + 2fs$$
$$0 = 20^2 + 2(-\tfrac{1}{6})s$$
$$s = \frac{20^2 \times 6}{2}$$
$$s = 1\,200 \text{ m as before.}$$

Example 2. A particle starts with a speed of 0·6 m/s and has an acceleration of magnitude −0·03 m/s² (i.e. a retardation of 0·03 m/s²). If it is travelling in a straight line, how long will it be before it comes to rest and how far will it then have travelled?

In this case $u = 0\cdot6$, $f = -0\cdot03$, $v = 0$. Using

$$v = u + ft,$$
$$0 = 0\cdot6 - 0\cdot03t$$
∴ $\quad t = 20$ s.

Also $\quad s = ut + \tfrac{1}{2}ft^2$

∴ $\quad s = 0\cdot6 \times 20 + \tfrac{1}{2}(-0\cdot03)20^2$
$\quad\quad = 6$ m.

Example 3. A particle is travelling in a straight line with constant acceleration. It covers 3·5 m in the third second and 4·1 m in the fourth second, find its acceleration and initial speed.

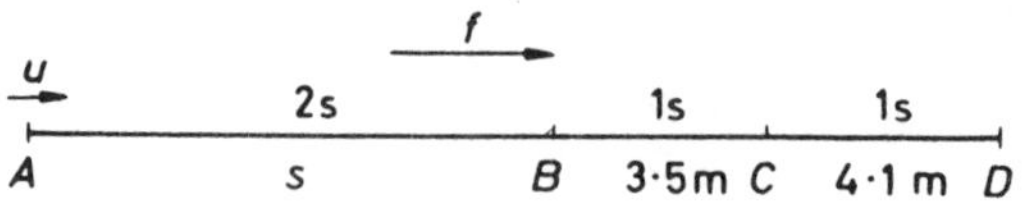

Figure 5.2

UNIFORM ACCELERATION

Applying $s = ut + \tfrac{1}{2}ft^2$ to:

section AB $\qquad s = 2u + \tfrac{1}{2}f4$ $\qquad$ ....(i)

section AC $\qquad 3\cdot5 + s = 3u + \tfrac{1}{2}f9$ $\qquad$ ....(ii)

section AD $\qquad 7\cdot6 + s = 4u + \tfrac{1}{2}f16.$ $\qquad$ ....(iii)

Subtracting (i) from (ii)
$$3\cdot5 = u + \tfrac{1}{2}f5. \qquad \text{....(a)}$$

Subtracting (ii) from (iii)
$$4\cdot1 = u + \tfrac{1}{2}f7. \qquad \text{....(b)}$$

Subtracting (a) from (b)
$$0\cdot6 = \tfrac{1}{2}f2.$$
$$\therefore \quad f = 0\cdot6 \text{ m/s}^2.$$

Substituting in equation (b)
$$u = 2 \text{ m/s.}$$

Example 4. A car moves a distance d in time T seconds. It starts from rest, and travels in a straight line, accelerating with a constant acceleration f_1, until the brakes are applied and the engine stopped. It is then subject to a constant retardation, f_2, until it comes to rest. Find an expression for T^2 and for V, the greatest speed attained, in terms of f_1, f_2 and d.

Let t be the time during which the car accelerates, therefore $T - t$ is the time during which it decelerates. Similarly, if s is the distance it travels while accelerating, $d - s$ is the distance travelled while decelerating.

First part of journey $\qquad\qquad$ Second part of journey

$V = 0 + f_1 t$ $\qquad\qquad\qquad\qquad 0 = V - f_2(T - t)$

$\therefore\ V = f_1 t$ $\qquad$ ....(i) $\qquad \therefore\ V = f_2(T - t)$ $\qquad$ ....(ii)

$V^2 = 2f_1 s$ $\qquad$ ....(iii) $\qquad 0 = V^2 - 2f_2(d - s)$ ....(iv)

From equations (i) and (ii)
$$t + T - t = \frac{V}{f_1} + \frac{V}{f_2}$$
$$\therefore \quad T = V\left(\frac{1}{f_1} + \frac{1}{f_2}\right) \qquad \text{....(a)}$$

89

ACCELERATION

From equations (iii) and (iv)

$$s + d - s = \frac{V^2}{2f_1} + \frac{V^2}{2f_2}$$

$$\therefore \quad d = \frac{V^2}{2}\left(\frac{1}{f_1} + \frac{1}{f_2}\right) \quad \ldots (b)$$

From equations (a) and (b)

$$\frac{T^2}{d} = \frac{V^2\left(\frac{1}{f_1} + \frac{1}{f_2}\right)^2}{\frac{V^2}{2}\left(\frac{1}{f_1} + \frac{1}{f_2}\right)}$$

$$\therefore \quad T^2 = 2d\left(\frac{1}{f_1} + \frac{1}{f_2}\right).$$

From equation (a)

$$V = T \Big/ \left(\frac{1}{f_1} + \frac{1}{f_2}\right)$$

$$\therefore \quad V = \sqrt{2d\left(\frac{1}{f_1} + \frac{1}{f_2}\right)} \Big/ \left(\frac{1}{f_1} + \frac{1}{f_2}\right)$$

$$= \sqrt{\frac{2d}{\left(\frac{1}{f_1} + \frac{1}{f_2}\right)}}.$$

Exercises 5a

1. In the following exercises, distance is measured in metres and time in seconds. f is a constant acceleration, u the initial speed, v is the final speed and t is the time.

(a) Given $u = 3, f = 4, t = 6$, find v and s.
(b) Given $u = 14, v = 32, s = 138$, find f and t.
(c) Given $f = -7, v = -7, s = 28$, find u and t.
(d) Given $u = 60, f = -\frac{3}{4}, t = 24$, find s and v.

2. A body starts from rest and moves with uniform acceleration in a straight line. It covers 30 m in the eighth second, find its acceleration.

3. A particle starts with a speed of 200 cm/s and moves in a straight line with a constant retardation of 4 cm/s^2. When will it come to rest, and how far will it have travelled?

4. A particle starting from rest has an acceleration of 4 m/s^2. Find its speed after 30 seconds and the distance s_1 covered in that time. It is then subject to a constant deceleration and comes to rest in 20 seconds. Find the distance s_2 covered while decelerating.

5. A body moves with constant acceleration in a straight line for 5 seconds during which time it covers 3·5 m. The acceleration then ceases and during the next 5 seconds it travels 6·5 m. Find its initial speed and its acceleration.

6. A train is subject to a uniform retardation while travelling in a straight line. It travels a distance of 100 m while its speed is reduced from 60 km/h to 20 km/h. Find how much further it will travel before coming to rest.

7. In two successive seconds a motor car moves through 13 m and 15 m respectively. Assuming that it is travelling with uniform acceleration, find its speed at the commencement of these two seconds and its acceleration.

8. A motorway runs directly from A to B. A car passes A at a constant speed of 36 km/h, and immediately another car starts from rest at A with a uniform acceleration of 2·5 m/s^2. When and where will the second car overtake the first?

9. In Question 8, what time elapses between the two cars arriving at B, if the second car stops accelerating as soon as it reaches the first car, the distance from A to B being 3 km.

10. A particle is moving in a straight line with uniform acceleration. It covers a distance a in the first n seconds, a total distance of b in the first $2n$ seconds. Find its acceleration and initial speed.

5.4. SPEED–TIME GRAPHS

For a variable acceleration which can be expressed as a function of the distance s, time t, or velocity v we have to solve the equation,

$$\text{Acceleration} = F(s) \quad \text{or} \quad G(t) \quad \text{or} \quad H(v)$$

and this will be dealt with in Section 5.5.

If there is no known functional relation between the acceleration and $s, t,$ or v a graphical method can sometimes be used.

Suppose that the acceleration varies but the speeds are known, or observed at different times, then a speed–time graph can be

ACCELERATION

obtained. Time is plotted along the horizontal axis and speed along the vertical axis. It is assumed that the variation is continuous and the successive points are joined by a smooth curve (refer to *Figure 5.3*). The slope of the curve (dv/dt) at any point approximates

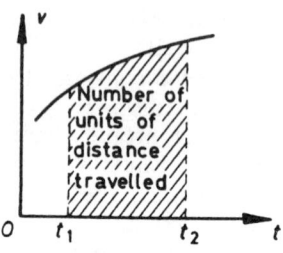

Figure 5.3

to the acceleration at that point. Also, the algebraic sum of the area under the curve is given by

$$\int_a^b v\, dt = \int_a^b \frac{ds}{dt}\, dt = \int_a^b ds = s_b - s_a,\text{ the displacement.}$$

Alternatively, if the areas above and below the axis are treated as positive, then their sum gives the distance travelled.

Example. In a journey of 14 km *the speed of a train is given by the following table:*

t minutes	0	1	2	3	4	5	6	7	8	9	10	11	12	13	14
v km/h	0	20	40	60	50	60	65	50	45	55	65	50	35	30	0

Draw a speed–time graph and estimate the distance travelled between the times t = 2 min and t = 7 min.

Referring to *Figure 5.4*, the distance travelled in kilometres is equal to the number of units of area in *ABCDE*. To estimate this we can construct the ordinates at the points $t = 2, 3, \ldots, 7$, and approximate to the area by the sum of the areas of the five trapezia so formed (converting the time to hours).

SPEED-TIME GRAPHS

$$\text{Area} = \tfrac{1}{2}(40 + 60) \cdot \tfrac{1}{60} + \tfrac{1}{2}(60 + 50)\tfrac{1}{60} + \tfrac{1}{2}(50 + 60) \cdot \tfrac{1}{60}$$
$$+ \tfrac{1}{2}(60 + 65) \cdot \tfrac{1}{60} + \tfrac{1}{2}(65 + 50) \cdot \tfrac{1}{60}$$
$$= \tfrac{1}{120}(40 + 60 + 60 + \cdots + 65 + 65 + 50)$$
$$= \tfrac{560}{120} = 4\tfrac{2}{3}.$$

The distance travelled between the second and seventh minute is $4\tfrac{2}{3}$ km.

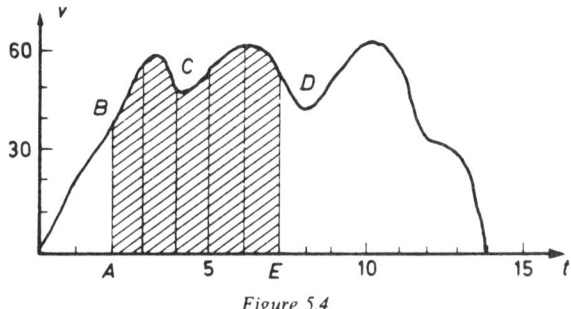

Figure 5.4

Exercises 5b

1. In a speed-time graph the velocities are measured in metres per second and the time in seconds. It consists of two straight lines AB, BC joining the points A (0, 50), B (10, 100); C (50, 0). Describe the motion of the particle and find the distance covered.

2. Draw a speed-time graph of a particle describing a straight line with uniform acceleration of 2 m/s^2, having started with an initial speed of 10 m/s. Plot the graph for values of t between 0 and 10 s and hence find the distance covered in the first 10 seconds.

3. Draw the speed-time curve of a particle moving in a straight line with initial velocity u, uniform acceleration f for t seconds. Deduce the formula $s = ut + \tfrac{1}{2}ft^2$.

4. A car is travelling along a straight road and its speeds taken at 10 second intervals over 2 minutes are as follows:

Time in s	0	10	20	30	40	50	60	70	80	90	100	110	120
Speed in m/s	20	25	28	30	31	27	25	20	18	16	14	8	0

Find the total distance covered in the 2 minutes and the car's speed and acceleration after 75 seconds.

5. A particle leaves a point O and travels in a straight line with a speed v m/s, given by the formula $v = 64 - t^2$, where t is the time

ACCELERATION

in seconds. Draw a speed–time curve from $t = 0$ to $t = 10$, and find from the graph how far the particle travelled in 10 seconds and how far it travelled in the fourth second. (If possible, check your answers by integration.) Is the distance of the car from O after 10 seconds the same as the distance travelled in 10 seconds?

5.5. VARIABLE ACCELERATION
(in a straight line)

When the acceleration varies, appropriate equations involving v, s or t can be found, provided the acceleration is given as a function of speed, distance or time.

If $f = F(t)$ we write $\dfrac{dv}{dt} = F(t)$ and integrate

if $f = G(s)$ we write $v\dfrac{dv}{ds} = G(s)$ and integrate

if $f = H(v)$ we write $\dfrac{dv}{dt} = H(v)$ or $v\dfrac{dv}{ds} = H(v)$

according to the information given and required, and the ease of integration.

Example 1. A particle travelling in a straight line has an acceleration of $(-v^2/200 - 32)$ m/s² where v is its speed at any time t. If its initial speed was 40 m/s, find the distance travelled before it comes to rest and the time taken.

$$\text{Acceleration} = -\frac{v^2}{200} - 32$$

$$\therefore \quad v\frac{dv}{ds} = -\frac{1}{200}(v^2 + 6400)$$

$$\therefore \quad \frac{v}{v^2 + 6400}\frac{dv}{ds} = -\frac{1}{200}$$

$$\therefore \quad \int\left(\frac{v}{v^2 + 80^2}\frac{dv}{ds}\right)ds = -\frac{1}{200}\int ds$$

$$\therefore \quad \int\frac{v\,dv}{v^2 + 80^2} = -\frac{1}{200}s$$

VARIABLE ACCELERATION

$$\therefore \quad \frac{1}{2}\log(v^2 + 80^2) + C = -\frac{1}{200}s.$$

Now when $s = 0$ $v = 40$.

$$\therefore \quad \tfrac{1}{2}\log_e(1\,600 + 6\,400) + C = 0$$

$$\therefore \quad C = -\tfrac{1}{2}\log_e(8\,000)$$

$$\therefore \quad -\frac{1}{200}s = +\tfrac{1}{2}\log_e(v^2 + 80^2) - \tfrac{1}{2}\log_e(8\,000)$$

$$\therefore \quad s = -100\log_e(v^2 + 80^2) + 100\log_e 8\,000$$

$$s = 100\log_e\left(\frac{8\,000}{v^2 + 80^2}\right).$$

The particle is at rest when $v = 0$

$$\therefore \quad s = 100\log_e \frac{8\,000}{6\,400}$$

$$= 100\log_e \tfrac{5}{4} \text{ m}.$$

To find the time taken we reconsider the initial equation, acceleration $= -v^2/200 - 32$,

i.e. $$\frac{dv}{dt} = -\frac{1}{200}(v^2 + 6\,400)$$

$$\therefore \quad \frac{1}{v^2 + 80^2}\frac{dv}{dt} = -\frac{1}{200}$$

$$\therefore \quad \int \frac{dv}{v^2 + 80^2} = -\frac{1}{200}\int dt$$

$$-\frac{1}{200}t = C + \frac{1}{80}\tan^{-1}\left(\frac{v}{80}\right).$$

When $t = 0$, $v = 40$

$$\therefore \quad 0 = C + \frac{1}{80}\tan^{-1}\frac{40}{80}$$

$$\therefore \quad -\frac{1}{200}t = -\frac{1}{80}\tan^{-1}\left(\frac{1}{2}\right) + \frac{1}{80}\tan^{-1}\frac{v}{80}$$

$$\therefore \quad t = \frac{5}{2}\left[-\tan^{-1}\left(\frac{v}{80}\right) + \tan^{-1}\left(\frac{1}{2}\right)\right] \text{ s}$$

ACCELERATION

and when $v = 0$

$$t = \frac{5}{2} \tan^{-1}\left(\frac{1}{2}\right) \text{ s.}$$

Example 2. A car is moving in a straight line and its acceleration is given by $k^2(a - s)$, where s is the distance moved in time t seconds and k and a are constants. If the car starts from rest, show that its speed v is given by $v^2 = k^2(2as - s^2)$. Hence verify that $s = a(1 - \cos kt)$.

$$\text{Acceleration} = k^2(a - s)$$

$$v\frac{dv}{ds} = k^2(a - s)$$

$$\therefore \int v \, dv = \int k^2(a - s) \, ds$$

$$\therefore \tfrac{1}{2}v^2 = C + k^2(as - s^2/2)$$

and since $v = 0$ when $s = 0$, $C = 0$

$$\therefore \quad v^2 = k^2(2as - s^2) \quad \dots \text{(i)}$$

If $\quad s = a(1 - \cos kt) \quad \dots \text{(ii)}$

$$v = \frac{ds}{dt} = ak \sin kt$$

$$\therefore \quad v^2 = a^2 k^2 \sin^2 kt$$

$$= k^2(a^2 - a^2 \cos^2 kt)$$

$$= k^2[a^2 - (a - s)^2] \quad \text{from (ii)}$$

$$= k^2[2as - s^2]$$

which agrees with equation (i).

Example 3. If, at time t, the speed of a particle is given by $v = 1/(A + Bt)$ (A, B positive), show that its acceleration is negative and proportional to the square of the speed. If, initially, the acceleration is -1 unit and $v = 40$, find A and B. Hence find s, the distance moved, in terms of t, the time, and find v in terms of s.

Since $v = \dfrac{1}{A + Bt}$,

$$\text{acceleration} = \frac{dv}{dt} = \frac{-B}{(A + Bt)^2}.$$

96

VARIABLE ACCELERATION

Combining these equations,

$$\text{Acceleration} = -Bv^2.$$

Thus the acceleration is negative and proportional to the square of the speed.

Initially $t = 0$ and $v = 40$,

$$\therefore \quad 40 = \frac{1}{A}; \quad A = \frac{1}{40}.$$

Also acceleration $= -1$ when $t = 0$

$$\therefore \quad -1 = \frac{-B}{A^2}$$

$$\therefore \quad B = A^2 = \left(\frac{1}{40}\right)^2 = \frac{1}{1\,600}$$

$$\therefore \quad v = \frac{1}{\dfrac{1}{40} + \dfrac{1}{1\,600}t} = \frac{1\,600}{40 + t}$$

$$\therefore \quad \frac{ds}{dt} = \frac{1\,600}{40 + t} \quad \quad \ldots\text{(i)}$$

$$\therefore \quad s = 1\,600 \log_e (40 + t) + C.$$

Now $s = 0$ when $t = 0$, $\therefore C = -1\,600 \log_e 40$,

$$\therefore \quad s = 1\,600 \log_e \left(\frac{40 + t}{40}\right). \quad \quad \ldots\text{(ii)}$$

From (i) $$v = \frac{1\,600}{40 + t}$$

$$\therefore \quad 40v + vt = 1\,600$$

$$\therefore \quad t = \frac{1\,600 - 40v}{v} \quad \quad \ldots\text{(iii)}$$

From (ii) $$e^{s/1600} = \frac{40 + t}{40}$$

$$\therefore \quad 40\,e^{s/1\,600} = 40 + t$$

$$t = 40\,e^{s/1\,600} - 40. \quad \quad \ldots\text{(iv)}$$

ACCELERATION

From (iii) and (iv)

$$t = \frac{1\,600 - 40v}{v} = 40e^{s/1\,600} - 40$$

∴ $\quad 40 - v = ve^{s/1\,600} - v$

∴ $\quad ve^{s/1\,600} = 40$

$$v = 40e^{-s/1\,600}.$$

Exercises 5c

1. A particle starts from rest and moves in a straight line with an acceleration of $(t^2 + 3t)$ m/s^2, where t is the time at any instant. Find its velocity and displacement at any time t, and after 6 seconds.

2. The acceleration of a particle moving in a straight line is $(12 - 3v^2)$ m/s^2, where v is its speed. If over a distance of 30 m, v increases from u to $2u$, find u.

3. A particle starts from rest and moves in a straight line with an acceleration of $(g - kv^2)$, where v is its speed at any time t and g and k are constants. Show that $v^2 = g(1 - e^{-2ks})/k$, where s is the distance travelled in time t. If the speed increases from V to $\tfrac{5}{4}V$ as s increases from d to $2d$, find an expression for k in terms of V and g.

4. A car is travelling in a straight line so that the distance s and its speed v are connected by the relation $s = \dfrac{5v}{75 - v}$. Show that the acceleration is $\dfrac{v(75 - v)^2}{375}$. Also show that $v = \dfrac{75s}{5 + s}$ and find an expression for t in terms of s.

5. The magnitude of the acceleration of a particle starting from rest and moving in a straight line is $16(1 - s/50)$, where s is the displacement in time t. Find its greatest speed and the distance travelled before coming to rest again.

6. A particle starts from rest at a point A and moves in a straight line towards a point B, distance 800 m away, with an acceleration of $1/(s - 800)^2$ m/s^2. If s is the distance of the particle from A, find its speed v in terms of s. Hence show that the time taken to reach the mid-point of AB is

$$20 \int_0^{400} \sqrt{\left(\frac{800 - s}{s}\right)}\, ds$$

and use the substitution $s = 800 \sin^2 \theta$ to evaluate this integral.

5.6. SIMPLE HARMONIC MOTION

An important case of variable acceleration is that of a particle moving with Simple Harmonic Motion.

Definition—When a particle moves so that its acceleration is proportional to the distance it has moved along its path from a fixed point O, and is directed towards O, the particle is said to move with *Simple Harmonic Motion* (S.H.M.).

It is to be noted that simple harmonic motion applies not only to motion in a straight line, but to any path the particle is traversing. For example, if the particle is moving along a curve and the acceleration along the curve is proportional to the arc length measured from some fixed point on the curve.

For simple harmonic motion if s is the displacement from a point O then,

$$\text{Acceleration} \propto -s$$

or
$$\frac{d^2s}{dt^2} = -n^2 s \quad \ldots \text{(i)}$$

where n is some constant. Note that n^2 is used not only to simplify subsequent formulae but also because n^2 is essentially positive and maintains the negative sign.

The solution of equation (i) is mathematically always of the same form whatever the path of the particle. For convenience we shall consider motion in a straight line.

Referring to *Figure 5.5*

$$\text{Acceleration} = -n^2 s$$

∴
$$v \frac{dv}{ds} = -n^2 s$$

∴
$$\int v \, dv = -n^2 \int s \, ds$$

∴
$$\tfrac{1}{2} v^2 = C - \tfrac{1}{2} n^2 s^2$$

$$v^2 = C' - n^2 s^2 \quad (C' = 2C).$$

If $s = a$, when $v = 0$, $C' = n^2 a^2$

∴
$$v^2 = n^2 a^2 - n^2 s^2$$

i.e.
$$v^2 = n^2(a^2 - s^2) \quad \ldots \text{(ii)}$$

ACCELERATION

From this equation, we see that $v = 0$ when $s^2 = a^2$, i.e. $s = +a$ or $-a$. Also, since $a^2 - s^2$ must be positive [otherwise we cannot find a square root of $a^2 - s^2$], the value of s must vary between $+a$ and $-a$. The particle oscillates between two extreme points A' and A, distance a on either side of the fixed point O [refer to *Figure 5.5*]. a is known as the *amplitude*.

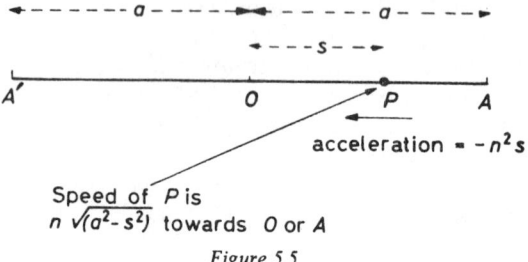

Figure 5.5

The maximum speed is that when $s = 0$, that is, at the point O and
$$v_{max} = n\sqrt{(a^2 - 0)} = na.$$
The acceleration has its greatest magnitude when $s = \pm a$, that is, at the extreme points A' and A and its magnitude is then na.

From equation (ii)
$$\frac{ds}{dt} = n\sqrt{(a^2 - s^2)}$$

$$\therefore \frac{dt}{ds} = \frac{1}{n\sqrt{(a^2 - s^2)}}.$$

$$\therefore t = \frac{1}{n}\int \frac{ds}{\sqrt{(a^2 - s^2)}}.$$

The solution of the integral is $\sin^{-1} s/a$ and therefore,
$$C + nt = \sin^{-1}\frac{s}{a}$$
where C is a constant to be found from the initial conditions. For example, if the particle starts from O, then $s = 0$ when $t = 0$, and therefore $C = 0$ and the equation becomes
$$s = a \sin nt.$$
If the time is measured from the time when the particle is passing

SIMPLE HARMONIC MOTION

through B, the mid-point of OA then $s = a/2$ when $t = 0$ and
$$C = \sin^{-1} \tfrac{1}{2}$$
$$\therefore \quad C = \frac{\pi}{6} \quad \text{or} \quad \frac{5\pi}{6}.$$

$C = \pi/6$ if the particle is at B but moving away from O, and $C = 5\pi/6$ if the particle is moving towards O.

In general, if when $t = 0$ the particle is at a distance d from O ($-a \leqslant d \leqslant a$), then $C = \sin^{-1} d/a = \varepsilon$ say

$$\therefore \quad s = a \sin(nt + \varepsilon). \quad \ldots \text{(iii)}$$

Note from this result that by differentiating

$$v = \frac{ds}{dt} = an \cos(nt + \varepsilon), \quad \ldots \text{(iv)}$$

which is an alternative expression for v to that given in equation (ii), it has the advantage of indicating the direction of the speed as well as its magnitude.

If in equation (iii), we increase t to $t + 2\pi/n$,

$$s = a \sin \left[n \left(t + \frac{2\pi}{n} \right) + \varepsilon \right]$$
$$= a \sin(nt + \varepsilon + 2\pi)$$
$$= a \sin(nt + \varepsilon)$$

we obtain the same value of s and similarly from equation (iv) v has the same value. This proves that after successive intervals of time $2\pi/n$, the particle passes through the same position with the same velocity,

i.e. $$T = \frac{2\pi}{n} \quad \ldots \text{(v)}$$

is the time for a complete oscillation, called the period of the motion. Summarizing we have

$$\begin{array}{|l|}
\hline
\ddot{s} = -n^2 s \quad (\ddot{s} = d^2s/dt^2) \\
v^2 = n^2(a^2 - s^2) \\
s = a \sin(nt + \varepsilon) \\
v = an \cos(nt + \varepsilon) \\
T = \dfrac{2\pi}{n} \\
\hline
\end{array}$$

ACCELERATION

In the solution of problems involving S.H.M. it is usually necessary to find n and a, and then solve the problem.

Example 1. *A particle is moving in a straight line with S.H.M. whose amplitude is 6 m and whose period is 10 s. Find the maximum speed and the speed when the particle is 3 m from the central position.*

The period is 10 s

$$\therefore \quad \frac{2\pi}{n} = 10$$

$$\therefore \quad n = \frac{2\pi}{10}.$$

Also the amplitude is given as 6 m

and $\quad v = n\sqrt{(a^2 - s^2)}$

$$\therefore \quad v = \frac{2\pi}{10}\sqrt{(36 - s^2)}.$$

The maximum speed is when $s = 0$.

$$\therefore \quad v_{max} = \frac{2\pi}{10}\sqrt{36}$$

$$= 3\cdot770 \text{ m/s}.$$

When the particle is 3 m from the centre $s = \pm 3$,

$$\therefore \quad v = \frac{2\pi}{10}\sqrt{[36 - (\pm 3)^2]}$$

$$= \frac{2\pi}{10} 3\sqrt{3}$$

$$= 3\cdot264 \text{ m/s}.$$

Example 2. *A particle moving with S.H.M. has speeds of 32 m/s and 24 m/s when its distances from the centre of oscillation are respectively 3 m and 4 m. Find the periodic time of the motion.*

$$v^2 = n^2(a^2 - s^2)$$

$\therefore \quad\quad\quad 32^2 = n^2(a^2 - 9)$ ....(i)

and $\quad\quad\quad 24^2 = n^2(a^2 - 16)$ ....(ii)

SIMPLE HARMONIC MOTION

dividing equations (i) and (ii) we have

$$\frac{32^2}{24^2} = \frac{a^2 - 9}{a^2 - 16}$$

i.e.
$$\frac{16}{9} = \frac{a^2 - 9}{a^2 - 16}$$

$$\therefore \quad 16a^2 - 256 = 9a^2 - 81$$

$$a^2 = 25$$

The amplitude is 5 m. Substituting we have

$$24^2 = n^2(25 - 16)$$
$$64 = n^2$$
$$8 = n$$

$\therefore$ periodic time is $2\pi/8$ or $\pi/4$ seconds.

Example 3. A particle is moving with S.H.M. of period $2\pi/n$, amplitude a, and centre of oscillation O. When $t = 0$ it is at a distance $a/\sqrt{2}$ from O, and when $t = \pi/24$ it is a distance $a\sqrt{3}/2$ from O. Find n (*four values*). Using the positive values of n, find the speed when $t = -\pi/24$ and show that there are two possible values of ε. Draw diagrams to illustrate the two cases.

For S.H.M. of period $2\pi/n$ and amplitude a

$$s = a \sin(nt + \varepsilon).$$

When $t = 0$, $\quad s = \dfrac{a}{\sqrt{2}}$

$$\therefore \quad \frac{a}{\sqrt{2}} = a \sin \varepsilon$$

$$\therefore \quad \sin \varepsilon = \frac{1}{\sqrt{2}}$$

$$\varepsilon = \frac{\pi}{4} \quad \text{or} \quad 3\frac{\pi}{4}.$$

Case $\varepsilon = \pi/4$

$$s = a \sin\left(nt + \frac{\pi}{4}\right)$$

ACCELERATION

but when $t = \dfrac{\pi}{24}$, $s = \dfrac{a\sqrt{3}}{2}$

$\therefore \quad \dfrac{a\sqrt{3}}{2} = a\sin\left(n\dfrac{\pi}{24} + \dfrac{\pi}{4}\right)$

$\therefore \quad \dfrac{\sqrt{3}}{2} = \sin\left(n\dfrac{\pi}{24} + \dfrac{\pi}{4}\right)$

$\therefore \quad n\dfrac{\pi}{24} + \dfrac{\pi}{4} = \dfrac{\pi}{3} \quad \text{or} \quad \dfrac{2\pi}{3}$

$\therefore \quad n\dfrac{\pi}{24} = \dfrac{\pi}{12} \quad \text{or} \quad \dfrac{5\pi}{12}$

$\therefore \quad n = 2 \quad \text{or} \quad 10.$

Case $\varepsilon = 3\pi/4$

$$s = a\sin\left(nt + \dfrac{3\pi}{4}\right)$$

when $t = \dfrac{\pi}{24}$, $s = \dfrac{a\sqrt{3}}{2}$

$\therefore \quad \dfrac{a\sqrt{3}}{2} = a\sin\left(n\dfrac{\pi}{24} + \dfrac{3\pi}{4}\right)$

$\therefore \quad \dfrac{\sqrt{3}}{2} = \sin\left(n\dfrac{\pi}{24} + \dfrac{3\pi}{4}\right)$

$\therefore \quad n\dfrac{\pi}{24} + \dfrac{3\pi}{4} = \dfrac{\pi}{3} \quad \text{or} \quad \dfrac{2\pi}{3}$

$\therefore \quad n\dfrac{\pi}{24} = -\dfrac{5\pi}{12} \quad \text{or} \quad -\dfrac{\pi}{12}$

$\therefore \quad n = -10 \quad \text{or} \quad -2.$

Only positive values of n are to be considered, therefore these two cases are inadmissible.

Two cases are left:

$$s = a\sin\left(2t + \dfrac{\pi}{4}\right) \quad \text{or} \quad a\sin\left(10t + \dfrac{\pi}{4}\right)$$

SIMPLE HARMONIC MOTION

when $t = -\pi/24$

$$s = a \sin\left(-\frac{2\pi}{24} + \frac{\pi}{4}\right), \quad \text{or} \quad a \sin\left(-\frac{10\pi}{24} + \frac{\pi}{4}\right)$$

$$= a \sin\frac{\pi}{6}, \quad \text{or} \quad a \sin\left(-\frac{\pi}{6}\right)$$

$$= \frac{a}{2}, \quad \text{or} \quad -\frac{a}{2}.$$

Similarly the velocities are given by

$$v = an \cos(nt + \varepsilon)$$

that is $\quad v = 2a \cos\left(2t + \frac{\pi}{4}\right) \quad \text{or} \quad 10a \cos\left(10t + \frac{\pi}{4}\right)$

solving when $\quad t = \frac{\pi}{24}$

$$v = a, \quad \text{or} \quad -5a$$

when $\quad t = -\frac{\pi}{24}$

$$v = \sqrt{3}a, \quad \text{or} \quad -5\sqrt{3}a.$$

Gathering the results into a tabular form we have

$n = 2$

t	s	v
$-\dfrac{\pi}{24}$	$\dfrac{a}{2}$	$\sqrt{3}a$
0	$\dfrac{a}{\sqrt{2}}$	$\sqrt{2}a$
$\dfrac{\pi}{24}$	$\dfrac{a\sqrt{3}}{2}$	a

$n = 10$

t	s	v
$-\dfrac{\pi}{24}$	$-\dfrac{a}{2}$	$-5\sqrt{3}a$
0	$\dfrac{a}{\sqrt{2}}$	$5\sqrt{2}a$
$\dfrac{\pi}{24}$	$a\dfrac{\sqrt{3}}{2}$	$-5a$

which are illustrated at the top of page 106.

ACCELERATION

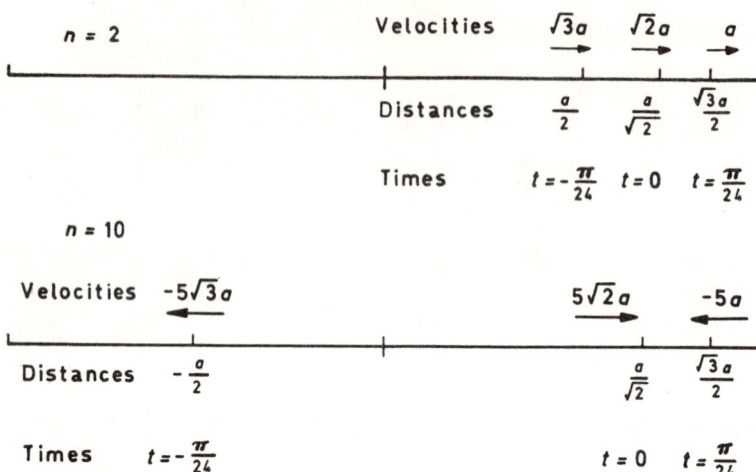

Example 4. A particle P describes a circle, radius a, with constant angular velocity ω rad/s AA' is any diameter of the circle and Q is the foot of the perpendicular from P to AA'. Show that the point Q moves with S.H.M. along AA' about O the centre of the circle.

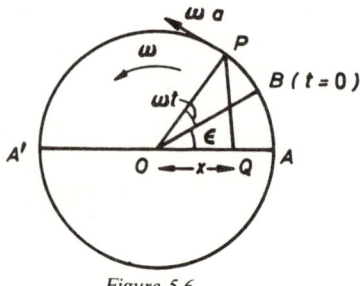

Figure 5.6

Let t be the time in seconds and $OQ = x$. Let B be the initial position of P when $t = 0$ and $\angle BOA = \varepsilon$ (refer to *Figure* 5.6).

Since P is moving with angular velocity ω the angle POB described in any time t is ωt.

$$\angle POA = \angle POB + \angle BOA$$
$$= \omega t + \varepsilon.$$

The velocity of P is of magnitude $a\omega$ tangential to the circle, that is, perpendicular to OP.

SIMPLE HARMONIC MOTION

∴ Speed of $Q(v)$ = Resolved part of $a\omega$ parallel to AOA'
$$= a\omega \sin P\hat{O}A'$$
$$= \omega PQ$$
∴ $\quad v = \omega\sqrt{(a^2 - x^2)}$ by Pythagoras
∴ $\quad v^2 = \omega^2(a^2 - x^2)$

and this shows that Q is moving with S.H.M. because differentiating this expression with respect to x.

$$2v\frac{dv}{dx} = -2\omega^2 x$$

or $\quad\quad$ acceleration $= -\omega^2 x \quad\quad\quad\quad$ ....(i)

i.e. $\quad\quad$ acceleration $\propto -x$.

Note. In this special case the general constant of proportionality n becomes ω the angular velocity of P. O is the centre of the motion and A and A', the limiting points.

Exercises 5d

1. A particle is moving with S.H.M. of amplitude 1·3 m and its periodic time is 5 s. It passes through its central position at time $t = 0$. Find its acceleration and speed when it is (a) 0·5 m (b) −1·2 m from O.

2. A particle is moving in a straight line with S.H.M. about a point O. Find the time of a complete oscillation when
(a) magnitude of its acceleration at 3 m from O is 6 m/s²,
(b) magnitude of its acceleration at 4 cm from O is 16 cm/s²,
(c) magnitude of its acceleration at 1 cm from O is 4 m/s².

3. A particle moving in a straight line with S.H.M. has speeds of magnitudes 8 m/s and 6 m/s, in the same sense at distances 3 m and 4 m, measured in the same direction, respectively, from its central position. Find the period, amplitude and maximum acceleration of the particle.

4. A particle is moving with S.H.M. of period $\pi/2$ seconds, and its maximum speed is 12 m/s. Find its amplitude, and its speed at a distance 1 m from the central position.

5. A point is moving in a straight line with S.H.M. Its speed when moving through its mean position is 12 m/s, and the magnitude of the acceleration at a point 3 m from the mean position is 6 m/s². Find its amplitude and periodic time.

ACCELERATION

6. The amplitude of a particle moving with S.H.M. is 15 m and the speed at a distance of 9 m from the mean position is 6 m/s. Find the maximum speed of the particle and its speed when 12 m from its mean position.

7. A particle P is moving with S.H.M. of amplitude a and periodic time $2\pi/p$. The time t is measured from the instant when P is at an extreme point. Show that the distance x from the mean position is given by $x = a \cos pt$.

8. Prove that a particle whose displacement, x, is given by $x = a \cos nt + b \sin nt$, is moving with S.H.M. [*Hint*: show that $x = a \cos nt + b \sin nt$ is a solution of the equation $\ddot{x} = -n^2 x$.]

If $a = 4$, $b = 3$ and $n = 1$, find the amplitude of the motion and the maximum speed.

9. A particle is moving in a straight line $A'POQA$ with S.H.M. where O is the mid-point of the line and P and Q are two points 6 cm on either side of O. The particle passes P, then 4 s later passes Q. It continues to A and returns to Q in 4 more seconds. Find the period and the amplitude of the oscillation.

10. A particle is moving with S.H.M. of period 12π seconds and amplitude 8 m. B is 4 m from the mean position. The particle passes through the mean position O in the direction OB at time $t = 0$ and makes 1 oscillation. Find the times at which the particle is at B during this oscillation. If the particle had been passing through O, in the direction away from B, at $t = 0$, what would have been the time at which the particle was next at B?

5.7. ANGULAR ACCELERATION

Suppose a particle P moves in a plane so that the line OP joining it to a fixed point O makes an angle θ with a fixed line OA at time t (*Figure 5.7*).

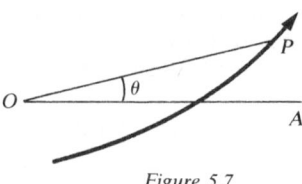

Figure 5.7

Then the angular speed, ω, of P about O has been defined (Section 4.5) as the rate at which $\angle POA$ increases.

EXERCISES

The rate of change of ω is called α its (scalar) angular acceleration and

$$\alpha = \frac{d\omega}{dt} = \frac{d^2\theta}{dt^2}.$$

In the special case of a particle moving in a circle of radius r we have that the arc distance

$$s = r\theta$$

$$\frac{ds}{dt} = r\frac{d\theta}{dt} \quad \text{(as in Section 4.5)}$$

and

$$\frac{d^2s}{dt^2} = r\frac{d^2\theta}{dt^2}$$

Thus the scalar acceleration of the particle along the tangent to the circle is given by the product of the radius and the angular acceleration.

EXERCISES 5*

1. A particle moves in a straight line so that at time t, measured in seconds, the acceleration, in m/s^2, of the particle is $(2 + 6t)$. The particle travels 11 m in the interval from $t = 1$ to $t = 2$. Calculate the speed of the particle when $t = 0$. (London, part)

2. At time t, the position vector of a point P is

$$(2 + \cos 2t)\mathbf{i} + (\sin 2t)\mathbf{j}.$$

Find the velocity and the acceleration of P at time t. Find also the times at which P is moving
(a) directly away from the origin,
(b) directly towards the origin. (London, part)

3. A body moves for 5 seconds with constant acceleration in which time it travels 95 cm. The acceleration then ceases and in the next 5 seconds it travels 170 cm, find its initial velocity and acceleration.

4. A particle moves so that its acceleration is equal to $-n^2x + C$, describe the motion n, C being constant and x its distance from a fixed point O. [Hint: let $y = x - C/n^2$.]

---

* Exercises marked thus, †, have been metricized, see Preface.

ACCELERATION

✗ 5. The acceleration of a train starting from rest is given by $(33/v - 1)$ m/s², where v is its speed in metres per second. If it is moving in a straight line, find an expression for the time t in terms of v. Calculate the greatest possible speed of the train and the time taken to acquire two-thirds of that speed.

6. A train moving in a straight line starts from A with uniform acceleration of 0·1 m/s². After it has attained full speed it moves uniformly for 10 min. It is brought to rest at B by the brakes, which apply a constant retardation of 0·8 m/s² for 20 s. Draw a speed–time curve and from it find the time of the journey and the distance from A to B.

7. The speed of a particle moving in a straight line is equal to $k/\sqrt{s}$ where k is constant and s is the displacement. Prove that the acceleration is equal to Cv^4 and find C in terms of k.

✗ 8. A motor car starts from rest with a uniform acceleration of f. After a certain time the acceleration ceases and it experiences a uniform retardation of $3f$, until brought to rest. If the total distance described is d, and the total time taken is T, prove that $d = 3fT^2/8$.

9. Two particles move along a straight line starting at the same time from the same point A. The first moves with a constant speed of 10 m/s and the second starts from rest with a constant acceleration of 2 m/s. They meet again at the point B. Draw a speed-time graph of the motion of the two particles and from it find the distance AB.

10. The speed of a body is observed at one second intervals to be as follows:

t seconds	0	1	2	3	4	5
v m/s	4	33	56	69	60	29

Draw a speed–time curve of the motion and estimate the total distance travelled and the average speed during the run.

11. A particle with a speed of 120 cm/s starts to decelerate at a constant rate. Ten seconds later another particle starts from rest with a constant acceleration. After the second particle has travelled 337·5 cm the speeds of the two particles are both 45 cm/s. Draw a speed–time graph of the two motions and find when the first particle is at rest and the velocity of the second particle at that instant.

12. The acceleration of a particle is equal to $(g \sin x)/a$. Find its maximum velocity if $v = 0$ when $x = 0$ and a is a constant.

13. A particle oscillates with S.H.M. between two points A and B which are 8 m apart. Its speed at O, the mid-point of AB is 14 m/s.

EXERCISES

What is its speed and the magnitude of its acceleration at a point 3 m from O?

14. A body floating in the sea oscillates up and down with the waves with S.H.M. It rises a vertical distance of 16 cm and falls a distance of 16 cm from its mean position in a periodic time of 4 seconds. Find its greatest speed and greatest acceleration.

15. A particle rotates with constant angular acceleration α (so that $d^2\theta/dt^2 = \alpha$ constant). Derive a set of equations for uniform angular acceleration similar to those already obtained for uniform linear acceleration.

A point moving in a circle with uniform angular acceleration describes angles θ_1 and θ_2 in successive time intervals t_1 and t_2. Show that the angular acceleration is $2(t_1\theta_2 - t_2\theta_1)/t_1t_2(t_1 + t_2)$ and find the angular speed at the beginning of the first interval. At what time before the start of the first interval was the particle at rest?

16. A flywheel is being brought to rest by a constant retarding torque. The flywheel is observed to do n_1 revolutions in the first half-minute and n_2 revolutions in the next half-minute. Find (a) the angular retardation of the wheel, (b) how many more revolutions the wheel does before coming to rest.   (London)
[*Note.* Constant torque produces constant angular acceleration. Also refer to previous question.]

17. Draw the speed–time curve of a particle moving in a straight line with initial velocity u, uniform acceleration f for t seconds. Deduce the formula $v^2 = u^2 + 2fs$.

18. A particle has speed v after travelling distance s. Explain briefly how the time taken from distance s_1 to distance s_2 could be found from the graph obtained by plotting $1/v$ against s.

The observed speeds of a vehicle when the front bumper is at different distances from a mark on a straight road are given by the following table:

Distance (m)	0	20	40	60	80	100	120
Speed (m/s)	16.7	17.9	19.05	20.4	22.2	24.1	26.3

Find the approximate time taken by the vehicle to travel the distance of 120 m.

State briefly how the acceleration at 100 m could be estimated from these data.   (London)†

19. Two railway stations A and B are 6 km apart. A train passes A at 40 km/h, maintains this speed for 5 km, then decelerates

uniformly coming to rest at B. A second train, starting from rest at A 2 min before the first passes A, accelerates uniformly for a time at 10 km h^{-1} min^{-1}, then decelerates uniformly coming to rest at B simultaneously with the arrival of the first train. Sketch the two velocity–time graphs, using the same axes for both.

Show the second train takes $12\frac{1}{2}$ min for the journey and find its maximum speed and its deceleration in kilometres per hour per minute. (London)†

20. A cyclist travelling at 8 m/s observes a motor car just beginning to accelerate uniformly away from rest 40 m distant. If the cyclist's speed remains constant, he can just catch the motor car. Find the acceleration of the car. If he had been cycling at a steady speed of 6 m/s, how near would he have got to the car?

21. A car X, moving with constant acceleration, is travelling at 30 km/h as it passes a fixed point A. After travelling a further quarter of a kilometre the car reaches a speed of v km/h and is thereafter driven at this speed. Six seconds after X passes A another car Y, travelling at 45 km/h in the same direction on the same road and accelerating at $\frac{5}{33}$ m/s^2, also passes A. On reaching a speed of 75 km/h, Y is thereafter driven at this speed. Y passes X 1 km beyond A. Prove that the time taken by Y to cover the kilometre beyond A is 59 seconds.

Calculate the value of v. (London)†

22. A particle falls from rest at a point A under gravity. After it has fallen a distance a, another particle is given a downward speed $\sqrt{8ga}$ from the same starting point A. Show that the two particles collide. Find the time and distance from A at which they collide.

23. Two cars start from rest at the same point and travel in the same direction. The first, accelerating uniformly, acquires a speed of 20 km/h in 4 seconds and then maintains this steady speed. The second car starts 2 seconds after the first and accelerates uniformly at $3\frac{1}{3}$ km h^{-1} s^{-1}. Draw the velocity–time graph for each car.

Find the time during which the first car is in motion before being overtaken by the second and the distance in metres then travelled by either car.

Find also the times, measured from the start of the first car, when the two cars have a relative speed of 12 km/h. (London)†

24. The maximum acceleration of a train is f_1, the maximum deceleration is f_2 and the maximum speed is v. Find the shortest time taken to travel a distance s from rest to rest

when $\qquad$ (a) $s < \dfrac{(f_1 + f_2)v^2}{2f_1 f_2}$,

EXERCISES

(b) $s > \dfrac{(f_1 + f_2)v^2}{2f_1 f_2}.$ (London)

25. A bus-stop is situated $13\frac{1}{3}$ m from the intersection of two straight roads at right angles. A bus passes the stop at a speed of 1 m/s and turns left at the junction, accelerating throughout at $\frac{2}{3}$ m/s². A boy, starting from the bus-stop as the bus turns the corner, runs at a uniform speed in a straight line and catches the bus at a point $46\frac{2}{3}$ m from the junction. Find his speed in metres per second to 3 significant figures.

With what uniform speed would he have had to run along the roads in order *just* to catch the bus? (London)†

26. A particle moves in a straight line against a resistance which varies as the cube of the speed. Prove that the distance travelled in any time is the same as if the particle were to move uniformly during this time with its speed at the mid-point of the distance. (Oxford)

27. A particle describing simple harmonic motion in a straight line reaches its maximum speed at a point O and is momentarily at rest at a point A distant $4a$ from O. B is a point on OA such that $OB = a$. C is a point on BA such that the speed at C is half the speed at B. Show that the acceleration at C is three and a half times that at B.

The least time taken from B to C is one second. Find the period of the motion correct to one-tenth of a second. (London)

28. Two engines, A and B, travelling on parallel rails in the same direction with uniform accelerations f and $\frac{3}{2}f$ respectively, simultaneously pass a signal box P with speeds u and $\frac{1}{2}u$ respectively. The engines are once again level when they pass the next signal box Q, and they immediately brake with uniform decelerations f and F respectively, to stop at the same station.

(a) Show that the greatest distance A gets ahead of B is $u^2/4f$.

(b) Find the speeds of the engines as they pass the signal box Q and the distance PQ.

(c) Show that $F:f = 49:36$. (WJEC)

29. A particle moves in a straight line in such a way that its acceleration is directed towards a fixed point O on the line and is equal to ω^2 times its distance from O. By considering the velocity, show that the motion is oscillatory and that the time of a complete oscillation is $2\pi/\omega$.

Such a particle moves so that its speed is 12 cm/s when its distance is 4 cm from O and 9 cm/s when its distance is 7 cm from O. Find

(a) the period, in seconds, of a complete oscillation,

ACCELERATION

(b) the speed, in centimetres per second, when the particle is 8 cm from O,

giving both results correct to three significant figures. (London)†

30. A point P describes a circle of centre O and radius r with uniform angular velocity ω. Q is the foot of the perpendicular from P upon a fixed diameter of the circle. Find the velocity and the acceleration of Q when it is a distance x from O.

A particle is describing simple harmonic motion in a straight line. Its speeds at distances 2, 3 and 5 m from a point A in the line are 0, 4 and 2 m/s respectively. Find the distance of the centre of the motion from A. Find also the amplitude, periodic time and maximum acceleration of the motion. (London)†

31. The position vector of a particle with respect to the origin at time t is $r = a \sin pt\,i + 2a \cos pt\,j + a \cos pt\,k$. Another particle is describing simple harmonic motion with period $2\pi/p$ between the points $\pm aj$, and when $t = 0$, its acceleration is $-ap^2 j$. Show that the magnitude of the relative velocity of the particles is greatest when they are closest together. (London, part)

32. A particle P is describing simple harmonic motion in the horizontal line $ADCB$, where $AD = DC = \frac{1}{2}CB$. The speed of P as it passes through C is 5 m/s and P is instantaneously at rest at A and B. Given that P performs 3 complete oscillations per second, calculate

  (i) the distance AB,
  (ii) the speed of P as it passes through D,
  (iii) the distance of P from C at an instant when the acceleration of P is 18π m/s²,
  (iv) the time taken by P to go directly from D to A. (AEB)

33. At time t the displacement from the origin of a particle moving along the x-axis under the action of a single force is given by $x = a \cos(\omega t + \alpha)$ where a, ω and α are constants. Find the acceleration of the particle in terms of its displacement from the origin and describe the force acting on the particle. What is the periodic time of this motion?

Two particles, each moving in simple harmonic motion, pass through their centres of oscillation at time $t = 0$. They are next at their greatest distances from their centres at times $t = 2$ s and $t = 3$ s respectively, having been at the same distance from their respective centres at time $t = 1$ s. Show that the amplitudes of their motions are in the ratio $1:\sqrt{2}$.

When will the particles next be at their centres, simultaneously? When will the initial conditions next be repeated?
(WJEC)

6
FORCE

6.1. MOMENTUM

Definition—The *momentum* of a particle is the product of its mass and its velocity, i.e. if a particle of mass m is moving with velocity v, its momentum is the vector quantity mv.

SI unit of momentum ... kg m/s

Example. Find the momentum of a particle of mass 5 kg *moving horizontally at* 20 m/s *in a north-easterly direction.*

The momentum of the particle is in the same direction as the velocity and of magnitude $5 \times 20 = 100$ kg m/s.

If i, j are horizontal unit vectors east and north respectively, then we could express this momentum in the form

$$mv = 5\left(\frac{20}{\sqrt{2}}i + \frac{20}{\sqrt{2}}j\right)$$
$$= 50\sqrt{2}(i + j) \text{ kg m/s}.$$

6.2. FORCE

In 1687, Sir Isaac Newton in his famous work *Principia* laid the foundations of Newtonian Mechanics.* The first two of his laws of motion can be stated in modern terms as follows:

1. Every body continues in its state of rest or with uniform momentum unless acted upon by an external force.

2. The rate of change of momentum of the body is directly proportional to the applied force and takes place in the direction of that force.

* When bodies move at speeds comparable with the speed of light, Newtonian Mechanics, used in this book, breaks down. A new system of mechanics, based on the *Theory of Relativity*, was developed at the beginning of this century. There is close agreement, however, between the two systems for bodies moving at 'ordinary' speeds.

FORCE

Thus Newton suggests that the force on a body is to be measured in terms of the rate at which its momentum changes.

Also it is implied that each force has its own effect. This is equivalent to stating that the effect of several forces is the same as that of their resultant found by successive application of the triangle rule. (That this is necessary, if we are to treat force as a vector quantity, was indicated in Section 2.1.) The justification for this assumption for forces is that it works in practice.

Definition—The resultant *force* acting on a particle is a vector quantity equal to the rate of change of momentum, i.e., the force P is given by

$$P = d(mv)/dt.$$

SI unit of force ... kg m/s^2 or newton (N).

6.3. FORCE AND ACCELERATION

The particles met with in this book are of constant mass. For such particles

$$P = d(mv)/dt$$
$$= m\, dv/dt$$

$\therefore$ $\boxed{P = mf.}$

This will be the basic equation used in determining the motion in many of our problems.

Example 1. A particle of mass 2 tonne *is moving in a straight line with an acceleration of* 12 m/s^2. *Find the resultant force acting on it.*

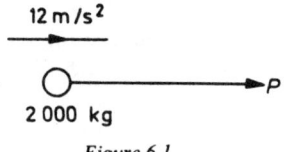

Figure 6.1

Since $P = mf$, the force will be in the same direction as the acceleration (refer to *Figure 6.1*), and

$$P = 2\,000 \times 12$$
$$= 24\,000 \text{ N}.$$

FORCE AND ACCELERATION

The resultant force acting on the particle is of magnitude 24 000 N and in the direction of the acceleration.

Example 2. *A constant force of* 60 N *is applied to a particle of mass* 5 kg. *Find the acceleration it produces.*

If the particle starts from rest, find the speed acquired in 10 *seconds.*

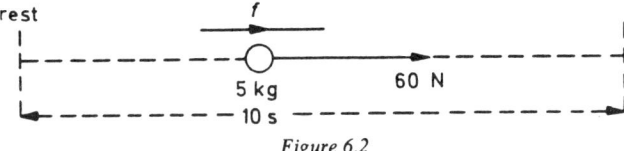

Figure 6.2

Since $P = mf$, the acceleration will be in the direction of the applied force (refer to *Figure 6.2*), and

$$60 = 5f.$$

$$\therefore \quad f = 12 \text{ m/s}^2.$$

The force being constant, the acceleration will be constant and

$$v = u + ft$$
$$v = 0 + 12 \times 10$$
$$= 120 \text{ m/s}.$$

The acceleration produced is of magnitude 12 m/s² and in the direction of the force. The particle acquires a speed of 120 m/s in 10 s.

Example 3. *A particle of mass* 2 kg *is moving under the action of two forces. They are of magnitude* 6 N *and* 4 N *acting at* 60 *degrees to each other. Find the acceleration of the particle.*

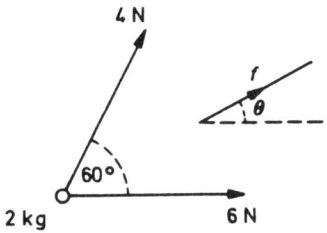

Figure 6.3

Let the resultant acceleration of the particle be of magnitude f making an angle θ with the 6 N force. (Refer to *Figure 6.3*.)
Since $P = mf$,

perpendicular to the 6 N force $\quad\quad 4 \sin 60° = 2 \times f \sin \theta$

parallel to the 6 N force $\quad\quad 6 + 4 \cos 60° = 2 \times f \cos \theta.$

From these two equations
$$f \sin \theta = \sqrt{3}, \quad\quad \ldots \text{(i)}$$
$$f \cos \theta = 4. \quad\quad \ldots \text{(ii)}$$

Squaring and adding (i) and (ii)
$$f^2 = (\sqrt{3})^2 + 4^2$$
$$= 19$$
$$\therefore \quad f = 4{\cdot}359 \text{ m/s}^2.$$

Dividing (i) by (ii)
$$\tan \theta = \sqrt{3}/4$$
$$= 0{\cdot}433 \quad \text{and } \sin \theta \text{ is positive}$$
$$\therefore \quad \theta = 23° \, 25'.$$

The acceleration of the particle is of magnitude $4{\cdot}36$ m/s² in a direction making an angle of 23° 25′ with the 6 N force. (Note that this result could equally well have been obtained by finding the resultant force and then applying $P = mf$.)

EXERCISES 6

1. A rifle bullet of mass 10 g is moving horizontally with an acceleration of $1{\cdot}6 \times 10^5$ m/s². Find the resultant force acting on it.

2. A particle of mass 160 kg is acted upon by a single force of magnitude 640 N. Find the acceleration produced.

3. The resultant force acting on a particle is 960 N. Its acceleration is 6 m/s². Find the mass of the particle.

4. A particle of mass 1 kg moves in a straight line under the action of a single force F. F is of magnitude $8s$ N where s metres is the distance of the particle from a fixed point O in the line, and is directed towards O. Find the acceleration of the particle when it is 8 cm from O.

How would you describe the motion of this particle?

EXERCISES

5. A 1 000 kg car moves along a road under the action of a constant net force of 5 000 N. Find its acceleration and the time taken to travel 3 m from rest.

6. A 1 000 kg trailer is being towed along a level road against horizontal resistances totalling 700 N. The tow rope, also horizontal, can withstand tensions of up to 3 000 N. If no other forces affect the horizontal motion, find the maximum acceleration that can be imparted to the trailer.

7. The only two forces acting on a particle are equal in magnitude and opposite in direction. What can be said about the motion of the particle?

8. A particle of mass 3 kg moves under the action of two forces of 2 N and 4 N which act at 60 degrees to each other. Find the components of the acceleration of the particle which are parallel and perpendicular to the 4 N force.

9. Three forces of 160, 320 and 416 N act in one plane on a particle. The angle between any two of them is 120 degrees. If the mass of the particle is 56 kg, find the acceleration these forces give to it.

10. A particle of mass 4 kg is acted upon by a force $32i + 64j$ N. Find the acceleration it produces.

11. A particle of mass m moves so that at time t its velocity is $t^2 a + b$, where a and b are constant vectors. Find the resultant force acting on it at time t.

12. A particle of mass m is moving under the action of a single force. Its position vector with respect to the origin at time t is

$$r = a \sin pt\, i + 2a \cos pt\, j + a \cos pt\, k.$$

Show that the particle is moving in a plane and find an expression for the force acting on it. (London, part)

13. A particle of mass 2 kg moves so that after t seconds it has a displacement of $(\cos 2t)i + (\sin 2t)j$ m from a fixed point. Find the magnitude of the resultant force acting on it at time t, and verify that the force is always at right angles to the direction of motion.

14. Two particles A and B, with masses 3 and 4 units respectively, move so that at time t they have position vectors r_A and r_B where

$$r_A = (t - 1)i + \sin \pi t\, j + (t^2 + 2)k,$$

$$r_B = \sin \frac{\pi t}{4} i + (t^3 - 8)j + 3t k.$$

Find
- *(i) the total kinetic energy of the particles at time t,
- (ii) the magnitude of the resultant force acting on A at time t,
- (iii) the cosine of the angle which the acceleration of the particle A makes with its path at $t = 1$.

Show that A and B eventually collide and find the position vector of the point of collision. (AEB)

---

* This part of the question to be omitted on first reading; Section 12.5 refers to kinetic energy.

7
PHYSICAL LAWS

7.1. INTRODUCTION

So far, quantities (such as speed, acceleration, force etc.) have been defined to describe the motion of a particle. Practical problems, however, may involve not only the description of its motion but the prediction of future behaviour. Given the mass of the particle and the initial conditions then, *provided the forces acting are known*, this prediction can be made. (*See*, for example, *Example 2*, Section 6.3.) So we need to know the forces which act in different situations.

Scientists provide this information in the form of physical laws which are the result of observation and experiment. Some of the laws relevant to the work in this book are discussed in the sections that follow.

7.2. GRAVITATION

From observations of astronomical bodies and of bodies on Earth, Newton deduced that every body attracts every other body with a force called gravitational attraction.* The force due to the gravitational attraction of the Earth is called the *weight* of the body. Near the Earth's surface the magnitude of this weight (W) is found to be directly proportional to the mass (m) of the body

$$W = g \times m.$$

The constant of proportionality g (the same for all bodies) is called the *local gravitational constant*. It varies a little over the Earth's surface but it is approximately $9 \cdot 8 \text{ m/s}^2$ (g has the dimensions of an acceleration and is, in fact, the acceleration of a particle in free fall near the Earth's surface; hence it is often called the *acceleration of gravity*).

* Newton's universal law of gravitation states that the magnitude of the gravitational attraction F between two bodies of mass, m_1, m_2, distance r apart, is given by $F = Gm_1m_2/r^2$ (G being the universal gravitation constant). In the case of a body on the Earth's surface its weight will be $W = GmE/R^2$ (E the mass of the earth, R its radius) $=.mg$ since GE/R^2 is approximately constant over the Earth's surface.

Example. A stone of mass 2 kg lies on the ground. With what force is it being pulled by the Earth?

The weight of the stone,
$$W = mg = 2 \times 9\cdot 8$$
$$= 19\cdot 6 \text{ N}.$$

This is the force exerted by the Earth on the stone.

7.3. HOOKE'S LAW
(for a spring or elastic thread)

When a spring is stretched, the spring pulls back with a force called the *tension* in the spring. Experiments show that the magnitude (T) of this tension is, within certain limits, directly proportional to the fractional extension of the spring.

i.e. $$T = \lambda \times x/l$$

where x is the extension of the spring, l its unstretched length.

The constant of proportionality λ has the dimensions of a force and is called the *modulus* of the spring. It is the same for different springs of the same construction but of different lengths.

Alternatively we can write $T = kx$. Here k ($=\lambda/l$) is called the force constant or stiffness of the spring and is different for each spring.

Example. An elastic thread of modulus 10 N is 2 m in length when unstretched. Find the tension in the thread when it is stretched to a length of 3 m.

The modulus of the thread is 10 N and the extension is 1 m. Hence by Hooke's law the tension

$$T = \lambda(x/l) = 10(\tfrac{1}{2})$$
$$\therefore \quad T = 5 \text{ N}.$$

Note that if a string is inelastic then Hooke's law does not apply, though there will of course be a tension in the string as long as it is tight.

If a string is light and passes over smooth contacts, then the tension will have the same magnitude throughout its length. For a

FRICTION

small length it will have negligible mass and, if T_1, T_2 are the tensions at its ends, $P = mf$ gives $T_1 - T_2 = 0$. This will hold whether the string is at rest or in motion.

7.4. FRICTION

When a body A is in contact with another body B, they exert forces on one another. In particular, B will push A with a force that can be regarded as made up of two components: a component along the common tangent of the surfaces in contact, called the *frictional reaction*, and a component perpendicular to the surfaces, called the *normal reaction*. (Refer to *Figure 7.1*.)

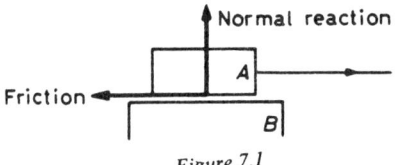

Figure 7.1

If no attempt is made to slide one surface over the other, or if the surfaces are smooth, the frictional component is of zero magnitude. If, however, such an attempt *is* made, the friction is not zero and its direction is such as to oppose the tendency to slide. Experiments show that the friction will increase to prevent sliding up to a maximum or "limiting" value and then sliding begins.

It is found that the magnitude (F_{max}) of this limiting frictional force is directly proportional to that of the normal reaction (N), i.e.

$$F_{max} = \mu \times N.$$

The constant of proportionality μ is dimensionless and is called the *coefficient of friction*. It has different values for different pairs of surfaces.

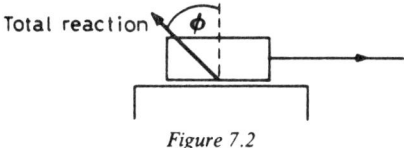

Figure 7.2

An alternative way of expressing this law is obtained by considering the behaviour of the resultant of the friction and the normal reaction. This force is called the total reaction between the two surfaces. (Refer to *Figure 7.2*.)

123

PHYSICAL LAWS

If ϕ is the angle between the total reaction and the normal to the surfaces, then as the friction increases, ϕ will increase up to a maximum λ (say) when sliding will begin. λ is called the *angle of friction*. (It is left to the reader to show that $\tan \lambda = \mu$.)

Note that it is an important point that friction opposes *relative* motion between the surfaces. For instance, as a conveyor belt starts up, a frictional force arises between the belt and a box on it so as to move the box along with the belt.

In practice, when sliding begins there is a slight drop in the magnitude of the frictional force. So there are two coefficients of friction: the coefficient of statical friction (μ_1 say) and the coefficient of sliding friction (μ_2) and $\mu_1 > \mu_2$. In problems in this book the reader should assume that the values of μ (or λ) quoted are the appropriate ones for the situations described.

7.5. NEWTON'S THIRD LAW OF MOTION

This can be stated in the form. Whenever a body A exerts a force on a body B, then B exerts an equal and opposite force on A.

The pairs of forces to which the law applies might, for example, be the forces between electrostatic charges, the gravitational attractions between bodies or the forces which arise between bodies which come into contact. Some examples follow. (It should be noted that the equal and opposite forces do not "cancel each other out"; indeed each force contributes to the motion of the particle to which it is applied.)

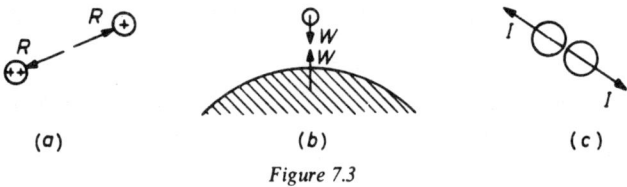

Figure 7.3

Figure 7.3a shows the repulsive forces between like electrostatic charges. The forces will be equal whatever the magnitude of the charges and will cause them to fly apart (provided no other forces intervene).

Figure 7.3b shows the gravitational forces between a particle and the Earth. The particle will, if free, fall towards the Earth under its weight W, and the Earth will be attracted by an equal force W.

Figure 7.3c shows the impulsive reactions between colliding particles. These will be equal whatever the masses of the particles.

NEWTON'S THIRD LAW OF MOTION

In *Figure 7.4a* the forces between a box and a table are shown. Since they are at rest the force N on the box is balanced by its weight, while the force N on the table, together with its weight, is balanced by reactions from the ground.

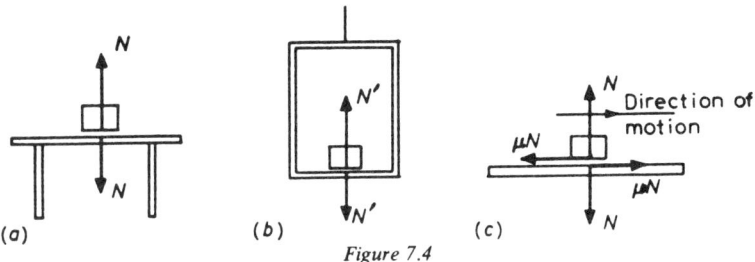

Figure 7.4

Figure 7.4b shows similar forces acting between a box and a lift. In this case, however, if the lift accelerates upwards the force N' on the box will be greater than its weight, while there will be a large tension in the lift rope to overcome the weight of the lift and the force N' acting downwards on it.

Figure 7.4c shows the forces between a box and a plank on which it is sliding. If the box is accelerating, it must be because it is being pulled by a force greater than the friction μN, while, if free to do so, the plank will move in the same direction under an equal and opposite force μN.

Summarizing we have that the forces that can act on a particle include its weight, tensions or thrusts in rods, strings etc. and the reactions between surfaces in contact. Concerning the magnitude of these forces we have seen that:

near the Earth's surface	$W = mg$,
in a spring or elastic thread	$T = \lambda(x/l)$,
between rough surfaces	$F \leqslant \mu N \quad (\phi \leqslant \lambda)$,
and if two bodies interact the forces between them are equal and opposite.	

With this information we can now decide what forces will act on a particle in a given situation.

PHYSICAL LAWS

Example 1. A particle of mass 7 kg is rotating in a horizontal circle on the end of an inelastic string. Indicate, on a diagram, the forces acting on the particle and any physical laws that apply.

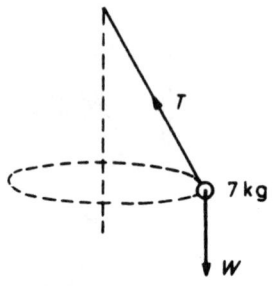

Figure 7.5

Refer to *Figure 7.5*.
For a particle near the Earth's surface
$$W = mg_0$$
$$\therefore \quad W = 7g \text{ N}.$$
(This is usually written directly on to the diagram.)

Example 2. A particle of mass 5 kg lies on a smooth plane inclined to the horizontal. It is attached to the end of an elastic string the other end of which is fastened to a point higher up the plane. The string, of modulus 100 N and natural length 2 m, is extended to a length of 3 m. Indicate, on a diagram, the forces acting on the particle and any physical laws which apply.

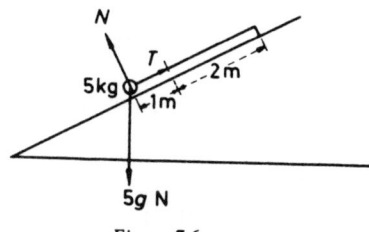

Figure 7.6

Refer to *Figure 7.6*.
The modulus of the thread is 100 N and its extension is 1 m.

Hence by Hooke's law
$$T = \lambda(x/l)$$
$$= 100(\tfrac{1}{2})$$
$$\therefore \quad T = 50 \text{ N}.$$

Example 3. A particle of mass 8 kg lies at rest on a rough inclined plane. Indicate the forces acting on the particle and any physical laws that apply.

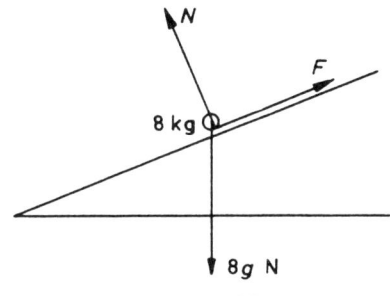

Figure 7.7

Refer to *Figure 7.7*.
As the particle is not about to slip $F < \mu N$.

Example 4. A particle of mass m slides down the rough inclined face of a wedge, which is free to move on a smooth horizontal surface. The mass of the wedge is M and the coefficient of friction between particle and wedge is μ. Indicate, on two diagrams, the forces acting on the particle and the forces acting on the wedge, giving any physical laws that apply.

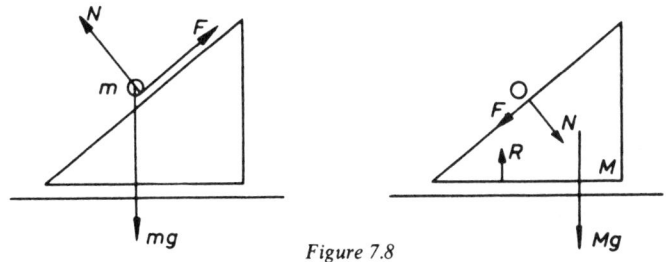

Figure 7.8

Refer to *Figure 7.8*.
Since the particle is sliding down the wedge $F = \mu N$.

127

PHYSICAL LAWS

Exercises 7a

Draw diagrams to illustrate each of the situations described as follows: Indicate clearly all the forces acting on the *particle* mentioned and any physical laws which apply.

1. A particle of mass 3 kg hangs on the end of a spring. The modulus of the spring is 6 N, its natural length is 4 m and it is extended by 2 m.

2. A particle of mass 5 kg is pulled along a smooth horizontal surface by a force of 13 N acting parallel to the surface.

3. A particle of mass 3 kg is describing horizontal circles on the inside of a smooth hemispherical bowl fixed with its rim horizontal.

4. A particle of mass 2 kg is about to slip on a rough horizontal plane when pulled by a horizontal force Q. The coefficient of (static) friction between the particle and the plane is $1/2$.

5. A particle of mass 8 kg has been projected horizontally so that it now slides along a rough horizontal plane. The angle of friction between the particle and plane is 30 degrees.

6. A particle of mass 2 kg describes horizontal circles on the end of an elastic string (modulus 980 N, unstretched length 50 cm). The string is inclined at 60 degrees to the vertical and is stretched to a length of 52 cm.

7. A particle of mass m is in the form of a small smooth ring attached to the end of an inextensible string of length l. The ring is threaded on a circular wire, of radius r, fixed in a vertical plane. The other end of the string is fastened to the topmost point of the wire and the ring is in equilibrium with the string taut ($l < 2r$).

8. A particle of mass m lies on a rough plane inclined at an angle θ to the horizontal. It is just prevented from slipping down the plane by a force Q acting parallel to the plane. The coefficient of friction between particle and plane is μ.

9. A particle of mass 3 g lies at rest on a rough horizontal table and is connected by a light inextensible string passing over a smooth pulley at the edge to a mass of 2 g hanging freely. (Show the forces acting on both particles.)

10. On a smooth fixed plane inclined at 45 degrees to the horizontal, a smooth wedge of mass 7 kg and angle 45 degrees is placed so that its upper face is horizontal. A particle of mass 3 kg is placed on this horizontal face and the wedge begins to move. (Show the forces acting on both particle and wedge.)

Further practice may be obtained by carrying out the same process with the situations described in the exercises of Chapters 8–11.

7.6. SUGGESTED APPROACH TO PROBLEM SOLVING

The following systematic approach can be of help in solving many problems concerning the motion of particles.

(a) Read the question carefully and draw a clear diagram of the situation described.
(b) Transfer all the information given to the diagram, making sure that all units belong to a coherent system (in our case SI units).
(c) Mark all the forces acting on the body being considered and note any physical laws that apply.
(d) Apply $P = mf$ to find the acceleration. (It may be necessary to apply it in two or three directions at right angles.)
(e) Use the appropriate equations for the kind of acceleration found, obtaining them by integration if necessary.

It is not suggested that this scheme should be rigidly adhered to or that it applies to all problems. Special techniques for individual types of problem are explained in succeeding chapters.

Example 1. A train of mass 200 *tonne is travelling at* 36 km/h *on a level track. The tractive force exerted by the engine is constant and of magnitude* 100 kN, *while the resistances (also constant) amount to* 40 N/t. *Find the time taken to reach a speed of* 72 km/h. *Find also the total normal reaction between the train and the track.*

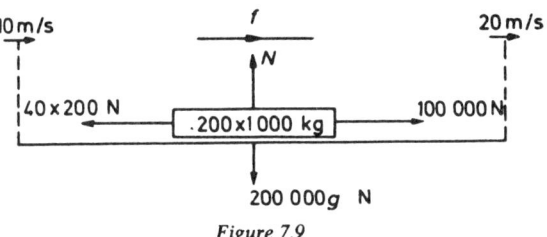

Figure 7.9

Since $P = mf$, referring to *Figure 7.9*, we have

$$\text{horizontally} \quad 100\,000 - 8\,000 = 200\,000 f \quad \ldots\text{(i)}$$
$$\text{vertically} \quad N - 200\,000\,g = 0 \quad \ldots\text{(ii)}$$

From (ii)
$$N = 200\,000\,g \text{ N}$$
$$= 1\,960 \text{ kN}.$$

From (i)
$$f = \frac{92\,000}{200\,000}$$
$$= 0{\cdot}46 \text{ m/s}^2.$$

Since the tractive force is constant, this acceleration will be uniform and hence the formula $v = u + ft$ applies.

∴ $\qquad 20 = 10 + 0{\cdot}46t$

giving $\qquad t = 21{\cdot}74$ s.

The normal reaction is 1 960 kN and the train takes a further 21·7 s to reach a speed of 72 km/h.

Example 2. A particle of mass 2 kg is suspended by an elastic thread of modulus 15 N and natural length 15 cm. If the particle is projected vertically downwards, find its acceleration when the thread is stretched to a total length of 25 cm.

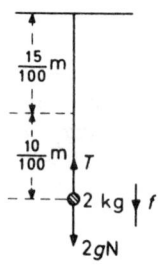

Figure 7.10

The modulus of the thread is 15 N and its extension 0·1 m. Hence by Hooke's law,
$$T = \lambda(x/l)$$
$$= 15(10/15)$$
$$= 10 \text{ N}.$$

Since **P** = m**f**, referring to *Figure 7.10*, we have vertically downwards
$$2g - T = 2f$$
∴ $\qquad 19{\cdot}6 - 10 = 2f$
∴ $\qquad f = 4{\cdot}8 \text{ m/s}^2.$

SUGGESTED APPROACH TO PROBLEM SOLVING

At the instant when the thread is 25 cm long, the particle has a downward acceleration of 4·8 m/s². (Note that this is the instantaneous acceleration for this position. To investigate the general motion of the particle it would be necessary to find the acceleration in a typical position say at a distance s from some fixed point.)

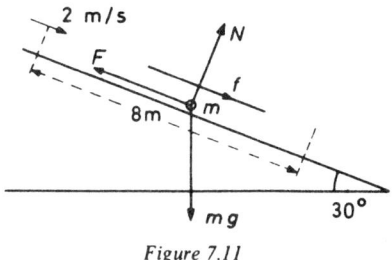

Figure 7.11

Example 3. A particle is projected at 2 m/s down a rough plane inclined at 30 degrees to the horizontal. If the coefficient of friction between particle and plane is $1/\sqrt{3}$, find the time taken to slide 8 m.

The particle is sliding. $\therefore$ $F = \mu N = N/\sqrt{3}$.
Since $P = mf$, referring to *Figure 7.11*, we have

perpendicular to the plane $\qquad N - mg \cos 30° = 0 \qquad$ .... (i)

down the plane $\qquad mg \sin 30° - N/\sqrt{3} = mf \qquad$ .... (ii)

From (i) $N = mg\sqrt{3}/2$ and substituting this value in (ii),

$$mg \cdot \frac{1}{2} - \left(mg\frac{\sqrt{3}}{2}\right) \cdot \frac{1}{\sqrt{3}} = mf$$

$\therefore \qquad\qquad\qquad f = 0.$

This means that the particle must be moving with constant speed, so

$$s = vt$$
$\therefore \qquad\qquad 8 = 2t$
$\therefore \qquad\qquad t = 4 \text{ s}.$

The particle slides 8 m in 4 s.

PHYSICAL LAWS

Example 4. A particle of mass m, initially at rest, moves on a smooth horizontal surface in a straight line under a variable force which has magnitude 2at + b after time t (a and b being constants). Find an expression for the speed of the particle at this time.

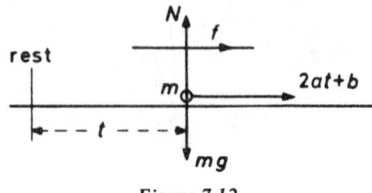

Figure 7.12

Figure 7.12 shows the situation at time t. Since $\mathbf{P} = m\mathbf{f}$,

horizontally $\quad 2at + b = mf$

vertically $\quad N - mg = 0$

∴ $\quad f = (2at + b)/m$

i.e. $\quad dv/dt = (2at + b)/m$.

Integrating with respect to t,

$$v = (at^2 + bt)/m + C.$$

But when $t = 0$, $v = 0$, so that $0 = 0 + C$, i.e. $C = 0$

∴ $\quad v = (at^2 + bt)/m$

and this is the required expression.

(Note that when no units are given it is assumed that all quantities quoted are already in a coherent system of units.)

It is recommended that the following questions should be studied in detail.

Exercises 7b

1. A particle is placed at the tóp of a smooth inclined plane and released. If the plane is 2 m long and of inclination 1 in 196, find the speed of the particle when it reaches the bottom.

2. An elastic string, of modulus 2g N and natural length 1 m, has one end fixed while the other end B is pulled vertically downwards until the extension is 0·5 m. While stationary in this position a particle of mass 1 kg is gently attached to B and is released. What happens?

SUGGESTED APPROACH TO PROBLEM SOLVING

3. A sled, which with its load has a mass of 50 kg, is moving horizontally at an instantaneous speed of 2 m/s. The motion is due to a horizontal force of 170 N acting against resistances totalling 20 N. Find the speed of the sled 3 seconds later.

4. A spring (modulus 16 N, unstretched length 2 m) is fastened at one end to a smooth horizontal table. To the other end a particle of mass 4 kg is attached. This particle is now pulled along the table to stretch the spring and is released. Find the acceleration of the particle when at a distance s metres from the unstretched position. Hence show that the motion is simple harmonic and find its period.

5. A mass of 1 tonne is pulled up a rough slope by a constant force of 17 kN. The slope is at 30 degrees to the horizontal and the coefficient of friction is $1/\sqrt{3}$. Find the time taken to travel 100 m from rest.

6. A particle of mass 2 kg is pulled from rest along a smooth horizontal surface by a force which has magnitude $(t^2 + 1)$ N after t seconds. Find its acceleration at this time and its speed when $t = 3$. Comment on *all* the forces acting.

7. The vertical descent of a lift-cage is undertaken in three stages. During the first stage the lift uniformly accelerates from rest at $5k$ m/s^2, during the second stage it moves at a constant speed of 10 m/s and during the third stage it uniformly retards at $2k$ m/s^2 until it comes to rest.
  (i) Express, in terms of k, the times taken during the first and third stages of the descent.
  (ii) Given that the total distance covered by the lift during the descent is 350 m and that this distance is covered in 40 s, calculate the value of k.

A tool-box of mass 5 kg was placed on the floor of the lift-cage before the descent.
  (iii) Find, in newtons, the force exerted by the floor of the lift-cage on the tool-box during each stage of the descent.

(Take the acceleration due to gravity to be 10 m/s^2.) (AEB)

8

MOTION IN A STRAIGHT LINE UNDER CONSTANT FORCES

8.1. GENERAL METHOD

IF all the forces acting on a body are constant, then (from $P = mf$) its acceleration will be constant. Hence, the formulae for uniform acceleration, developed in Section 5.3, can be applied. They are

$$v = u + ft$$
$$v^2 = u^2 + 2fs$$
$$s = ut + \tfrac{1}{2}ft^2.$$

Illustrative examples are given below but the reader should first refer again to Examples 1, 2, 3 and 4 of Section 5.3.

Example 1. A nail of mass 4 g is hammered horizontally into a fixed block of wood. The first blow of the hammer imparts a speed of 180 m/s to the nail and it moves 6 mm into the wood. Find the resistance (assumed constant) that it encounters.

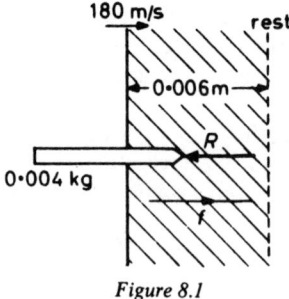

Figure 8.1

Referring to *Figure 8.1* and applying $P = mf$ horizontally,

$$-R = 0{\cdot}004f,$$

∴
$$f = -250\,R.$$

GENERAL METHOD

Since R is constant, f is constant and hence the formula $v^2 = u^2 + 2fs$ applies.

$$\therefore \qquad 0 = 180^2 + 2(-250\,R)\,.\,0{\cdot}006$$

giving $\qquad R = 10\,800\text{ N}.$

The resistance to the nail is equivalent to a constant force of 10·8 kN.

Example 2. A stone is thrown vertically downwards with a speed of 21 m/s from the top of a building of height 180 m. Another stone is thrown vertically upwards from the bottom of the building with the same speed. Find the time taken by the first stone to reach the ground and show that in this special case the second stone takes the same time. (Neglect the effect of air resistance.)

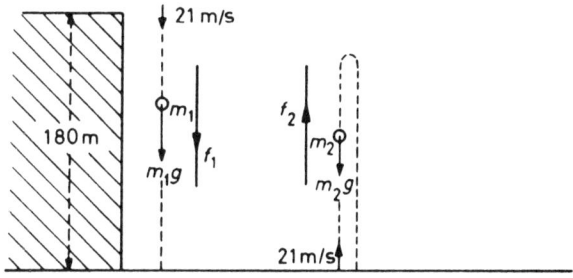

Figure 8.2

Let the stone thrown downwards have mass m_1. Then the only force acting on it is its weight $m_1 g$, and from $P = mf$, its downward acceleration $f_1 = g$ (refer to *Figure 8.2*). This acceleration is constant, so $s = ut + \tfrac{1}{2}ft^2$ applies.

$$\therefore \qquad 180 = 21t + \tfrac{1}{2} 9{\cdot}8 t^2$$

giving $\qquad t = 4{\cdot}29 \quad \text{or} \quad -8.57.$

Hence the first stone takes 4·29 s to reach the ground. (The second root of our equation is inadmissible.)

Now consider the stone thrown upwards (of mass m_2 say). Again the only force acting on it is its weight $m_2 g$, so its upward acceleration $f_2 = -g$. This acceleration is constant, so $s = ut + \tfrac{1}{2}ft^2$,

$\therefore \quad$ at the ground $\qquad 0 = 21t - \tfrac{1}{2} 9{\cdot}8 t^2,$

giving $\qquad t = 4{\cdot}29 \quad \text{or} \quad t = 0.$

135

MOTION IN A STRAIGHT LINE UNDER CONSTANT FORCES

Hence, the second stone takes 4·29 s to reach the ground, the same as the first stone. (The second root of our equation refers to the moment of projection.)

It is important to note that in these formulae s stands for the distance from some origin O, *not* for the distance travelled. Hence the formula $s = ut + \frac{1}{2}ft^2$ will give us the two times when the particle is distant s from O, provided the acceleration remains constant in magnitude and direction throughout the motion.

Example 3. Two particles of mass 1 kg and 2 kg are connected by a light inelastic string. The particles are held with the 1 kg mass on a rough horizontal table and the 2 kg mass hanging vertically. The string is taut and passes over a smooth pulley at the edge of the table. The system is now released from rest. If the coefficient of friction between the 1 kg mass and the table is $\frac{1}{2}$, find the tension in the string and the speed acquired by each particle in 1 second. (Assume that in this time, the 1 kg mass does not reach the edge of the table, nor the 2 kg mass the floor.)

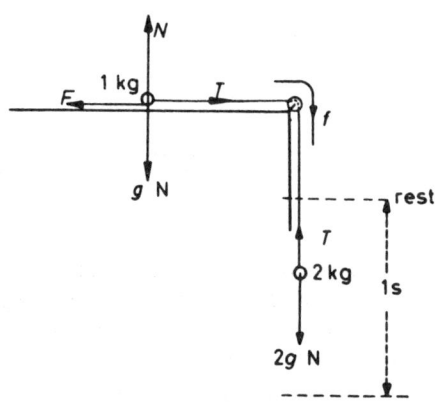

Figure 8.3

Refer to *Figure 8.3*.
When the particles are in motion $F = \mu N$,
$$\therefore \qquad F = \tfrac{1}{2}N.$$
Consider the 1 kg mass. Since $P = mf$,

$$\text{horizontally} \quad T - \tfrac{1}{2}N = 1f \qquad \ldots\text{(i)}$$
$$\text{vertically} \quad N - 1g = 0 \qquad \ldots\text{(ii)}$$

GENERAL METHOD

Consider the 2 kg mass. Since $P = mf$,

$$\text{vertically} \quad 2g - T = 2f \quad \ldots \text{(iii)}$$

From (ii) $N = g$, and substituting this value in (i) gives

$$T - \tfrac{1}{2}g = f \quad \ldots \text{(a)}$$

Multiplying (a) by 2 and subtracting (iii), $3T - 3g = 0$.

$$\therefore \quad T = g$$
$$\therefore \quad T = 9 \cdot 8 \text{ N}.$$

Adding (iii) and (a) $2g - \tfrac{1}{2}g = 3f$

$$\therefore \quad f = g/2$$
$$= 4 \cdot 9 \text{ m/s}^2.$$

Since this acceleration is uniform, $v = u + ft$,

$\therefore$ after one second $\quad v = 0 + 4 \cdot 9 \times 1$

$\therefore \quad\quad\quad\quad\quad\quad\quad v = 4 \cdot 9$ m/s.

The tension in the string is 9·8 N and in 1 second each particle acquires a speed of 4·9 m/s.

Exercises 8a

1. A particle of mass 8 kg is pulled from rest along a smooth horizontal surface by a constant horizontal force of magnitude 320 N. Find the distance travelled in 2 seconds.

2. A stone is thrown vertically downwards with a speed of 100 cm/s. Neglecting the effect of air resistance, find how far it falls in one-tenth of a second and the speed acquired.

3. A car of mass 0·5 tonne moves along a level road against a constant resistance of 30 N. If the engine exerts a constant tractive force of 1 kN, find the time taken to increase speed from 18 km/h to 54 km/h.

4. A particle is projected at 14 m/s along a rough horizontal surface. The coefficient of friction between particle and surface is $\tfrac{1}{5}$. How far does the particle travel before coming to rest?

5. A bullet of mass 10 g, travelling horizontally at 600 m/s, enters a fixed block of wood. If it penetrates the wood to a depth of 15 cm, find the resistance (assumed constant) that it encounters.

6. A particle slides along a rough horizontal surface pulled by a horizontal force equal in magnitude to its own weight. The coefficient of friction between particle and surface is μ ($\mu < 1$).

MOTION IN A STRAIGHT LINE UNDER CONSTANT FORCES

Show that, if t is the time taken to travel a distance s from rest, then $gt^2(1-\mu) = 2s$.

7. A 4 kg mass on a smooth horizontal table is connected by a light inextensible string, which passes over the edge of the table, to a 10 kg mass hanging freely. The 4 kg mass is held at a distance of 50 cm from the edge of the table and then released. Find the time it takes to reach the edge of the table and the speed acquired.

8. A particle slides from rest down a rough plane inclined at $\tan^{-1}\frac{3}{4}$ to the horizontal. If it covers the first 7 m in 5 seconds, find the coefficient of friction between the particle and the plane.

9. A particle is projected vertically upwards at 280 cm/s. At what times will it be at a height of 30 cm above the point of projection? Find its speed at those times.

10. Consider again the situation described in Example 2. Show that for a building of height h, the two stones will take equal times if each is thrown with speed $\frac{1}{2}\sqrt{gh}$.

11. A particle is projected directly up a rough inclined plane with speed u. The plane makes an angle α with the horizontal and the coefficient of friction between the particle and plane is μ. If after travelling a distance s, the particle is moving with speed v, find a formula for s in terms of u, v, μ and α.

12. A particle, of mass 4 kg, on a rough plane inclined at 30 degrees to the horizontal is attached to one end of a light inelastic string. The string passes over a small smooth pulley at the top and carries at its other end a particle of mass 5 kg hanging freely. The coefficient of friction between the 4 kg mass and the plane is $1/2\sqrt{3}$. Find the acceleration of the system when released and the speed acquired in 0·5 second.

13. A particle of mass 3 kg on a smooth horizontal plane is acted upon by a force F. $F = 9i + 6j$ N, where i, j are horizontal unit vectors at right angles to each other. If, initially, the particle is at rest, show that after 2 seconds its displacement is $6i + 4j$ metres.

14. i, j are horizontal unit vectors in the directions east and north respectively. Forces F_1, F_2 act on a 2 kg particle at rest on a smooth horizontal table $F_1 = 2i + 4j$ N and $F_2 = 3j - 9i$ N. Find the velocity of the particle, in magnitude and direction, after 3 seconds.

15. A particle of mass m moves from A to B under the action of constant forces. A has position vector $a - b$, B has position vector $a + b$ and the particle takes time t to travel from A to B. If it starts from rest at A, show that the resultant force R acting on the particle is given by

$$t^2 R = 4mb.$$

COMPOUND PROBLEMS

8.2. COMPOUND PROBLEMS

If the forces acting change suddenly during the motion of a particle, each part of the motion should be considered separately. The speed of the particle at the "change-over" forms the link between the two parts of the motion and must usually be found.

Also, in this kind of uniform acceleration problem it may sometimes be an advantage to use a speed-time curve rather than the standard equations. This method is used in one of the examples that follow.

Example 1. Two particles of mass m and 2m are connected by a light inextensible string passing over a smooth pulley. The system is held with the mass m on the ground and the string taut so that the mass 2m hangs at a height h above the ground. It is then released. Find the maximum height reached by the mass m and the time taken to get there.

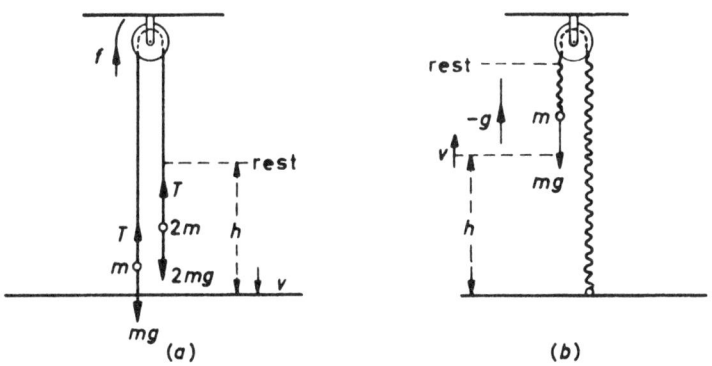

Figure 8.4

Consider the motion with the string taut. (Refer to *Figure 8.4a.*)
Since $P = mf$,

$$\text{for mass } m \quad T - mg = mf \quad \ldots (i)$$

$$\text{for mass } 2m \quad 2mg - T = 2mf. \quad \ldots (ii)$$

Adding (i) and (ii) $2mg - mg = 3mf$

$$\therefore \quad f = \tfrac{1}{3}g.$$

This is a uniform acceleration so, $v^2 = u^2 + 2fs$. Hence, when $2m$

MOTION IN A STRAIGHT LINE UNDER CONSTANT FORCES

mass strikes the ground,

$$v^2 = 0 + 2\tfrac{g}{3}h$$

or
$$v = \sqrt{\tfrac{2}{3}gh}.$$

Also
$$s = ut + \tfrac{1}{2}ft^2$$

∴
$$h = 0 + \tfrac{1}{2}\tfrac{g}{3}t^2$$

∴
$$t = \pm\sqrt{\frac{6h}{g}}.$$

That is, the 2m mass hits the ground after a time $\sqrt{6h/g}$ moving with speed $\sqrt{\tfrac{2}{3}gh}$. Then the string goes slack (refer to *Figure 8.4b*) and the mass m rises with uniform acceleration $-g$ and initial speed $\sqrt{\tfrac{2}{3}gh}$.

Again for uniform acceleration $v^2 = u^2 + 2fs$,

∴ at maximum height $\quad 0 = \tfrac{2}{3}gh - 2gs$

∴ $\quad s = \tfrac{1}{3}h.$

Hence $\quad$ total height above ground $= h + \tfrac{1}{3}h$

$$= \tfrac{4}{3}h.$$

Also, since $v = u + ft$

$$0 = \sqrt{\tfrac{2}{3}gh} - gt$$

∴
$$t = \sqrt{\frac{2h}{3g}}.$$

∴ $\quad$ total time taken to reach max. height $= \sqrt{\dfrac{6h}{g}} + \sqrt{\dfrac{2h}{3g}}$

$$= 4\sqrt{\frac{2h}{3g}}.$$

The mass m reaches a maximum height $\tfrac{4}{3}h$ after a time $4\sqrt{2h/3g}$.

Example 2. An engine pulls, from rest, a train of mass 250 tonne with a constant tractive force of 100 kN until it reaches a speed of 54 km/h. The engine is then cut-off and the brakes are applied producing a force

COMPOUND PROBLEMS

of 200 kN *which brings the train to rest. If the resistance to motion is* 80 N/t *throughout, find the total distance travelled.*

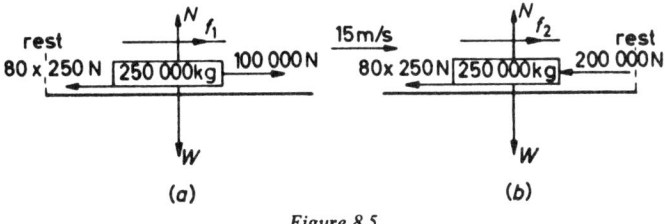

(a)   (b)

Figure 8.5

Consider the train as it accelerates. (Refer to *Figure 8.5a*.)
Since $P = mf$,

horizontally $\quad 100\,000 - 80 \times 250 = 250\,000 f_1$

giving $\quad\quad\quad\quad\quad\quad\quad f_1 = 8/25 \text{ m/s}^2.$

Consider the train as it decelerates. (Refer to *Figure 8.5b*.)
Since $P = mf$,

horizontally $\quad -200\,000 - 80 \times 250 = 250\,000 f_2,$

giving $\quad\quad\quad\quad\quad\quad\quad f_2 = -22/25 \text{ m/s}^2.$

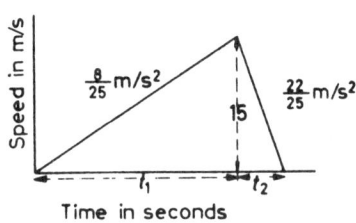

Figure 8.6

Sketching a speed–time curve we have *Figure 8.6*.

Slope of speed–time curve = acceleration

∴ $\quad \dfrac{15}{t_1} = \dfrac{8}{25} \quad \ldots\text{(i)} \quad$ and $\quad -\dfrac{15}{t_2} = -\dfrac{22}{25} \quad \ldots\text{(ii)}$

Area under curve = distance travelled

∴ $\quad \tfrac{1}{2}(t_1 + t_2)15 = s$ (the total distance travelled by
the train). $\quad\ldots\text{(iii)}$

141

MOTION IN A STRAIGHT LINE UNDER CONSTANT FORCES

From (i) $t_1 = \dfrac{25 \times 15}{8}$ and from (ii) $t_2 = \dfrac{25 \times 15}{22}$, $\therefore$ substituting these values in (iii),

$$s = \tfrac{1}{2}\left(\dfrac{25 \times 15}{8} + \dfrac{25 \times 15}{22}\right)15$$

$$= 479{\cdot}4 \text{ m.}$$

The total distance moved by the train is 479 m.

Exercises 8b

1. A car accelerates from rest under a constant tractive force of 100 N. After travelling 300 m the engine stalls and the car slows down to a stop, without the application of the brakes. If, throughout the motion, there is a constant resistance of 30 N, find the total distance travelled.

2. A particle of mass 2 kg, on a smooth plane inclined at 30 degrees to the horizontal, is attached to one end of a light inelastic string. The string passes over the top of the plane and carries at its other end a 3 kg mass hanging freely. The system is released from rest in a position where the 2 kg mass is 3 m from the top of the plane and the 3 kg mass is 2·5 m from the ground. Find the speed of the 2 kg mass when it reaches the top.

3. A particle of mass 3 g falls from a height of 40 cm onto a soft material into which it is found to have penetrated 2·42 cm. Find the constant resistance that could account for this. (Neglect the effect of air resistance.)

4. A light inextensible string passes around a smooth pulley and carries masses of 500 g and 1 500 g at its ends. The system is released from rest with the string taut and the 1 500 g mass 2 m above the floor. Find the time taken until the string jerks tight for the first time.

5. A horizontal surface is part smooth part rough. A particle is projected along the smooth part with speed u. After travelling a distance d, it meets the rough section and is shortly brought to rest. If μ is the coefficient between particle and surface, find the further distance travelled.

Show also that if t is the total time in motion, then $\mu g(ut - d) = u^2$.

6. A light inextensible string passes over a smooth pulley and carries 7 g masses at either end. A further 2 g mass is added to one side and the system is released from rest. After travelling 10 cm the double particle passes through a ring and the 2 g mass is lifted off. The ring is 14 cm from the floor. How long will it be before the string goes slack?

LINKED ACCELERATIONS

7. A car of mass 500 kg travelling on the level is approaching an incline of 1 in 5. When 200 m from the foot of the incline its speed is 36 km/h. If the maximum tractive force the car can produce is 400 N, how far up the hill can it go? (Assume a constant resistance of 40 N throughout.)

8. An engine, exerting a constant tractive force of 100 kN, pulls from rest a train of total mass 250 tonne. After travelling 200 m, several coaches are uncoupled so that the mass is reduced to 150 tonne. At the same time the engine is shut off. If there is a constant resistance of 40 N/t, how far will the train go before coming to rest?

8.3. LINKED ACCELERATIONS

In pulley systems where a single string passes round movable pulleys, the relationships between the various accelerations must first be established.

Example 1. A light inextensible string fastened to a point in the ceiling passes under a smooth movable pulley of mass 5 kg, over a smooth fixed pulley and carries a 4 kg mass hanging freely. All parts of the string which are not touching the pulleys are vertical. If the system is released from rest, find the speed acquired by the movable pulley in 2 seconds.

Suppose the movable pulley rises a distance d in time t. Then $2d$ of string is released so that the 4 kg mass falls a distance $2d$ in the same time. Thus the distances moved are in the ratio of $1:2$ and, differentiating twice with respect to t, the accelerations will be in the same ratio.

i.e. $\qquad F = 2f \qquad$ (refer to *Figure 8.7*).

(Alternatively, at any instant let the pulley be distant x below the line LM and the 4 kg mass distant y. Then the length of the string

$$l = 2x + y + k \qquad \text{(a constant)}.$$

Again differentiating twice with respect to time

$$0 = 2\frac{d^2x}{dt^2} + \frac{d^2y}{dt^2} + 0$$

or $\qquad \dfrac{d^2y}{dt^2} = -2\dfrac{d^2x}{dt^2}.$

(Hence $F = 2f$ as before.)

MOTION IN A STRAIGHT LINE UNDER CONSTANT FORCES

Now since $P = mf$,

for 4 kg mass $\quad 4g - T = 4.2f \quad$ ....(i)

for pulley $\quad 2T - 5g = 5.f. \quad$ ....(ii)

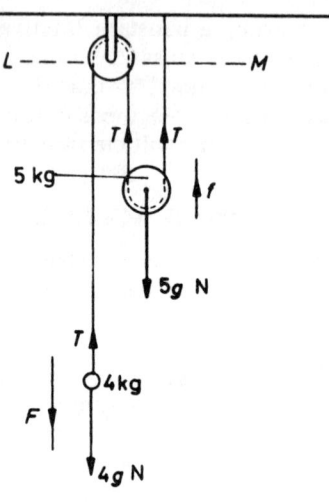

Figure 8.7

Multiplying (i) by 2 and adding (ii)

$$8g - 5g = 21f$$
$$f = \frac{3g}{21}$$
$$= 1\cdot 4 \text{ m/s}^2.$$

This is a uniform acceleration so

$$v = u + ft$$

∴ after 2 s from rest $\quad v = 0 + 1\cdot 4 \times 2$

$$= 2\cdot 8 \text{ m/s}.$$

The pulley acquires an upward velocity of 280 cm/s in 2 s.

In pulley systems with more than one string, and in other situations where one body (free to move) is carried with another, relative accelerations can be used.

Example 2. A light inextensible string passes over a smooth fixed pulley and carries at its ends a particle of mass 4 g and another smooth

144

LINKED ACCELERATIONS

pulley A of mass 5 g. A second string carrying 3 g and 2 g masses at its ends hangs round A. Find the acceleration of A when the system is released.

Let f be the downward acceleration of the 3 g mass relative to the pulley A. Then the resultant accelerations of the various bodies will be as shown in *Figure 8.8*.

Since $\boldsymbol{P} = m\boldsymbol{f}$,

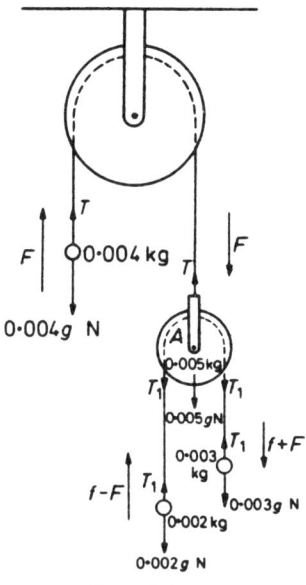

Figure 8.8

for 4 g mass upwards	$T - 0\cdot004g = 0\cdot004F$	.... (i)
for A downwards	$2T_1 + 0\cdot005g - T = 0\cdot005F$	.... (ii)
for 2 g mass upwards	$T_1 - 0\cdot002g = 0\cdot002(f - F)$	.... (iii)
for 3 g mass downwards	$0\cdot003g - T_1 = 0\cdot003(f + F)$	.... (iv)

Adding (i) and (ii),

$$2T_1 + 0\cdot001g = 0\cdot009F. \qquad \ldots\text{(a)}$$

Multiplying (iii) by 3, (iv) by 2 and subtracting,

$$5T_1 - 0\cdot012g = -0\cdot012F. \qquad \ldots\text{(b)}$$

MOTION IN A STRAIGHT LINE UNDER CONSTANT FORCES

Multiplying (a) by 5, (b) by 2 and subtracting,

$$0\cdot029g = 0\cdot069F$$

$$\therefore \quad F = \tfrac{29}{69}g$$

$$= 4\cdot119 \text{ m/s}^2.$$

Pulley A moves with a downward acceleration of $4\cdot12$ m/s².

Example 3. *A particle of mass m is placed on the rough inclined face of a wedge which stands on a smooth horizontal table. The mass of the wedge is 2m and the coefficient of friction between particle and wedge is μ, μ being less than $\tan \alpha$ where α is the inclination of the face to the horizontal. The system is released from rest. Show that the wedge begins to move with an acceleration f given by*

$$2f = (g \cos \alpha - f \sin \alpha)(\sin \alpha - \mu \cos \alpha).$$

If $\alpha = 45$ degrees and $\mu = \tfrac{1}{2}$, find the distance moved by the particle relative to the wedge in 1 second.

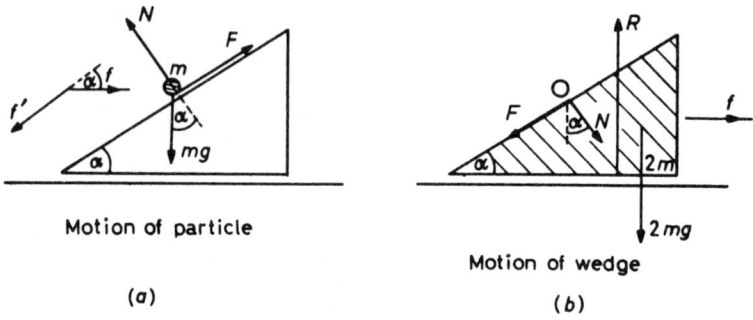

Motion of particle
(a)

Motion of wedge
(b)

Figure 8.9

Since $\mu < \tan \alpha$, the particle will slip down the plane and hence $F = \mu N$.

Consider the motion of the particle (refer to *Figure 8.9a*) and let f' be the acceleration of the particle relative to the wedge. Then its resultant acceleration will have components as shown in the diagram.

Since $P = mf$,

down the wedge $\qquad mg \sin \alpha - \mu N = m(f' - f \cos \alpha)$

$$\dots \text{(i)}$$

146

LINKED ACCELERATIONS

perpendicular to the wedge $\quad mg \cos \alpha - N = mf \sin \alpha \quad \ldots$ (ii)

Consider the motion of the wedge. (Refer to *Figure 8.9b*.) Applying $P = mf$ horizontally,

$$N \sin \alpha - \mu N \cos \alpha = 2mf. \quad \ldots \text{(iii)}$$

From (ii)

$$N = m(g \cos \alpha - f \sin \alpha) \quad \ldots \text{(a)}$$

and substituting this value in (iii),

$$m(g \cos \alpha - f \sin \alpha)(\sin \alpha - \mu \cos \alpha) = 2mf,$$

or $\quad (g \cos \alpha - f \sin \alpha)(\sin \alpha - \mu \cos \alpha) = 2f \quad \ldots$ (b)

as required.

Now putting $\alpha = 45$ degrees, $\mu = \frac{1}{2}$ in (b) and then (a), we have,

$$f = \tfrac{1}{9}g \quad \text{and} \quad N = 8/9\sqrt{2}\,mg.$$

Substituting these values in equation (i),

$$\frac{mg}{\sqrt{2}} - \frac{1}{2}\left(\frac{8}{9\sqrt{2}}mg\right) = m\left(f' - \frac{1}{9\sqrt{2}}g\right)$$

$$\therefore \quad g - \tfrac{4}{9}g = \sqrt{2}f' - \tfrac{1}{9}g$$

$$\therefore \quad f' = \frac{\sqrt{2}}{3}g$$

$$= 4\cdot62 \text{ m/s}^2 \quad \text{(in SI units)}.$$

This relative acceleration is uniform so that the relative distance is given by $s = ut + \tfrac{1}{2}ft^2$.

$\therefore \quad$ after 1 s from rest $\quad s = 0 + \tfrac{1}{2}(4\cdot62)1^2$

$$= 2\cdot31 \text{ m}.$$

The particle moves 2·31 m relative to the wedge in 1 second.

(Note that in these, and indeed in all, problems care should be taken in choosing directions in which to resolve. In this particular case f' was not required initially. By resolving for the particle along and perpendicular to the wedge face, f' was isolated in equation (i) and its elimination was not necessary.)

MOTION IN A STRAIGHT LINE UNDER CONSTANT FORCES

Exercises 8c

1. A particle of mass M on a smooth plane inclined at an angle α to the horizontal is attached to one end of a light inelastic string. The string passes over the upper edge of the plane, under a smooth movable pulley of mass m and is attached to a fixed point. The string on either side of the pulley is vertical. The system is released from rest. Show that the acceleration of the mass M down the plane is $2g(2M \sin \alpha - m)/(4M + m)$.

If $\alpha = 30$ degrees, $M = 10$ g and $m = 8$ g, show that the 10 g mass will move down the plane and find its speed when it has moved 30 cm.

2. A 7 kg particle is attached to one end of a light inextensible string which passes over a smooth fixed pulley, under a smooth movable pulley and over another smooth fixed pulley to another 7 kg particle hanging freely. The movable pulley has a mass of 15 kg and the strings on either side of it are vertical. This pulley is given an upward velocity of 8 m/s. Find the time taken until it reverses direction.

3. A particle of mass 5 kg on a smooth horizontal table is connected by a light inelastic string passing over the edge of the table to a smooth pulley of mass 1 kg. A second string passes round this pulley and carries masses of 1 kg and 3 kg at its ends. If the system is released from rest with the 5 kg mass 1 m from the edge of the table, show that it will reach the edge in less than a second and find its speed as it does so.

4. A particle of mass 9 kg hangs on the end of a light inextensible string which passes over a smooth fixed pulley. The other end carries a *light* smooth pulley, over which hangs a second string carrying masses of 3 kg and 1 kg. Find the accelerations of all three masses when the system is released.

5. A wedge of mass 3 kg is placed on a smooth horizontal surface so that one smooth face is inclined at 30 degrees to the horizontal. A particle of mass 1 kg is placed on that face and the system begins to move. Find the magnitude of the resultant acceleration of that particle.

6. A plank of mass 7 kg rests on a smooth floor with its rough face uppermost. A particle of mass 3 kg is projected along this face. If the coefficient of friction between particle and plank is $\frac{1}{2}$, find the acceleration of the particle relative to the plank.

If, in the subsequent motion, there is a moment when the particle is moving at 10 m/s and the plank at 3 m/s, how much longer will it be before the particle and plank are moving together at the same speed.

EXERCISES

EXERCISES 8*

1. A particle of mass 8 kg on a smooth horizontal surface is pulled by a horizontal force T of magnitude 64 N. If initially, when T is applied, the particle is at O moving at 10 m/s in a direction opposite to T, find its position 3 seconds later.

2. A particle of mass 5 kg, at rest on a smooth horizontal surface, has applied to it three forces P, Q and R. $P = i + 7j$ N, $Q = 6i - 3j$ N and $R = 6j - 2i$ N, where i, j are perpendicular horizontal unit vectors. Find the velocity and displacement of the particle after 4 seconds.

3. A particle of mass $2m$ is held on a smooth plane inclined at an angle α to the horizontal. It is connected by a light inextensible string passing over the top of the plane to a mass m hanging freely. The system is released from rest. Show that the $2m$ mass will begin to move up the plane provided that $\alpha <$ 30 degrees.

If $\sin \alpha = \frac{1}{4}$, find the speed acquired by the particles in 3 seconds.

4. A particle is projected upwards at 9·8 m/s from the top of a building 14·7 m high. How long does it take to reach the ground? (Neglect air resistance.)

5. A car "free wheels" down a hill, inclined at an angle α to the horizontal, with constant speed u. When it reaches the bottom, the brakes are applied producing a constant force equal to $\frac{1}{50}$ of the weight of the car. The car is thus brought to rest in a distance s. If the resistance to the motion is constant throughout show that

$$s = \frac{25u^2}{g(1 + 50 \sin \alpha)}.$$

6. A light inextensible string lies on a smooth horizontal table with its ends passing over opposite edges. To these ends are attached masses m and $2m$ respectively, while a mass M is tied to the mid-point of the string on the table. The system is released and when the strings are tight the particles move with acceleration f. Show that

$$f = g\left(\frac{m}{3m + M}\right),$$

and find the tensions in the strings.

7. i, j, k are unit vectors horizontally east, horizontally north and vertically upwards respectively. To a particle of mass 5 kg at rest on a smooth horizontal surface, forces F_1, F_2 are applied. $F_1 = 3i - j + 2k$ N and $F_2 = 2i + j - 2k$ N. In what direction does the particle begin to move and how far does it go in 2 seconds?

* Exercises marked thus, †, have been metricized, *see* Preface.

8. A nail of mass 10 g is being hammered vertically downwards into a fixed block of wood. The first blow of the hammer imparts a velocity of 140 cm/s to the nail which penetrates 1 cm into the wood. Find the constant resistance which would account for this.

9. A particle is projected vertically upwards at a speed of 14 m/s. When it reaches its greatest height another particle is projected upwards at 28 m/s in the same vertical line. Find the height at which they collide.

10. A small lift of mass 48 kg contains a load of mass 2 kg. When ascending, its speed increases uniformly from 2 m/s to 3 m/s in 1 second. Find the constant tension in the rope and the reaction between the load and the lift.

When descending, it again increases its speed from 2 m/s to 3 m/s in 1 second. Find, in this case, the tension and reaction.

11. A block of mass 20 kg is held at rest on a fixed plane of inclination 30 degrees to the horizontal. A light inextensible string attached to the block runs up the plane parallel to a line of greatest slope, passes over a smooth fixed pulley at the top of the plane and is attached to a particle of mass 13 kg hanging freely. When the system is released, contact between the block and the plane is smooth for the first 1 m of the motion. Thereafter the contact is rough, the coefficient of friction being $\frac{1}{5}$. Assuming that the block does not reach the pulley, find the total distance it travels before coming to rest.

Show that, when the block meets the rough portion of the plane, the tension in the string suddenly increases by an amount $\frac{26}{33}\sqrt{3g}$ N.
(London)†

12. A particle of mass 2 kg, at a point whose displacement is $2i + 4j + 10k$ m from a fixed point O, is acted upon by three forces F_1, F_2 and F_3. The particle is initially at rest and $F_1 = i + 2j - 3k$, $F_2 = 2i - j - 4k$, $F_3 = i - 2j + 2k$ N. Find the displacement of the particle from O after 2 seconds.

13. A particle is pulled along a rough horizontal table by a string which is inclined at an angle θ to the table. The tension in the string is constant and equal to the weight of the particle. If λ is the angle of friction, show that the particle moves with acceleration f, where

$$f \cos \lambda = g\{\cos(\theta - \lambda) - \sin \lambda\}$$

If the inclination of the string may be varied (but not the tension in it) and the coefficient of friction is $\frac{3}{4}$, show that the greatest distance that can be covered from rest in 1 second is 245 cm.

14. Two particles A and B each of mass 6 kg are connected by a light inelastic string which passes over a smooth fixed pulley.

EXERCISES

A mass of 3 kg is attached to A. Find the acceleration with which the system begins to move.

After descending 6 m from rest, A reaches the floor and B continues to rise. Find how far B rises above its original level and its velocity just before the string next becomes taut.

15. A particle of mass $5m$ lies on a rough horizontal table, the coefficient of friction being μ. The particle is connected by a light smooth inextensible string, passing over a small fixed pulley at the edge of the table, to a small light pulley hanging freely. Another light smooth inextensible string passes over the movable pulley and carries particles of masses $3m$ and $2m$ at its ends. The system is released from rest. Show that the acceleration of the particle on the table is $(24 - 25\mu)g/49$, provided that $\mu < 24/25$.

Determine the tension in the second (lower) string. (London)

16. A particle slides down a smooth plane inclined at an angle α to the horizontal. After it has covered a distance s, its speed is found to be v. Show that the time taken (t) satisfies the equation

$$g \sin \alpha \, t^2 - 2vt + 2s = 0.$$

Discuss the meaning of both roots of this equation.

17. A wedge of mass m lies on a smooth horizontal plane and a particle of mass m is in contact with a smooth inclined face of the wedge. The inclination of this face to the horizontal is 30 degrees. The system is released from rest with the particle at a vertical height h above the horizontal plane. Find (a) the acceleration of the wedge, (b) the time taken for the particle to reach the horizontal plane.

(London)

18. Two particles of mass m_1 and m_2 are connected by a long light inextensible string which passes over a smooth pulley. When they are moving with speed u, the first particle upwards and the second downwards, the second particle is instantaneously stopped and released. Prove that, just as the string becomes taut again, the first particle is at rest. (Oxford, part)

19. A 2 kg particle starts from rest at a point O moving under the action of two forces $3i + 6j$ and $2i + 6j$ N. Simultaneously, an identical particle passes through a point P moving under the action of a single force $4i - 3j$ N. The two particles collide after 2 seconds with the second particle moving at 5 m/s. Find the speed with which it passed through P and the displacement of P relative to O.

20. A wedge of triangular cross-section lies on a rough horizontal table and its faces are each inclined at an angle of 45 degrees to the horizontal. Two particles of masses $3m$ and m lie on the smooth inclined faces connected by a light inextensible string passing over

MOTION IN A STRAIGHT LINE UNDER CONSTANT FORCES

a small smooth pulley at the top of the wedge. The string and the particles lie in a vertical plane perpendicular to the top edge of the wedge. If the wedge does not move, find the acceleration of the particles.

Find also the pressure exerted by the string on the wedge and show that the coefficient of friction between wedge and table must be greater than $2m/(2M + 7m)$.

21. A train is travelling on a level track at 72 km/h. Assuming there is no air or track resistance, find the constant brake resistance in newtons per tonne required to stop the train in half a kilometre.

The train descends an incline at 72 km/h and when the brakes are applied it comes to rest in three-quarters of a kilometre. Assuming the air, track and brake resistances are unchanged, find the time taken to stop and the gradient of the incline.

[Take g as 9.8 m/s^2.]   (London)†

22. Two particles of masses 3 kg and 4 kg respectively, lie on a horizontal table with which their coefficients of friction are each $\frac{1}{2}$. They are connected by a smooth inextensible thread which passes through a hole in the table and carries in its loop a pulley of mass M kg, the thread being entirely in one vertical plane and the hanging portions of the thread being vertical. Show that there will be no motion unless $M > 3$.

If $M = 5$, show that the particles and the pulley all move and find their accelerations.   (London)†

23. A light inextensible string passes over a smooth fixed peg and carries at one end a particle A of mass $5m$ and at the other a light smooth pulley; over the latter passes a second light inextensible string carrying particles of masses $2m$ and $3m$ at its ends. Find the acceleration of A when the particles move vertically under gravity.

Find by how much the mass of A must be reduced in order that A can remain at rest while the other two particles are in motion.

(JMB)

24. A smooth plank is fixed at an angle $\tan^{-1}\frac{3}{4}$ with horizontal ground. On it is placed a wedge of mass 40 g whose shape is such that its smooth upper face is horizontal. On this face a 25 g particle is placed 13 cm from the plank. In what direction (relative to the ground) will the particle move?

Find the acceleration of the wedge and the time taken for the particle to reach the plank.

25. A bullet of mass m is fired with speed u into a fixed block of wood and emerges with speed $2u/3$. When the experiment is repeated with the block free to move, the bullet emerges with speed $u/2$ *relative* to the block. Assuming the same constant resistance to

penetration in both cases, find the mass and the final speed of the block in the second case.

(Neglect the effect of gravity throughout.) (London)

26. A particle of mass $3m$ is connected by a light inextensible string passing over a smooth pulley to a smooth pulley A of mass M. Over this pulley passes a similar string carrying masses m and $2m$ at its ends. The system is released from rest. Show that the acceleration of A is $g(3M - m)/(3M + 17m)$.

If A is replaced by a pulley of negligible mass, find the new acceleration of A and the tensions in the strings.

27. A car of mass 700 kg accelerates from rest under a constant tractive force of 250 N exerted by its engine. After 40 seconds, when its speed is 12 m/s, it ceases to accelerate and continues at constant speed for 2 minutes. The engine is then switched off and the car brought to rest with a constant brake force, the total time for the journey being 3 minutes. If the resistance to motion was the same throughout, find,

(a) the tractive force during the middle period of the motion,
(b) the force exerted by the brakes,
(c) the total distance travelled.

28. A particle of mass m is attached to one end of a smooth light inextensible string. The string passes over a smooth fixed pulley, under a movable pulley A of mass M, back over another smooth fixed pulley to be fastened to the axle of A. The pulleys are arranged so that all free parts of the string are vertical. If the system is released from rest, show that the particle will descend provided $m > \frac{1}{3}M$.

If, when $M = 2m$, the pulley A is given a downward velocity u, find the time it takes to return to that position.

29. A smooth wedge, of mass M and triangular cross-section ABC, is placed with the face through AB on a horizontal plane; the angles CAB and ABC are α and $\pi/2$ respectively. A particle of mass m is placed on the inclined face, directly above the centre of gravity G of the wedge. A constant horizontal thrust P is then applied perpendicular to the vertical face and towards G. If the wedge has acceleration F in the direction of the thrust, show that

$$P = mg \sin \alpha \cos \alpha + (M + m \sin^2 \alpha)F,$$

and that the reaction of the horizontal plane on the wedge is

$$(M + m \cos^2 \alpha)g + mF \sin \alpha \cos \alpha.$$

(JMB)

30. A smooth wedge, whose central cross-section is a triangle ABC, right-angled at C, rests with the face containing AB on a

smooth horizontal plane. When the wedge is held fixed, a particle, released from rest, takes a time t_1 to slide the full length of CA. The corresponding time for CB is t_2. Show that $\tan A = t_2/t_1$ and find AB in terms of t_1 and t_2.

If the mass of the wedge is n times that of the particle and the wedge is free to move, show that the time of sliding down CA becomes.

$$t_1\left[1 - \frac{t_1^2}{(n+1)(t_1^2 + t_2^2)}\right]^{1/2}.$$

(London)

31. A light inextensible string passes round a fixed smooth pulley and carries at each end a smooth pulley of mass 1 kg. Over each of these pulleys a string hangs, one carrying masses of 1 kg and 2 kg at its ends, the other masses of 1 kg and 3 kg. The system is set in motion. Find the acceleration of the pulleys and the tension in the string to which they are attached.

32. Two bodies A and B of masses 6 kg and x kg ($x < 6$) respectively are connected by a light inextensible string which passes over a smooth fixed pulley. Show that the acceleration of the system when released is

$$(6-x)g/(6+x).$$

After descending 8 m from rest, A strikes the floor which is inelastic and B continues to rise a further 4 m. Find the value of x.

(WJEC, part)†

33. Two points A and B are on a rough horizontal table at a distance $2d$ apart. A particle Q leaves B with initial speed $2u$ and moves along the line AB produced until it comes to rest at the point C. The coefficient of friction between Q and the table is $1/3$.

 (i) Show that $BC = 6u^2/g$ and find the time taken by Q to travel from B to C.

At the same instant as Q leaves B, a second particle P leaves A with initial speed $5u$ and moves towards B. The coefficient of friction between P and the table is $1/3$ and Q comes to rest at C before P collides with Q.

 (ii) Show that $d > 9u^2/g$.

Given that $d = 12u^2/g$, calculate

 (iii) the speed of P when it collides with Q,
 (iv) the time taken by P to move from A to C.

(AEB)

9

MOTION UNDER FORCES CAUSING SIMPLE HARMONIC MOTION

9.1. BASIC METHODS

CONSIDER a particle moving along a straight line at a distance s from a fixed point O in the line. If the forces acting on the particle are such that their resultant is towards O and of magnitude proportional to s, then (from $P = mf$) we can show that $d^2s/dt^2 = -n^2s$ and the motion is simple harmonic (n^2 being any positive constant determined from $P = mf$).

For simple harmonic motion the following equations apply (refer to Section 5.6):

$$v^2 = n^2(a^2 - s^2)$$
$$s = a \sin(nt + \varepsilon)$$
$$v = an \cos(nt + \varepsilon)$$

and the period $\qquad T = 2\pi/n.$

The amplitude a is the value of s when $v = 0$, determined usually from the initial conditions. ε is determined from the value of s when $t = 0$. In particular, if timed from the centre, $\varepsilon = 0$, if timed from an extreme position, $\varepsilon = \pi/2$.

Example 1. A light elastic string, of natural length 2 m and modulus 5 N, is stretched between two points 3 m apart on a smooth horizontal surface. A particle of mass 100 g is fastened to its mid-point, pulled 3 cm in the direction of the string and released. Find the time taken to travel the first centimetre.

When the particle is fastened to the centre of the string, it divides it into two parts. Each part is of natural length 1 m and extended by $\frac{1}{2}$ m. Since the particle is pulled aside only 3 cm, these strings will remain taut throughout the motion.

Consider the motion of the particle when at a distance s from O (refer to *Figure 9.1*). Then from Hooke's law $T = \lambda(x/l)$ and the modulus $\lambda = 5$ N. Hence $T_1 = 5(\frac{1}{2} - s)$ and $T_2 = 5(\frac{1}{2} + s)$.

MOTION UNDER FORCES CAUSING SIMPLE HARMONIC MOTION

Since $P = mf$,

horizontally
$$T_1 - T_2 = 0 \cdot 1 \frac{d^2s}{dt^2}$$

$$\therefore \quad 5(\tfrac{1}{2} - s) - 5(\tfrac{1}{2} + s) = 0 \cdot 1 \frac{d^2s}{dt^2}$$

$$\therefore \quad -10s = 0 \cdot 1 \frac{d^2s}{dt^2}$$

or
$$\frac{d^2s}{dt^2} = -100s.$$

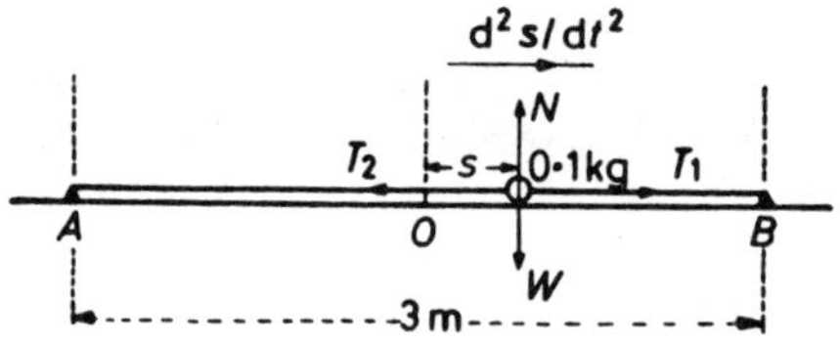

Figure 9.1

This is of the form $d^2s/dt^2 = -n^2s$ and hence the particle is performing S.H.M., about O, for which $n = 10$.

From the initial conditions, the speed $v = 0$ when $s = 0.03$ and hence the amplitude is 0·03 m.

Timing from the initial position, $s = a \sin(nt + \pi/2)$, so when the particle has travelled 1 cm, $s = 0.03 - 0.01$ and

$$0 \cdot 02 = 0 \cdot 03 \cos 10t$$

$$\therefore \quad \cos 10t = \tfrac{2}{3}$$

$$\therefore \quad 10t = 0 \cdot 8410 \text{ (radians)}$$

and
$$t = 0 \cdot 08410 \text{ s.}$$

The time taken to travel the first centimetre is 0·0841 s.

Note that to prove that a motion is simple harmonic it is important to consider the particle in a typical position (distance s say from some fixed point), *not* in its initial position.

In the above problem it was clear that if the particle proved to be moving in S.H.M., then the centre of the motion would be the centre O of the line AB. Hence the distance s was marked from that point. In general, the centre of the motion will be the equilibrium position for the system.

BASIC METHODS

However, if this point is not easily determined, s can be measured from any fixed point in the line of motion. Then we note that an equation of the form $d^2s/dt^2 = -n^2(s + k)$ denotes S.H.M. about the point where $s + k = 0$. For it we put $x = s + k$, then $d^2x/dt^2 = d^2s/dt^2$ and our equation becomes $d^2x/dt^2 = -n^2x$.

Example 2. A and B are two points 16 m apart. A particle P of mass 2 kg moves under the action of two forces $5\overrightarrow{PB}$ and $3\overrightarrow{PA}$ N. If the particle is projected from A towards B at 6 m/s, show that it reaches B and find its speed there.

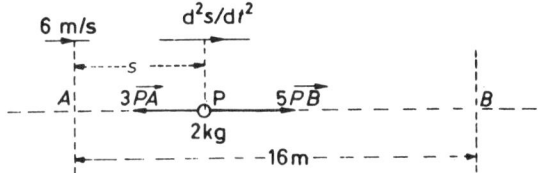

Figure 9.2

Consider the particle when at a distance s from A (refer to Figure 9.2).
Since $P = mf$, in the direction of $\overrightarrow{AB}$,

$$5PB - 3PA = 2\,d^2s/dt^2$$

∴ $$5(16 - s) - 3s = 2\,d^2s/dt^2$$

giving $$d^2s/dt^2 = -4s + 40$$
$$= -4(s - 10).$$

So that if we put $x = s - 10$, $d^2x/dt^2 = d^2s/dt^2$ and
$$d^2x/dt^2 = -4x.$$

This is of the form $d^2x/dt^2 = -n^2x$ and hence the particle is performing S.H.M., about the point $s = 10$ ($x = 0$), for which $n = 2$.
For S.H.M.
$$v^2 = n^2(a^2 - x^2)$$
∴ at A $$6^2 = 4[a^2 - (-10)^2]$$
∴ $$a^2 = 109$$
∴ $$a = \sqrt{109} \text{ m}.$$

MOTION UNDER FORCES CAUSING SIMPLE HARMONIC MOTION

Since this amplitude is greater than the value of x at B, the particle will reach B. Now applying the same equation again, at B

$$v^2 = 4(109 - 6^2)$$
$$= 4 \times 73$$
$$\therefore \quad v = 2\sqrt{73} \text{ m/s}.$$

The particle passes through B with a speed of $2\sqrt{73}$ m/s.

Example 3. *A particle of mass m is hung on the end of a spring of modulus 2mg and natural length l. The particle is pulled vertically downwards until the string is of length 2l and then released. Find the period of the subsequent motion and the maximum speed attained.*

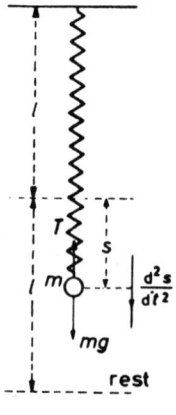

Figure 9.3

Consider the particle in motion distant s from the unstretched position (*Figure 9.3*).
By Hooke's law

$$T = \lambda x/l$$
$$\therefore \quad T = 2mgs/l.$$

Applying $P = mf$ downwards,

$$mg - 2mgs/l = m\, d^2s/dt^2$$
$$\therefore \quad d^2s/dt^2 = -2gs/l + g$$

158

BASIC METHODS

or
$$\frac{d^2s}{dt^2} = -\frac{2g}{l}(s - \tfrac{1}{2}l).$$

Putting $x = s - \tfrac{1}{2}l$, this becomes
$$\frac{d^2x}{dt^2} = -\frac{2g}{l}x.$$

This is of the form $d^2x/dt^2 = -n^2x$ and hence the motion is simple harmonic, about the point where $s = \tfrac{1}{2}l$, for which $n = \sqrt{2g/l}$. Initially, $v = 0$ when $s = l$ ($x = \tfrac{1}{2}l$), so that the amplitude $a' = \tfrac{1}{2}l$.
For S.H.M. the period
$$T = 2\pi/n$$

∴
$$T = 2\pi\sqrt{\frac{l}{2g}} = \pi\sqrt{\frac{2l}{g}}.$$

For S.H.M. also $v^2 = n^2(a^2 - x^2)$, and hence v is a maximum when $x = 0$

∴
$$v_{max} = na$$
$$= \sqrt{\frac{2g}{l}} \cdot \tfrac{1}{2}l$$

∴
$$v_{max} = \tfrac{1}{2}\sqrt{2gl}.$$

The particle moves in simple harmonic motion of period $\pi\sqrt{2l/g}$ and its maximum speed is $\tfrac{1}{2}\sqrt{2gl}$.

Exercises 9a

1. A spring, of modulus 1·5 N and natural length 30 cm, has one end fastened to a smooth horizontal table. To the other end a 40 g particle is attached and pulled out horizontally so that the spring is extended by 30 cm. If the particle is now released, show that it moves in simple harmonic motion about the unstretched position as centre, and find the speed of the particle when 20 cm from that position.

2. A simple pendulum consists of a small particle, of mass m, swinging freely in a vertical plane on the end of a string of length l. Show that, when the particle is at an arc distance s from its equilibrium position, the tangential component of the resultant force on it is $mg \sin(s/l)$.

MOTION UNDER FORCES CAUSING SIMPLE HARMONIC MOTION

Hence show that for small displacements the motion of the particle is simple harmonic and find its period. If the string is 20 cm long, how many swings does it make per minute?

3. A particle of mass 4 kg moves between two points A and B on a smooth horizontal surface under the action of two forces such that, when it is at a point P, the forces are $2\overrightarrow{PA}$ and $2\overrightarrow{PB}$ N. If the particle is released from rest at A, find the time it takes to travel a quarter of the way from A to B.

4. A 100 g particle hangs freely at rest on the end of a spring of modulus 5 N and natural length 50 cm. If this particle is projected upwards with a speed of 2 m/s, find the time taken till it first comes to rest and the distance travelled.

5. An elastic thread, of modulus 19·6 mN and unstretched length 20 cm, is stretched between two points 28 cm apart on a smooth horizontal table. A particle of mass 8 g is fastened to the centre of the thread, displaced a little in the direction of the string (less than 4 cm) and released. Show that the particle performs S.H.M. of period $2\pi/7$ seconds.

If, during this motion, it passes through the centre moving at 10 cm/s, find its speed one-tenth of a second later.

6. A light helical spring of natural length 10 cm is fixed upright on a table and a small platform of mass 10 g placed on top of it. When the platform is depressed and released, it vibrates vertically completing 5 oscillations in 1 second. Find the modulus of the spring.

7. When a particle of mass m is hung on an elastic string, it is extended by a distance x. Find the ratio of the modulus to the unstretched length of the string. (This is sometimes called the *force constant* or *stiffness* (see page 122) of the string.)

One end of the string is now fastened at A to a smooth horizontal surface and the particle pulled out horizontally until the extension of the string is again x. The particle is then projected towards A with speed u. If v is its speed when the extension is halved, show that $4(v^2 - u^2) = 3gx$.

8. Two particles of equal mass hang on the end of a spring of natural length l and whose modulus is twice the weight of one of the particles. Find the extension of the string.

If one of the particles is suddenly removed, find the time taken for the remaining particle to reach maximum speed and the magnitude of that speed.

9. One end of an elastic thread of natural length l is fastened to a point O on a rough horizontal table. A particle is attached to the other end and is held on the table at a distance $2l$ from O. The

FURTHER EXAMPLES

modulus of the thread is equal to the weight of the particle, and the coefficient of friction between particle and table is μ ($\mu < 1$). If the particle is now released show that after a time t it will have travelled a distance $l(1 - \mu)(1 - \cos\sqrt{(g/l)}t)$.

10. A particle P, of mass m, moves along a line AB, where A and B are fixed points distant $3a$ apart. The particle moves under the action of two forces $2(mg/a)\overrightarrow{PA}$, $(mg/a)\overrightarrow{PB}$. If it is projected from A towards B with speed $3\sqrt{ga}$, show that it moves in S.H.M. of period $(2\pi/3)\sqrt{3a/g}$ and amplitude $2a$.

Show also that the particle reaches B after a time equivalent to one-third of a complete oscillation.

9.2. FURTHER EXAMPLES

Example 1. When a particle is hung from a fixed point by a light elastic string it extends the string by a length e. If, now, it is projected vertically upwards with speed $\sqrt{2ge}$, show that the particle will come to instantaneous rest after a time, $\sqrt{e/g}(\pi/4 + 1)$. (Assume that the string does not tighten again before the particle reaches this position.)

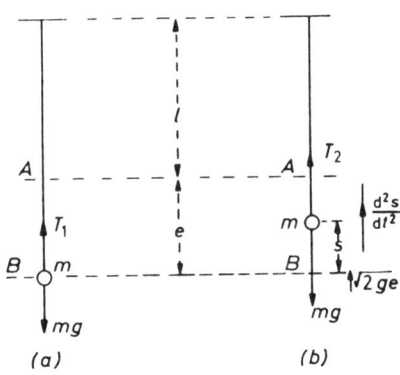

Figure 9.4

(a) Consider the particle when hanging at rest (refer to *Figure 9.4a*).
By Hooke's law
$$T = \lambda(x/l)$$
$$\therefore\quad T_1 = \lambda(e/l).$$

MOTION UNDER FORCES CAUSING SIMPLE HARMONIC MOTION

And since the particle is at rest,
$$T_1 - mg = 0$$
$$\therefore \quad \lambda(e/l) - mg = 0$$
$$\therefore \quad \lambda = mgl/e.$$

(b) Consider the particle in motion between A and B, and at a distance s from B (refer to *Figure 9.4b*).

By Hooke's law
$$T = \lambda(x/l)$$
$$\therefore \quad T_2 = \frac{mgl}{e}\left(\frac{e-s}{l}\right)$$
$$= \frac{mg}{l}(e-s).$$

Since $P = mf$, vertically upwards
$$T_2 - mg = m\,d^2s/dt^2$$
$$\therefore \quad \frac{mg}{e}(e-s) - mg = m\,d^2s/dt^2$$

giving
$$\frac{d^2s}{dt^2} = -\frac{g}{e}s.$$

This is of the form $d^2s/dt^2 = -n^2s$ and the motion is therefore simple harmonic about B with $n = \sqrt{g/e}$.

For S.H.M. $\quad v^2 = n^2(a^2 - s^2)$

$\therefore$ at B $\quad 2ge = \frac{g}{e}(a^2 - 0)$

$\therefore \quad a = e\sqrt{2}.$

This amplitude is greater than e so the particle will reach A.

For S.H.M. $\quad s = a\sin(nt + \varepsilon)$

$\therefore$ at A $\quad e = e\sqrt{2}\sin t\sqrt{\dfrac{g}{e}}$

(timing from the centre B of the motion)

FURTHER EXAMPLES

giving $\quad t = \dfrac{\pi}{4}\sqrt{\dfrac{e}{g}}.$

This is the time taken to move from B to A.

Also $\quad v^2 = n^2(a^2 - s^2)$

$\therefore$ at $A \quad v^2 = \dfrac{g}{e}(2e^2 - e^2)$

$\therefore \quad v = \sqrt{ge}.$

(c) Consider the motion of the particle above A with the string slack. Its initial speed will be $\sqrt{ge}$ and it will rise with acceleration $-g$. Then applying $v = u + ft$ at the point where it comes to instantaneous rest,

$$0 = \sqrt{ge} - gt$$

$$t = \sqrt{\dfrac{e}{g}}.$$

This is the time taken to reach this point from A. Hence the total time from B to this position is

$$\dfrac{\pi}{4}\sqrt{\dfrac{e}{g}} + \sqrt{\dfrac{e}{g}} = \sqrt{\dfrac{e}{g}}\left(\dfrac{\pi}{4} + 1\right).$$

Example 2. A light elastic string of natural length 2 m is stretched between two points A and B, 4 m apart, on a smooth horizontal surface. A particle is fastened to the mid-point of the string whose modulus is equal to the weight of the particle. If this particle is now pulled aside to A and released, find the time taken for it to reach the mid-point of AB again.

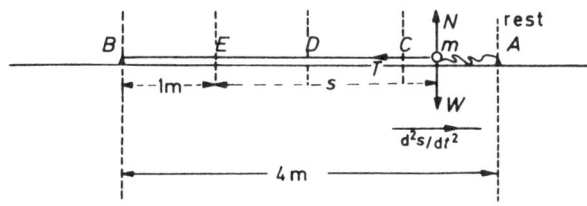

Figure 9.5(a)

MOTION UNDER FORCES CAUSING SIMPLE HARMONIC MOTION

The particle divides the string into two parts each of modulus mg (where m is the mass of the particle) and unstretched length 1 m.

(a) Consider the motion between A and C with one string slack and the particle at a distance s from E, the unstretched position for the remaining string. (Refer to *Figure 9.5a*.)

By Hooke's law $\qquad T = \lambda(x/l)$

$\therefore \qquad T = mg(s/1)$

Hence applying $P = mf$ towards A,

$$-mgs = m\,d^2s/dt^2$$

$\therefore \qquad \dfrac{d^2s}{dt^2} = -gs.$

This is of form $d^2s/dt^2 = -n^2s$ and indicates S.H.M. about E with $n = \sqrt{g} = 3\cdot 13$.

At A, $v = 0$ and $s = 3$, hence the amplitude is 3 m.

For S.H.M. $\qquad s = a \sin(nt + \varepsilon)$.

$\therefore$ at C $\qquad 2 = 3 \sin(3\cdot13t + \pi/2)$

timing from extreme position A.

$\therefore \qquad 2 = 3 \cos(3\cdot13t)$

$\therefore \qquad 3\cdot13t = 0\cdot8409 \quad \text{(radians)}$

$\therefore \qquad t = 0\cdot2687 \text{ s}.$

This is the time taken to travel from A to C.
Also $\qquad v^2 = n^2(a^2 - s^2)$

$\therefore$ at C $\qquad v^2 = 9\cdot8(9 - 4)$

$\therefore \qquad v = 7 \text{ m/s}.$

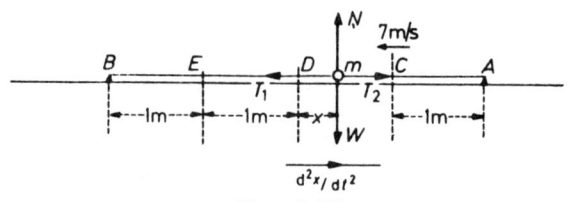

Figure 9.5(b)

(b) Consider the motion between C and D with both strings tight and the particle distant x from D. (Refer to *Figure 9.5b*.)

FURTHER EXAMPLES

By Hooke's law $\quad T = \lambda(x/l)$

$\therefore \quad T_1 = mg(1 + x) \quad \text{and} \quad T_2 = mg(1 - x)$

Hence applying $P = mf$ towards A

$$mg(1 - x) - mg(1 + x) = m\, d^2x/dt^2$$

$\therefore \quad d^2x/dt^2 = -2gx$

indicating S.H.M. about D with $n = \sqrt{2g} = 4\cdot427$.

For S.H.M. $\quad v^2 = n^2(a^2 - x^2)$

$\therefore$ at $C \quad 7^2 = 19\cdot6(a^2 - 1^2)$

giving $\quad a = 1\cdot871$ m.

This is the amplitude of this part of the motion.

Also $\quad x = a \sin(nt + \varepsilon)$

$\therefore$ at $C \quad 1 = 1\cdot871 \sin(4\cdot427t + 0) \quad$ timing from centre D

$\therefore \quad \sin(4\cdot427t) = 0\cdot5345$

$\therefore \quad 4\cdot427t = 0\cdot5638 \quad$ (radians)

$\therefore \quad t = 0\cdot1274$ s.

This is the time taken from C to D.

Hence the total time taken to travel from A to D is

$$0\cdot2687 + 0\cdot1274 \approx 0\cdot396 \text{ s.}$$

Example 3. ABC is an equilateral triangle, of side 5 m, marked out on a smooth horizontal surface. A particle P of mass 5 kg moves on this surface under the action of two forces $3\overrightarrow{PB}$ and $2\overrightarrow{PC}$ N. Show that if the particle is released from A it will move in a straight line cutting BC in D where $BD:DC = 2:3$, and find its speed when it reaches D.

When the particle comes to instantaneous rest again at A', the force towards B ceases to act (so that it moves now towards C). Find the total time taken to travel from A to A' and to C.

(a) Consider the motion while both forces are acting (refer to Figure 9.6a).

By a standard result obtained in Section 3.3

$$\lambda\overrightarrow{OA} + \mu\overrightarrow{OB} = (\lambda + \mu)\overrightarrow{OC}$$

MOTION UNDER FORCES CAUSING SIMPLE HARMONIC MOTION

where C divides AB in the ratio $\mu:\lambda$. Hence in this case
$$3\overrightarrow{PB} + 2\overrightarrow{PC} = 5\overrightarrow{PD}$$
where D divides BC in the ratio $2:3$, i.e., the resultant of the two forces on the particle is $5\overrightarrow{PD}$, and if the particle starts from A, it will begin to move towards D and continue along the line AD.

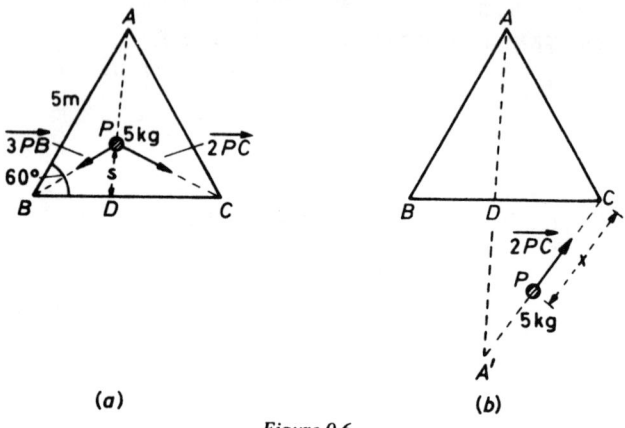

Figure 9.6

Then applying $P = mf$ towards A when the particle is at a distance s from D,
$$-5PD = 5\, d^2s/dt^2$$
$$\therefore \quad -5s = 5\, d^2s/dt^2$$
$$\therefore \quad \frac{d^2s}{dt^2} = -s.$$

This is the form $d^2s/dt^2 = -n^2s$ and the motion is S.H.M. about D as centre, with $n = 1$. At A the speed is zero and therefore AD is the amplitude of the motion. Applying the cosine rule to triangle ABD,
$$AD^2 = 5^2 + 2^2 - 2 \times 5 \times 2 \cos 60°$$
$$\therefore \quad AD = \sqrt{19} \text{ m}.$$
For S.H.M. $v^2 = n^2(a^2 - s^2)$
$$\therefore \quad \text{at } D \text{ where } s = 0 \qquad v^2 = 1(19 - 0)$$
$$\therefore \qquad\qquad\qquad v = \sqrt{19} \text{ m/s}.$$

FURTHER EXAMPLES

The particle will next come to instantaneous rest at its other extreme position A' after half a period (measured from A), i.e.

$$t_1 = \frac{1}{2}\left(\frac{2\pi}{n}\right) = \pi \text{ s}.$$

(b) Consider now the motion while only the force towards C acts (refer to *Figure 9.6b*).

Since the particle is at rest at A', this force $2\overrightarrow{PC}$ will cause it to move directly towards C. We shall consider the instant when the particle is at a distance x from C.

Since $P = mf$, towards A'

$$-2PC = 5 \, d^2x/dt^2$$
$$\therefore \quad -2x = 5 \, d^2x/dt^2$$
$$\therefore \quad \frac{d^2x}{dt^2} = -\frac{2}{5}x.$$

This indicates S.H.M. about C with $n = \sqrt{\frac{2}{5}}$. Hence the time taken from A' to C will be a quarter of a period, i.e.,

$$t_2 = \frac{1}{4}\left(2\pi\sqrt{\frac{5}{2}}\right) = \frac{\pi}{4}\sqrt{10} \text{ s}.$$

$$\therefore \quad \text{total time from } A \text{ to } C = t_1 + t_2$$
$$= \pi + \frac{\pi}{4}\sqrt{10}$$
$$= \frac{\pi}{4}(4 + \sqrt{10}) \text{ s}.$$

The particle passes through D at $\sqrt{19}$ m/s and reaches C after a time $\pi(4 + \sqrt{10})/4$ seconds.

Exercises 9b

1. An elastic string, of unstretched length 1 m and modulus 1 N, has a particle of mass 20 g fastened to one end. The other end is fixed to a point O on a smooth horizontal table. The particle is held on the table at a distance of 2 m from O. Find the time it takes to reach O when released.

2. When a particle is hung on an elastic thread of natural length l it extends the thread by $\frac{1}{2}l$. Show that the modulus of the thread is twice the weight of the particle.

If the particle is now pulled down a further distance l and released, show that it will rise to within $\frac{1}{4}l$ from the point of suspension in time

$$\sqrt{\frac{2l}{g}}\left(\frac{\pi}{3} + \frac{\sqrt{3}}{2}\right).$$

3. P is any point in the plane of a triangle ABC. Show that the resultant of vectors represented by $\overrightarrow{PA}, \overrightarrow{PB}$ and $\overrightarrow{PC}$ is a vector $3\overrightarrow{PG}$, where G is the centroid of the triangle ABC.

Three points A, B, C on a smooth horizontal table form an equilateral triangle of side a. When a particle of mass m is at any point P on the table, it is simultaneously attracted to A by a force $\frac{mg}{a}\overrightarrow{PA}$, to B by a force $\frac{mg}{a}\overrightarrow{PB}$ and to C by a force $\frac{mg}{a}\overrightarrow{PC}$. Initially it is placed at A and released. Find how long it takes to reach the side BC and its speed as it crosses that side. (London)

4. An elastic string, of modulus λ and natural length l, has one end fastened to a point O on a smooth horizontal surface. A particle of mass m is attached to the other end and is projected from O horizontally at speed u. Show that the greatest extension of the string in the subsequent motion is $u\sqrt{ml/\lambda}$ and that the time taken to return to O is $2l/u + \pi\sqrt{ml/\lambda}$.

5. A particle has a mass such that when it is hung on an elastic string of natural length 30 cm, it increases its length to 90 cm. If, now, the particle is lifted to the point of suspension and released, find the time taken till it again comes to rest.

6. A light helical spring of natural length 15 cm and modulus 1·5g mN is fixed in a vertical position on a table. A scale-pan of mass 10 g containing a 30 g mass is placed on top of the spring, depressed until the spring measures only 5 cm, and then released. Find the position at which the mass loses contact with the scale-pan and the height above the table to which it rises.

7. An elastic string, of modulus λ and natural length $2a$, is stretched between two points P and Q, distant $3a$ apart on a smooth horizontal surface. A particle of mass m is then attached to the mid-point of PQ, pulled to one side to P and released. Show that it will just reach Q and that this will happen after a time

$$2\sqrt{\frac{ma}{\lambda}}\left(\frac{\pi}{3} + \frac{1}{\sqrt{2}}\sin^{-1}\frac{1}{\sqrt{7}}\right).$$

EXERCISES 9

1. O is a fixed point and a particle P of mass m moves under a net force $k\overrightarrow{PO}$. If the particle is instantaneously at rest when at a distance d from O, find its speed as it passes through O. (k is a positive constant.)

2. A particle moves in simple harmonic motion of period 2 seconds and amplitude 100 cm. If the mass of the particle is 10 g, find the greatest and least force that it experiences.

3. A small ball bearing is released from the edge of a smooth spherical watch glass so that it slides back and forth across it. Show that, if the radius of the sphere is large compared with the diameter of the watch glass, then the ball bearing will perform simple harmonic motion. Find the period of the motion.

4. A light spring, of modulus λ and natural length l, has one end fixed to a point O on a smooth horizontal surface. The other end carries a mass m which is held on the surface with the string just taut. If this mass is now projected away from O with speed u, find its furthest distance from O in the subsequent motion.

Find also the time taken to reach this position for the first time.

5. A rough plank moves horizontally in simple harmonic motion of amplitude a and period T. Show that a particle on the plank will not slip provided that the coefficient of friction is greater than $4\pi^2 a/(gT^2)$.

6. A particle, of mass m, is attached to the mid-point of an elastic string stretched between two points on a smooth horizontal surface distant $2a$ apart. If this particle is slightly displaced a distance x at right angles to the line joining the two points, show that, neglecting $(x/a)^2$ and higher powers, the tension in the string remains constant.

Hence show that, if released to vibrate in this direction the mass performs S.H.M. If the magnitude of the tension is T, find the period of the motion.

7. When a 5 kg mass is hung from a spring, it extends the spring by 20 cm. If it is then pulled down a further 20 cm and released, find the period of the subsequent motion and the maximum speed.

8. Some kitchen "scales" consist of a helical spring of natural length l on top of which is placed a scale pan of mass m. When a further mass M is added to the pan, the total compression of the spring is a. Find the modulus of the spring.

The pan is now depressed a further distance b and released. Find the reaction between mass and pan (a) immediately on release, (b) after travelling a distance b, (c) when the spring is its natural length.

MOTION UNDER FORCES CAUSING SIMPLE HARMONIC MOTION

9. An elastic thread of natural length 20 cm and modulus 0·196 mN is stretched between two points A, B on a smooth horizontal table 30 cm apart. A particle of mass 2 g, fastened to the mid-point of the string is released from rest at a point 20 cm from A. Find the period and amplitude of the subsequent motion.

What will be its speed when 17 cm from A?

10. A and B are two points 30 m apart on a smooth horizontal table. A particle P of mass 3 000 g moves under the action of two forces $2\overrightarrow{PA}$ and $\overrightarrow{PB}$ newton. The particle is released from rest at C, a point which divides AB internally in the ratio of 2:1. Find the position and velocity (in magnitude and direction) of the particle after $\pi/4$ seconds.

11. To the mid-point of an elastic string, of natural length l, a particle is attached whose weight is twice the modulus of the string. The two free ends of the string are attached to a fixed point O and the particle is pulled vertically downwards. If the particle is released from rest at a point $\tfrac{3}{2}l$ vertically below O, find the time taken to rise a distance $\tfrac{1}{4}l$.

12. A particle P of mass m lies on a smooth horizontal table and is attached by two light elastic strings, of natural lengths $3a$, $2a$ and moduli λ, 2λ to two fixed points, S, T respectively, on the table. If $ST = 7a$, show that, when the particle is in equilibrium, $SP = 9a/2$.

The particle is held at rest at the point in the line ST where $SP = 5a$, and then released. Show that the subsequent motion of the particle is simple harmonic of period $\pi\sqrt{(3ma/\lambda)}$. Find the maximum speed of the particle during this motion.   (JMB)

13. A small bead, of weight mg, moves along a smooth horizontal wire under the action of a force which is directed always towards a fixed point A at a distance h vertically below the wire; the magnitude of this force is $mn^2 r$, where n is a constant and r denotes the distance of the bead from A.

Show that the bead performs simple harmonic motion of period $2\pi/n$. Find the amplitude of this motion if the speed of the bead is v, when it is at the point O on the wire which is vertically above A.

Given that $v > hn$, obtain an expression for the time required for the bead to move from O to a point P on the wire such that $\angle OAP = \tfrac{1}{4}\pi$.   (Oxford)

14. A particle of mass m is attached to one end of an elastic thread of natural length l and modulus λ. The other end is fixed to a point on a smooth horizontal table. If the particle is released from a point on the table distant $a + l$ from O, show that the subsequent

motion of the particle is periodic and of period

$$\sqrt{\frac{ml}{\lambda}}\left(2\pi + \frac{4l}{a}\right).$$

15. A, B and C are the vertices of an equilateral triangle of side $5a$. A particle of mass m moves in the plane of the triangle under the action of two forces, one directed towards A of magnitude $2(PA/a)\,mg$ and the other directed towards B of magnitude $3(PB/a)mg$, where P is the position of the particle at any instant. Initially the particle is placed at C and then released.

(a) Show that the particle moves in a straight line and find where this line meets AB.

(b) Find the time that elapses before the particle first returns to C.

(c) Find the velocity and acceleration of the particle at the moment it crosses the line AB. (London)

16. A particle is suspended from a fixed point O by means of a light elastic string of natural length a and hangs at rest, its extension being c. If it is given a downward vertical velocity v when in this position, show that the ensuing motion is simple harmonic provided $v^2 \leqslant gc$.

If $v^2 = 2gc$, show that the time taken from the lowest point of its path to the highest is

$$\sqrt{\frac{c}{g}}\left(1 + \frac{3\pi}{4}\right).$$ (WJEC)

17. A particle of mass m is being pulled up vertically with uniform speed U at the end of an elastic thread of natural length l and modulus mg. If the other end of the thread is suddenly fixed, show that the string will remain taut provided $U^2 < gl$. Find the period and amplitude of the motion that takes place when this condition is satisfied.

18. A light elastic string, of natural length 2 m and modulus 10 N, is stretched between two points L and M on a smooth horizontal table 4 m apart. A particle of mass 0·1 kg is fastened to the mid-point of the string, then pulled aside to M and released. Show that it returns to M after a time $\frac{\sqrt{2}}{5}(\sqrt{2}\cos^{-1}\frac{2}{3} + \sin^{-1}\sqrt{\frac{2}{7}})$ seconds.

19. Two light springs AB, BC joined at B are fixed with AC in a vertical line. A small particle is attached at B and falls a distance d vertically. Show that if now pulled down and released it will perform S.H.M. of period $2\pi\sqrt{d/g}$.

20. State clearly the relation between the acceleration and the displacement in simple harmonic motion.

MOTION UNDER FORCES CAUSING SIMPLE HARMONIC MOTION

A man of mass M stands on a horizontal platform which performs a vertical simple harmonic motion of period T and amplitude a. Find the force he exerts on the platform when the latter is at a height x above its mean position; give your answer in terms of M, g, x and T. Deduce that he maintains contact with the platform provided that $T \geqslant T_0$, where $T_0 = 2\pi\sqrt{(a/g)}$.

If $T = nT_0$, where $n > 1$, show that the greatest and least forces exerted by the man on the platform are in the ratio $(n^2 + 1)/(n^2 - 1)$.

(JMB)

21. A particle is fastened to one end of an elastic thread of natural length 0·2 m and modulus equal to the weight of the particle. The other end is fixed at a point O. If the particle is released from a point 0·8 m below O, show that it will rise to a height of 0·1 m above O. Find also the time that elapses before it returns to its initial position.

22. ABC is a right-angled isosceles triangle in which BC, the hypotenuse, is of length $6a$. A particle P, of mass m, moves in the plane of the triangle under the action of three forces $2\lambda\dfrac{mg}{a}\overrightarrow{PA}$, $\lambda\dfrac{mg}{a}\overrightarrow{PB}$ and $\lambda\dfrac{mg}{a}\overrightarrow{PC}$, λ being a positive constant. The particle is released from rest at A. Show that it will just reach the mid-point D of BC in time $\tfrac{1}{2}\pi\sqrt{a/\lambda g}$.

If, when it reaches D, the force towards B ceases to act, find the speed with which it subsequently crosses the line AC.

23. A light elastic string LXM of natural length $3l/2$ has a mass m attached at X, where the unstretched length of $LX = l$. If the system is freely suspended from L, the mass hangs in equilibrium at a distance $2l$ below L. The system is placed on a smooth horizontal table with L and M fixed at a distance $3l$ apart. Find the position of the mass when the system is in equilibrium.

If the mass is placed midway between the fixed points and released, find the amplitude and period of the motion of the mass. (London)

24. A particle, of mass m, is attached to a point O by a light elastic string and unstretched length l. It is released from rest at O and falls a distance $2l$ before coming to instantaneous rest. Show that the time taken to fall to the lowest point is

$$\left[\tfrac{1}{4}\pi + \sqrt{2} + \tfrac{1}{2}\sin^{-1}\left(\tfrac{1}{3}\right)\right]\sqrt{\tfrac{l}{g}}.$$

25. A light elastic string, of natural length a and modulus $2mg$, is attached at one end to a fixed point A on a smooth horizontal table and at the other end to a particle P of mass $2m$. A second light string

EXERCISES

which is *inextensible* is attached to P and has at its other end a second particle Q of mass m. The distance from A to the edge of the table is $2a$. Initially the string PQ passes over the edge, the plane APQ is perpendicular to the edge, and Q hangs below the edge. P is held on the table with the elastic string just taut and then released. Write down the equations of motion for P and Q when the distance AP is $\frac{3}{2}a + x$. Hence find the differential equation of the motion, and the time taken for P to reach the edge of the table.

Find also the tension in each string just as P reaches the edge of the table. (JMB)

26. An elastic thread of unstretched length $2a$ is stretched between two points A and B. $AB = 2\sqrt{3}a$ and is horizontal. When a particle is attached to the mid-point M of the thread it hangs at rest with AM inclined at 30 degrees to the horizontal. Show that the modulus of the thread is equal to the weight of the particle.

The particle is now pulled down a further small distance s. Show that in this position, if $(s/a)^2$ and higher powers are neglected, $AM = 2a + \frac{1}{2}s$ and $\cos\theta = \frac{1}{2}(1 + \frac{3}{4}s/a)$ where θ is the angle between AM and the vertical. Deduce the period of small vertical vibrations if the particle is now released.

27. One end of a light elastic string, of modulus mg and natural length l, is fixed to a point O of a rough horizontal table. A particle of mass m is fixed to the other end of the string, and lies on the table with the string just taut. The particle is then struck so that it begins to move along the line of the string away from O with velocity u.

If μ is the coefficient of friction, show that the particle comes to rest for the first time after travelling a distance $l[\sqrt{(\mu^2 + u^2/gl)} - \mu]$ and after a time

$$\sqrt{\left(\frac{l}{g}\right)} \tan^{-1} \frac{u}{\mu\sqrt{(gl)}} \qquad \text{(Oxford)}$$

28. A particle P of mass m is attached to the ends of two light elastic strings of natural lengths l and $2l$, and having moduli $4mg$ and mg respectively. The other ends of the two strings are fixed at two points A and B respectively on a smooth horizontal table such that $AB = 4l$. The particle is released from rest at A. Prove that when $0 < AP < l$ and when $AP > l$ the particle executes a different simple harmonic motion in each case, and find the period of each motion.

Prove that the string attached at B never becomes slack during the motion, and that the greatest distance of the particle from A is $\frac{1}{9}l(10 + 2\sqrt{7})$. (London)

MOTION UNDER FORCES CAUSING SIMPLE HARMONIC MOTION

29. Two light elastic strings AB and CD have the same modulus of elasticity λ. The ends A and D are fixed at two points on a smooth horizontal table. The strings are joined at the ends B and C to a particle of mass m which lies between A and D. The particle is in equilibrium when the strings are stretched with $AB = a$ and $CD = b$ and the tension in each string is P.

Show that the natural length of the string AB is $\lambda a/(P + \lambda)$ and find the natural length of the string CD.

The particle is given a small displacement along AB and released. Prove that the particle moves with simple harmonic motion and that the period of oscillation is

$$2\pi \left[\frac{mab}{(P+\lambda)(a+b)} \right]^{1/2}.$$

Prove also that for a given value of $a + b$ the period is longest when $a = b$. (AEB)

30. A light elastic string has natural length 0·5 m and modulus of elasticity 140 N. The string is placed on a smooth horizontal table with one end tied at a fixed point A. A particle of mass 0·7 kg is attached at the other end of the string. Calculate the work done in stretching this string from its natural length to the position where the particle is held at rest at the point B on the table, where $AB = 0·8$ m. State the tension in the string in this position.

If the particle is released from rest at B, calculate

(i) the speed of the particle at the instant when it passes through A,

(ii) the time taken by the particle to reach A from B for the first time. (AEB)

31. A light elastic string of natural length a hangs from a fixed point O, and when a heavy particle at the other end is hanging in equilibrium, the extension of the string is b. The particle is pulled down a further distance c and released from rest. Given that $c^2 = 2ab + b^2$, prove that

(i) the particle will just rise to O,

(ii) the time taken to do this

$$\left(\frac{2a}{g}\right)^{1/2} + \left(\frac{b}{g}\right)^{1/2} \left\{ \pi - \cos^{-1}\left(\frac{b}{c}\right) \right\} \qquad \text{(Oxford)}$$

10

MOTION IN A STRAIGHT LINE UNDER VARIABLE FORCES

SINCE $P = mf$, when the forces acting on a particle vary, so will its acceleration. This chapter is concerned only with variable motion in a straight line. In this case, if the acceleration is a suitable function of time, displacement or velocity, then the methods of Section 5.5 can be applied.

As an introductory example, the reader is advised to study again Example 1 of Section 5.5.

Example 1. A train of mass 200 t moves on the horizontal against constant resistances of 75 N/t. Initially, the train is at rest and it accelerates under a tractive force that increases steadily from 20 kN to 120 kN in 50 seconds. Find the speed acquired (in km/h) and the distance travelled.

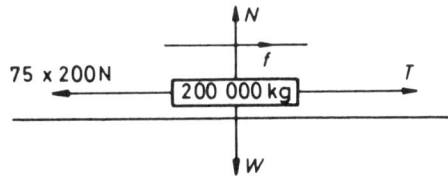

Figure 10.1

The tractive force T increases steadily at the rate of

$$\frac{120 - 20}{50} = 2 \text{ kN/s}$$

Hence, after t seconds

$$T = (20 + 2t) \text{ kN}$$
$$= (20 + 2t)1\,000 \text{ N}.$$

MOTION IN A STRAIGHT LINE UNDER VARIABLE FORCES

Considering the particle at this instant and applying $P = mf$ horizontally

$$T - 75 \times 200 = 200\,000 f$$

(*refer to Figure 10.1*)

$$\therefore \quad (20 + 2t)1\,000 - 15\,000 = 200\,000 f$$

giving
$$f = (5 + 2t)/200$$

i.e.
$$\frac{dv}{dt} = \frac{5 + 2t}{200}.$$

Integrating, $\quad v = \frac{1}{200}(5t + t^2) + C$

Now when $t = 0$, $\quad v = 0 \quad \therefore \quad 0 = 0 + 0 + C$

so
$$v = \tfrac{1}{200}(5t + t^2) \qquad \ldots (i)$$

$\therefore$ after 50 s $\quad v = \tfrac{1}{200}(5 \times 50 + 50^2)$

$$= \frac{50 \times 55}{200} \text{ m/s}$$

$$= \frac{50 \times 55}{200} \times \frac{18}{5} \text{ km/h}$$

$$= 49\cdot 5 \text{ km/h}.$$

Equation (i) may be written as

$$\frac{ds}{dt} = \tfrac{1}{200}(5t + t^2).$$

Integrating, $\quad s = \tfrac{1}{200}\left(5\tfrac{t^2}{2} + \tfrac{t^3}{3}\right) + A.$

Taking $s = 0$, when $t = 0$, $0 = 0 + 0 + A$

$\therefore \quad s = \tfrac{1}{200}(\tfrac{5}{2}t^2 + \tfrac{1}{3}t^3)$

$\therefore$ after 50 s $\quad s = \tfrac{1}{200}(\tfrac{5}{2} \times 50^2 + \tfrac{1}{3} \times 50^3)$

giving $\quad s = 240\cdot 6$ m.

After 50 seconds the train is travelling at 49·5 km/h and is 241 m from its starting point.

Example 2. Two particles in space are given electrostatic charges, one a positive charge and the other a negative charge. As a result, they are

MOTION IN A STRAIGHT LINE UNDER VARIABLE FORCES

attracted to one another with a force μ/s^2, s being their distance apart.

The first particle is fixed at a point O and the other (of mass m) is released a distance 2a from O. Show that when distant s from O, its speed is given by

$$v^2 = \frac{2\mu}{m}\left(\frac{1}{s} - \frac{1}{2a}\right)$$

and find the time taken to travel halfway to O.

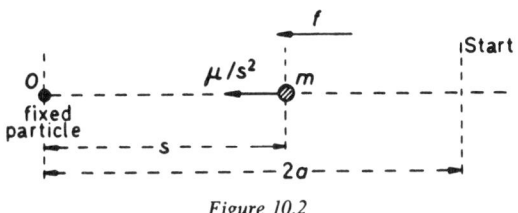

Figure 10.2

Consider the mass m when at a distance s from the fixed particle at O (refer to Figure 10.2).

Since $P = mf$, towards O,

$$\mu/s^2 = mf$$
$$f = \mu/ms^2$$

or since s increases away from O

$$-v\frac{dv}{ds} = \frac{\mu}{ms^2}$$

∴
$$\int v\, dv = -\frac{\mu}{m}\int \frac{ds}{s^2}$$

$$\frac{v^2}{2} = \frac{\mu}{ms} + C.$$

When $s = 2a$, $v = 0$,

∴
$$0 = \frac{\mu}{m2a} + C$$

∴
$$C = -\frac{\mu}{2am}$$

177

MOTION IN A STRAIGHT LINE UNDER VARIABLE FORCES

$$\therefore \quad v^2 = \frac{2\mu}{m}\left(\frac{1}{s} - \frac{1}{2a}\right)$$

as required.
Hence

$$v = \frac{ds}{dt} = -\sqrt{\left[\frac{2\mu(2a-s)}{2mas}\right]},$$

the negative sign being used since the particle moves towards O.

$$\therefore \quad -\int \sqrt{\left[\frac{mas}{\mu(2a-s)}\right]}\, ds = \int dt.$$

Making the substitution, $s = 2a\cos^2\theta$ on the left-hand side,

$$-\int \sqrt{\left[\frac{2ma^2\cos^2\theta}{2a\mu\sin^2\theta}\right]}(-4a\cos\theta\sin\theta)\,d\theta = t + A$$

$$\therefore \quad 4\sqrt{\frac{ma^3}{\mu}}\int \cos^2\theta\, d\theta = t + A$$

$$\therefore \quad 2\sqrt{\frac{ma^3}{\mu}}\int (1 + \cos 2\theta)\, d\theta = t + A$$

$$\therefore \quad 2\sqrt{\frac{ma^3}{\mu}}\left(\theta + \frac{\sin 2\theta}{2}\right) = t + B.$$

When $t = 0$, $s = 2a$, i.e. $2a\cos^2\theta = 2a$, giving $\theta = 0$. Substituting these values in the equation $B = 0$.

$$\therefore \quad t = 2\sqrt{\frac{ma^3}{\mu}}\left(\theta + \frac{1}{2}\sin 2\theta\right).$$

When the particle is halfway to O, $s = a$, i.e., $2a\cos^2\theta = a$, giving $\theta = \pi/4$. Substituting these values in the equation,

$$t = 2\sqrt{\frac{ma^3}{\mu}}\left(\frac{\pi}{4} + \frac{1}{2}\right).$$

Example 3. A particle falls from rest under gravity. Taking the air resistance as proportional to the square of the speed, find a formula expressing the speed v as a function of time t, and a formula expressing v as a function of the distance s fallen.

MOTION IN A STRAIGHT LINE UNDER VARIABLE FORCES

Let the particle have mass m and since the resistance is proportional to v^2, its magnitude may be expressed as mkv^2 (refer to Figure 10.3).

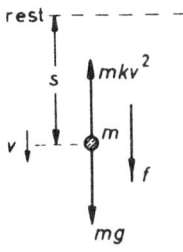

Figure 10.3

Since $P = mf$,

vertically downwards $mg - mkv^2 = mf$

$$\therefore \quad f = g - kv^2 \quad \ldots \text{(i)}$$

i.e.
$$\frac{dv}{dt} = g - kv^2.$$

Integrating,
$$\int \frac{dv}{g - kv^2} = \int dt,$$

giving
$$\frac{1}{2\sqrt{gk}} \log_e \left(\frac{\sqrt{g} + v\sqrt{k}}{\sqrt{g} - v\sqrt{k}} \right) = t + C.$$

The integral on the left-hand side can be obtained either directly or by splitting into partial fractions. When $t = 0, v = 0$.

$$\therefore \quad 0 = 0 + C$$

$$\log_e \left(\frac{\sqrt{g} + v\sqrt{k}}{\sqrt{g} - v\sqrt{k}} \right) = 2\sqrt{gk}\, t$$

$$\frac{\sqrt{g} + v\sqrt{k}}{\sqrt{g} - v\sqrt{k}} = e^{2\sqrt{gk}\, t}$$

giving
$$v = \sqrt{\frac{g}{k} \left(\frac{e^{2\sqrt{gk}\, t} - 1}{e^{2\sqrt{gk}\, t} + 1} \right)}$$

which expresses v as a function of t.

MOTION IN A STRAIGHT LINE UNDER VARIABLE FORCES

To obtain v as a function of s, equation (i) may be written

$$v\frac{dv}{ds} = g - kv^2.$$

Integrating
$$\int \frac{v\,dv}{g - kv^2} = \int ds$$

$\therefore \quad -\frac{1}{2k}\log_e(g - kv^2) = s + A$

when $v = 0, s = 0$

$\therefore \quad -\frac{1}{2k}\log_e g = 0 + A$

$\therefore \quad -\frac{1}{2k}\log_e(g - kv^2) = s - \frac{1}{2k}\log_e g$

$\therefore \quad -\frac{1}{2k}\log_e\left(\frac{g - kv^2}{g}\right) = s$

$\therefore \quad \frac{g - kv^2}{g} = e^{-2ks}$

giving
$$v^2 = \frac{g}{k}(1 - e^{-2ks})$$

$\therefore \quad v = \sqrt{\frac{g}{k}(1 - e^{-2ks})}$

which expresses v as a function of s.

Note that as s increases e^{-2ks} tends rapidly to zero. Hence v approaches the value $\sqrt{g/k}$ called the terminal velocity of the particle.

EXERCISES 10

1. A particle, of mass 2 kg, moves from rest on a smooth horizontal surface under a horizontal force which after t seconds has a magnitude of $(4 + 6t^2)$ N. Find the speed acquired by the particle in 2 seconds.

2. A particle, of mass m, moves in a straight line under the action of a single force ks, where s is its displacement from a fixed point O in the line and k is a positive constant. Find an expression for the speed of the particle in this position, assuming that it passed through O with speed u.

EXERCISES

3. A particle, of mass m, moves on a smooth horizontal surface under the action of a horizontal force of magnitude $mk\sqrt{v}$, where v is the speed of the particle and k a constant. If, at some instant, the particle is moving with speed u in the same direction as the force, show that this speed will be doubled in time $2(\sqrt{2} - 1)u^{1/2}/k$.

4. A 200 g mass moving on a smooth horizontal surface encounters a resistance of $14v^3/30$ N when its speed is v m/s. Show that its speed is reduced from $3\frac{1}{3}$ m/s to 1 m/s in a distance of 30 cm.

5. A car of mass 500 kg travelling at 36 km/h is slowed down by a gentle application of the brakes. If the force exerted by the brakes increases steadily from zero to 100 N in 10 s, find the distance travelled in this time and the speed attained. (Assume resistances to be negligible.)

6. A load of 60 kg is pulled up a smooth plane inclined at 30 degrees to the horizontal by a rope parallel to the plane. Initially the load is at rest and the initial tension applied in the rope is 414 N. Then, as the load moves, the tension increases uniformly at the rate of 120 N per metre travelled. Find the speed acquired in 1 m of travel from rest.

7. A particle of mass m moves in a straight line with initial speed u against a resisting force R, this being the only force acting on the particle. Obtain formulae for its speed v after time t, and its distance x from the starting point after that time, in each of the cases

(a) $R = mk_1 t$, (b) $R = mk_2 v$, (c) $R = mk_3 x$,

where k_1, k_2 and k_3 are constants.

Show that one of these motions is periodic and determine the period. (London)

8. A 60 g particle falls from rest through a medium in which the resistance is gkv N where v is the speed of the particle. Show that its speed approaches a terminal value.

If this terminal speed is 12 cm/s, find the value of k.

9. A particle, of mass m, moves along a line under the action of a force $k \sin(px)$, where x is its distance from a fixed point O in the line, k and p are constants, and the force is directed towards O. Given that the particle is instantaneously at rest when $x = a$, show that the maximum speed that it can attain is $2\sqrt{k/mp} \sin(\frac{1}{2}pa)$.

Find also the three positions nearest to O where this occurs.

10. A 200 tonne train travels on the level against resistances totalling 50 kN. The tractive force exerted by the engine is $1\,000/v$ kN when its speed is v m/s. Show that it accelerates from 10 m/s to 15 m/s in about 35 seconds.

MOTION IN A STRAIGHT LINE UNDER VARIABLE FORCES

11. A mass m falls from rest in a medium whose resistance is mkv when the speed of the mass is v. Show that when it has reached a speed V, it will have fallen a distance

$$\frac{g}{k^2}\log_e\left(\frac{g}{g-kV}\right) - \frac{V}{k}.$$

12. A particle moves from rest, at a point O, under the action of a single force whose magnitude is inversely proportional to $(1 + t)$ where t is the time taken. The force is directed away from O. If, after time T, its speed is U, show that after a time t it will be a distance s from O given by

$$s\log_e(1 + T) = U[(1 + t)\log_e(1 + t) - t].$$

13. A particle, of mass m, moves along a smooth horizontal surface under the action of a horizontal force of magnitude mk/s, where s is its distance from a fixed point O. The force is directed away from O. Initially the particle was projected with speed u, away from O, at a point distant a from O. If v is its speed when distant s from O, express s as a function of v.

14. A particle, of mass m, is acted on by a force $mk(u - v)^2$ where u is a constant and v is its speed at time t. If the particle is at rest when $t = 0$, find v as a function of t.

Find also the distance travelled in time t.

15. A particle P is moving on the axis of x, and at time t its acceleration and velocity in the direction of the positive axis of x are a and v respectively. Show that $a = v(dv/dx)$, where x is the displacement of P from the origin.

At time $t = 0$, P is at the origin O and $v = u(>0)$. Throughout the subsequent motion the particle is subject to a retardation kv^3, where k is a positive constant. Show that

$$v = u/(1 + ukx), \qquad t = \frac{x}{2u}(2 + ukx). \quad \text{(WJEC, part)}$$

16. A locomotive, of mass m, travels horizontally against resistances which can be expressed in the form mbv where v is its speed. If the tractive force exerted by the engine is of magnitude ma/v, find the time taken to increase the speed from u_1 to u_2. (a and b are constants.)

17. A particle of unit mass moves along a smooth horizontal surface under the action of a horizontal force of magnitude $(5 - k\sqrt{s})$ N, where s is its distance (in metres) from a fixed point O on the surface. The force is directed always away from O. If it starts from rest at O and comes to instantaneous rest again 400 m from O, find the value of the constant k.

EXERCISES

18. One end of a light inextensible string is attached to a particle A of mass $2m$ which lies on a smooth horizontal table. The string passes over a small smooth pulley P at the edge of the table. To the other end of the string is attached a particle B of mass m. At time $t = 0$, the system being at rest with the string taut and PB vertical, a variable force F is applied to A in the direction PA. If $F = mg(1 + \cos \omega t)$ at time t and ω is a constant, show that in the subsequent motion the displacement x of A from its initial position in the direction PA satisfies the equation

$$d^2x/dt^2 = \tfrac{1}{3}g \cos \omega t.$$

Find x in terms of g, ω and t. Show that the least value of the tension in the string is $\tfrac{2}{3}mg$. (JMB, part)

19. A particle, of mass m, moves in a straight line under a variable force such that when it is distant s from a fixed point in the line its speed v is given by $av^2 = s(b - v)$. Show that the resultant force acting on it at this instant is

$$\frac{m(b-v)^2}{a(2b-v)}.$$

Find the time taken for the particle to increase its speed from 0 to $\tfrac{1}{2}b$.

20. A particle moving in a straight line passes through a point O on it with a speed of 6 cm/s. It is acted on by a force which increases in direct proportion to the square root of the time taken from O, and is zero at O. The force is directed away from O. If the particle is $6\tfrac{1}{2}$ cm from O after 1 second, how far from O will it be after 4 seconds.

21. A particle, of mass m, is projected vertically upwards under gravity with initial speed u. The resistance to motion is mkv^2 where v is the speed at time t. Find the greatest height attained and the time taken to get there.

22. A particle, of mass m, moves along a straight line under the action of a force mk^2s, where s is its displacement from a point O in the line. If the particle starts from rest at a point distant a from O, show that

$$\frac{ds}{dt} = k\sqrt{s^2 - a^2}$$

Verify that $s = \tfrac{1}{2}a(e^{kt} + e^{-kt})$ is a solution of this differential equation satisfying the initial conditions. Find the time taken for the particle to double its initial distance from O.

23. A particle P is moving along the axis of x and at time t its acceleration and velocity, in the direction of the positive axis of x,

MOTION IN A STRAIGHT LINE UNDER VARIABLE FORCES

are a and v respectively. Show that $a = v\, dv/dx$ where x is the displacement of P from the origin.

A particle moves with velocity v in a horizontal straight line against a resistance $v^{3/2}$ per unit mass. Its initial velocity is V, and after a time t this has been reduced to kV; its distance from the initial position is then x. Prove that

$$x = k^{1/2} Vt.$$

Show also that the particle is never brought to rest and that its distance from the initial position is always less than $2V^{1/2}$.

(WJEC)

24. The only force acting on a particle moving in a straight line is a resistance $mk(c^2 + v^2)$ acting in that line; m is the mass of the particle, v its velocity and k, c are positive constants. The particle starts to move with velocity U and comes to rest in a distance s; its speed is $\tfrac{1}{3}U$ when it has moved a distance $\tfrac{1}{2}s$. Show that $63c^2 = U^2$. Show also that when the distance moved is x

$$63 \frac{v^2}{U^2} = 64\, e^{-2kx} - 1.$$

(JMB)

25. A particle, of mass m, is projected from a point O with velocity u. Thereafter it moves under the action of a force F, where $F = mau \sin(nt)/u$ at any time t (a, n are constants and $u = |u|$). Find the distance of the particle from O when $t = \pi$.

26. A particle of unit mass lies on a rough horizontal table and is repelled from a fixed point O on the table by a force kr^{-2}, where r is the distance from O. The particle starts from rest when $r = a$ and comes to rest again when $r = na$. Show that $n = k/\mu g a^2$, where μ is the coefficient of friction. (London, part)

• 27. A particle is projected vertically upwards with velocity u. The air resistance is proportional to the velocity, and is initially equal to λ times the weight of the particle. Prove that the particle reaches a height

$$\frac{u^2}{\lambda^2 g} [\lambda - \log_e (1 + \lambda)]$$

and that the time taken to reach this height is

$$\frac{u}{\lambda g} \log_e (1 + \lambda).$$

(Oxford)

28. Two particles of masses M and m ($M > m$) are connected by a light string which passes over a smooth fixed peg and hangs vertically

EXERCISES

on each side. The air resistance on each particle is kv^2 when the speed is v. Show that, if the resisted motion under gravity could continue indefinitely, the speed would tend to a limiting value V, where

$$V^2 = \frac{(M-m)g}{2k}.$$

Show that the acceleration of either particle, at speed v, has magnitude

$$\frac{(M-m)(V^2-v^2)}{(M+m)V^2}g.$$

Find the distance moved by either particle in attaining a speed of $\tfrac{1}{2}V$ from rest. (JMB)

29. A particle of unit mass starts from rest at a point A and is attracted towards a fixed point O by a force k/x^2, where x is the distance from O and k is constant. If $OA = a$, find the speed of the particle on reaching a point between O and A at a distance b from O and show that the time taken to this point is

$$\sqrt{\frac{a}{2k}}\int_b^a \sqrt{\frac{x}{a-x}}\,dx.$$

Using the substitution $x = a\sin^2\theta$, or otherwise, show that, if T is the time taken to reach the mid-point of OA,

$$k = \frac{(\pi+2)^2 a^3}{32T^2}.$$ (London)

30. A particle is projected along a rough horizontal surface with speed u. μ is the coefficient of friction and the air resistance per unit mass is gkv^2, when v is the speed. Find the distance travelled before it comes to rest and the time taken.

31. A particle of mass m is set in motion with speed u. Subsequently the only force acting upon the particle directly opposes its motion and is of magnitude $k(1+v^2)$, where v is its speed at time t and k is constant.

Show that the particle is brought to rest after a time $\dfrac{m}{k}\tan^{-1}u$ and find an expression in terms of m, k and u for the distance travelled by the particle in this time. (JMB)

MOTION IN A STRAIGHT LINE UNDER VARIABLE FORCES

32. (a) The position vector of a particle P at time t is given by
$$r = t \sin t\, i + t \cos t\, j$$
where i and j are constant orthogonal unit vectors. Calculate the velocity and acceleration vectors of P, and show that its speed is $\sqrt{(1 + t^2)}$.

(b) A particle is projected vertically upwards with speed u from a point O. The air resistance is proportional to the fourth power of the velocity and is initially equal to the weight of the particle. If v is the speed of the particle when it is at a height z above O, show that
$$\frac{d}{dz}(\tfrac{1}{2}v^2) + g\left(1 + \frac{v^4}{u^4}\right) = 0.$$

Solve this equation by writing w for v^2 and so prove that the particle is instantaneously at rest when its height above O is $\pi u^2/8g$.
(WJEB)

33. With the usual notation, prove that $\dfrac{dv}{dt} = v\dfrac{dv}{ds}$.

A particle P of mass m moves in a straight line and starts from a point O with velocity u. When $OP = x$, where $x \geq 0$, the velocity v of P is given by $v = u + \dfrac{x}{T}$, where T is a positive constant. Show that, at any instant, the force acting on P is proportional to v.

Given that the velocity of P at the point A is $3u$, calculate
  (i) the distance OA in terms of u and T,
  (ii) the time, in terms of T, taken by P to move from O to A,
  (iii) the work done, in terms of m and u, in moving P from O to A.
(AEB)

34. Two bodies, each of mass M, start simultaneously from rest and move in horizontal straight lines. The first moves under the action of a constant force of magnitude F. State the acceleration of this body.

The second body moves under the action of a force whose magnitude is variable but which does work at a constant rate K. Write down a differential equation which describes the motion of this body. If at a particular instant both bodies are moving at speed V, prove that $FV = 2K$.

Show further that at this instant the first particle has covered a distance which is $\tfrac{3}{4}$ of that which has been covered by the second particle.
(AEB)

35. A particle moving in a straight line is subject to a resisting force which produces a retardation kv^3 where v is the speed and

EXERCISES

k is a constant. If u is the initial speed, s is the distance moved in time t, and v is the speed at time t, find equations
  (i) connecting the variables v and t,
  (ii) connecting the variables v and s, and deduce that
$$ks^2 = 2t - 2s/u.$$

A bullet is fired horizontally from a rifle at a target 3 000 m away. The bullet is observed to take 1 second to travel the first 1 000 m and $1\frac{1}{4}$ seconds to travel the next 1 000 m. Assuming that the air resistance is proportional to v^3, and neglecting gravity, find the time taken to travel the last 1 000 m and show that the bullet reaches the target with speed $8\,000/13$ m s^{-1}. (JMB)

36. An aeroplane of mass 5 000 kg lands on an aircraft carrier at a relative velocity of 32 m s^{-1} parallel to the deck. It is immediately caught by an arrester gear which exerts a decelerating force directly proportional to the distance the aeroplane travels along the deck, the constant of proportionality being 3 000 N/m. When it has been slowed to 8 m s^{-1} the arrester gear releases the aeroplane. The pilot then brings it to rest with a constant braking force of 10 000 N.

Find the total landing distance measured from the point of contact with the arrester gear.

Show also that the time during which the arrester gear is operating is
$$\left(\frac{5}{3}\right)^{1/2} \sin^{-1}\left(\frac{\sqrt{15}}{4}\right) \text{ seconds.}$$

(The mass of the carrier is assumed to be very large compared with that of the aeroplane.) (JMB)

37. A particle of unit mass moves on a stright line subject to a force $k^2/2x^2$ towards a point O on the line, where x is the distance of the particle from O and k is a positive constant.

The particle is projected from the point $x = \frac{1}{4}a (>0)$ away from O with a speed $(3k^2/a)^{1/2}$. Show that during the motion
$$\left(\frac{dx}{dt}\right)^2 = k^2\left(\frac{1}{x} - \frac{1}{a}\right)$$
and that the particle comes to rest at the point $x = a$.

Express the time T for the particle to reach the point $x = a$ as a definite integral. By evaluating this integral, using the substitution $x = a\sin^2\theta$ or otherwise, show that
$$T = \left(\frac{\pi}{3} + \frac{\sqrt{3}}{4}\right)\frac{a^{3/2}}{k}.$$ (JMB)

11

EQUILIBRIUM OF A PARTICLE

11.1. RESOLUTION OF FORCES

A particle is said to be in equilibrium if it is at rest or moving with uniform velocity. Consider a particle in equilibrium under the action of forces $F_1, F_2, F_3 \ldots$ The particle has zero acceleration hence from $P = mf$ we have

$$\sum F = 0 \qquad \ldots \text{(i)}$$

The vector sum of the forces on a particle in equilibrium will be zero whatever the units in which the forces are measured, provided they are all measured in the same units. It is convenient in equilibrium problems to measure the forces in terms of the weight of unit mass of the particle. Thus a kilogramme weight (kg wt.) is the gravitational attraction on a mass of 1 kg placed in the position of the particle. Answers to problems have been given in these units. If it is required to convert to SI force units, we note that 1 kg wt. $= g$ newton where g is the magnitude of the local gravitational constant.

If each force F is resolved in three non-coplanar directions determined by the unit vectors $\hat{a}, \hat{b}, \hat{c}$, so that

$$F = F_1 \hat{a} + F_2 \hat{b} + F_3 \hat{c},$$

then equation (i) becomes

$$\sum (F_1 \hat{a} + F_2 \hat{b} + F_3 \hat{c}) = 0$$

or $\qquad \sum F_1 = 0, \qquad \sum F_2 = 0, \qquad \sum F_3 = 0.$

That is, if a particle is in equilibrium, the sums of the resolved parts of the forces in three non-coplanar directions are separately zero. The converse is also true, that if the resolved parts of the forces in three non-coplanar directions are separately zero, then the particle is in equilibrium. [In the case of *coplanar* forces it is sufficient that the sums in any *two* non-parallel directions are zero.] In practice it is usually desirable to resolve in mutually perpendicular directions.

RESOLUTION OF FORCES

Example 1. A particle of mass 2 kg is placed on a rough plane which is inclined at 30 degrees to the horizontal. A force of 2 kg wt. acts on the particle in a direction parallel to and up the plane. If the particle is just about to move up the plane, show that $\mu = 1/\sqrt{3}$ where μ is the coefficient of friction.

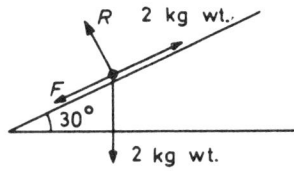

Figure 11.1

Let R be the normal reaction of the plane on the particle. Since the particle is about to slip up the plane, the frictional force F acts *down* the plane. Since the particle is about to slip, $F = \mu R$.

Resolving along the plane (refer to *Figure 11.1*)

$$\mu R + 2 \sin 30° - 2 = 0$$
$$\therefore \quad \mu R + 1 = 2 \quad \ldots (i)$$

Resolving perpendicular to the plane

$$R - 2 \cos 30° = 0$$
$$\therefore \quad R = \sqrt{3} \quad \ldots (ii)$$

Substituting from equation (ii) in equation (i)

$$\mu\sqrt{3} + 1 = 2$$
$$\mu\sqrt{3} = 1$$
$$\mu = 1/\sqrt{3}.$$

Example 2. A light inextensible string whose length is greater than AB is attached at its ends to two fixed points A, B on the same horizontal level. The string is threaded through a small smooth ring C of weight 4 W which can move freely on the string. The ring is pulled aside by a horizontal force 3 W in the vertical plane through AB. In equilibrium $\angle ABC = 80$ degrees and the vertical through C passes between A and B. Find the inclination of AC to the horizontal and the tension in the string.

EQUILIBRIUM OF A PARTICLE

Let $\angle BAC = \phi$ and T be the tension in the string. Since the ring C is smooth, the tension in each portion of the string AC and BC is T.

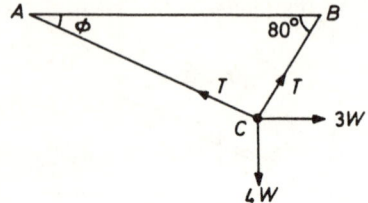

Figure 11.2

Resolving horizontally (refer to Figure 11.2),

$$T \cos \phi - T \cos 80° - 3W = 0.$$

$\therefore \qquad T \cos \phi - T \cos 80° = 3W \qquad \ldots \text{(i)}$

Resolving vertically

$$T \sin \phi + T \sin 80° - 4W = 0.$$

$\therefore \qquad T \sin \phi + T \sin 80° = 4W \qquad \ldots \text{(ii)}$

Dividing equation (i) by equation (ii),

$$\frac{\cos \phi - 0.1736}{\sin \phi + 0.9848} = \frac{3}{4}$$

$\therefore \qquad 4 \cos \phi - 0.6944 = 3 \sin \phi + 2.9544$

$\therefore \qquad 4 \cos \phi - 3 \sin \phi = 3.6488$

$\therefore \qquad 5 \cos (\phi + \varepsilon) = 3.6488$

where $5 \cos \varepsilon = 4$ and $5 \sin \varepsilon = 3$, hence $\varepsilon = \tan^{-1} \frac{3}{4}$ ($0 \leqslant \varepsilon \leqslant \pi/2$) and $\varepsilon = 36° 52'$.

$\therefore \qquad 5 \cos (\phi + 36° 52') = 3.6488$

$$\cos (\phi + 36° 52') = 0.7298$$

$$\phi + 36° 52' = 43° 8'$$

$$\phi = 6° 16'.$$

190

THREE-FORCE PROBLEMS

Substituting in equation (i)

$$T\cos 6° 16' - T\cos 80 = 3W$$
$$T\,0·9941 - T\,0·1736 = 3W$$
$$T = \frac{3W}{0·8205}$$
$$T = 3·66W.$$

The tension in the string is $3·66W$ and AC is inclined at $6° 16'$ to the horizontal.

11.2. THREE-FORCE PROBLEMS

Problems about a particle in equilibrium under the action of three forces can, of course, be solved by the general method of resolving described in the previous section. However, in the three-force case, calculations using a vector diagram may be more convenient.

We have seen that, when a particle is in equilibrium, the vector sum of the forces acting is zero. Hence they can be represented by the sides of a closed polygon taken in order. If there are only three forces, this polygon is a triangle which is particularly suitable for trigonometrical calculation.

So, instead of resolving, we may *if convenient* obtain our two equations from the trigonometry of the vector triangle.

Example 1. A particle of mass 2 kg hangs vertically on the end of an elastic string of modulus 4 kg wt. *and natural length* 0.5 m. *The particle is pulled aside by a horizontal force **F** until the string is* 1 m *long. Find the inclination of the string to the vertical and the magnitude of **F**.*

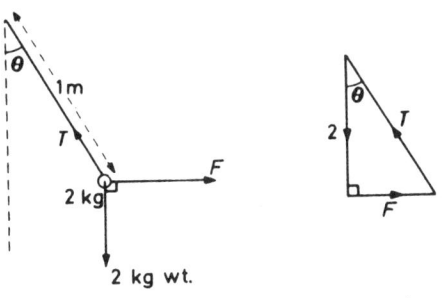

Space diagram   Vector diagram

Figure 11.3

EQUILIBRIUM OF A PARTICLE

Since the string is elastic

$$T = \lambda x / l$$
$$= 4 \times 0 \cdot 5 / 0 \cdot 5$$
$$= 4 \text{ kg wt.}$$

The vector diagram is a right-angled triangle (refer to *Figure 11.3*). Hence we may apply Pythagoras' theorem and

$$F = \sqrt{T^2 - 2^2}$$
$$= \sqrt{4^2 - 2^2}$$

$\therefore \qquad F = 2\sqrt{3} \text{ kg wt.}$

Also, from the vector diagram

$$\cos \theta = 2/T$$
$$= 2/4$$
$$= \tfrac{1}{2}.$$

$\therefore \qquad \theta = 60°.$

The string is inclined at 60 degrees to the downward vertical and the horizontal force is of magnitude $2\sqrt{3}$ kg wt.

Example 2. This is an alternative approach to Example 1 of Section 11.1, in which the number of forces is reduced to three. This is done by replacing the normal and frictional reactions by their resultant S, the total reaction of the plane on the particle.

Since the particle is about to slip, λ is the angle of friction (refer to Section 7.4) and $\tan \lambda = \mu$.

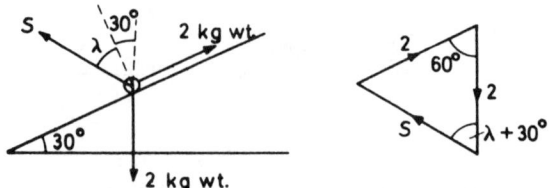

Figure 11.4

THREE-FORCE PROBLEMS

The vector triangle, having two sides equal, is isosceles. Hence the base angles are equal; (referring to *Figure 11.4*) we have

$$2(\lambda + 30°) + 60° = 180°$$

giving $\lambda = 30°$

$$\therefore \quad \mu = \tan 30° = 1/\sqrt{3}.$$

Example 3. A particle of mass 3 kg hangs in equilibrium supported by two strings inclined at 25 degrees and 50 degrees, respectively, to the horizontal. Find the tension in each string.

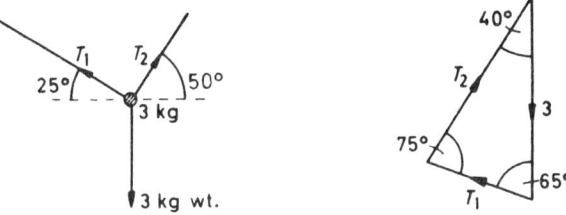

Figure 11.5

Applying the sine rule to the vector triangle (refer to *Figure 11.5*),

$$\frac{T_1}{\sin 40°} = \frac{T_2}{\sin 65°} = \frac{3}{\sin 75°}$$

$$\therefore \quad T_1 = \frac{3 \sin 40°}{\sin 75°} = 1·997 \text{ kg wt.}$$

and $$T_2 = \frac{3 \sin 65°}{\sin 75°} = 2·816 \text{ kg wt.}$$

The tension in the string inclined at 25 degrees to the horizontal is 2·00 kg wt., and in the other string 2·82 kg wt.

Exercises 11a

(At this stage the reader is recommended to solve three-force problems both by resolving and by means of the vector triangle.)

1. A particle of mass 9 kg hangs from a point on a light inextensible string. It is pulled by a horizontal force of $3\sqrt{3}$ kg wt. so that it is in equilibrium with the string inclined at an angle θ with the downward vertical. Find θ and the tension in the string.

2. A particle rests in equilibrium on a rough plane inclined at

193

an angle α to the horizontal. If λ is the angle of friction between the particle and plane, show that $\alpha \leq \lambda$.

3. A and B are two points at the same horizontal level 80 cm apart. A light elastic string of natural length 40 cm is stretched between A and B. When a mass of 90 g is attached to the mid-point of the string, it falls and eventually comes to rest 30 cm below AB. Find the modulus of the string.

4. A small ring of weight W can slide on a smooth circular wire of radius r fixed in a vertical plane. It is tied to the topmost point of the wire by a light inelastic string of length $4r/3$. Find the tension in the string and the reaction between ring and wire in the equilibrium position.

5. Three forces P, Q, R act on a particle which is in equilibrium. The angle between P and Q is 150 degrees, between Q and R is 120 degrees. If $P = 3\sqrt{3}$ kg wt., find Q and R.

6. A particle of mass 70 kg is placed on a smooth plane which is inclined at an angle of 30 degrees to the horizontal. The particle is kept in equilibrium by a horizontal force F. Find the magnitude of F and of the reaction of the plane.

7. If in Question 6 the force F acted parallel to the plane, find its magnitude.

8. A particle of weight 13 kg is suspended by two strings of length 1·2 m and 0·5 m from two points, A and B, on the same horizontal level. If A and B are 1·3 m apart, find the tension in the strings.

9. A light inextensible string has one end A attached to a fixed point and the other end B to a small smooth ring of negligible weight. Another similar string CBD has one end attached to a fixed point C on the same horizontal level as A, passes through the smooth ring B and supports a freely hanging weight W at its other end. In equilibrium, AB is inclined at 65 degrees to the vertical, find the tension in AB.

10. Prove Lami's Theorem: If a particle is in equilibrium under the action of three forces, the magnitudes of each of them are proportional to the sine of the angle between the other two.

11. A particle of mass 100 g rests on a rough plane which is inclined at an angle of 60 degrees to the horizontal. A force F is applied up the plane. If the coefficient of friction between the particle and the plane is 0·75, find the magnitude of F when the particle is on the point of slipping up the plane.

12. A particle of weight w rests on a rough plane which is inclined at an angle α to the horizontal, $\alpha > \lambda$ where λ is the angle of friction between particle and plane. A force F parallel to the plane supports

the particle which is just about to move *down* the plane, show that $F = w \sin(\alpha - \lambda) \sec \lambda$.

13. If, in Question 12, the magnitude of the force F is increased until the particle is just about to move *up* the plane, find the new magnitude of F in terms of ω, α and λ.

14. A light inextensible string of length 0·5 m has one end firmly attached to a point A of a rough horizontal rail. The other end is attached to a small ring, B, of weight W, which can slide on the horizontal rail. A weight $4W$ is attached to the mid-point of the string. If the coefficient of friction between the rail and the ring is $\frac{3}{4}$, show that for the system to be in equilibrium $AB \leq 0.36$ m.

15. A string of length a has one end A tied to a fixed point and the other end B to a small smooth ring. A second string CBD has one end attached to a fixed point C, where AC is horizontal and $AC = 6a$, passes through the ring and supports a weight W hanging freely from the other end D. Show that in equilibrium CB and BD are equally inclined to AB, and that AB is inclined to the vertical at an angle $\sin^{-1}(\frac{3}{4})$. Find the tension in AB. (London)

11.3. HARDER EXAMPLES

When a system of particles is connected by light inextensible strings or light rods, the forces on the particles can often be obtained by considering the equilibrium of each particle separately.

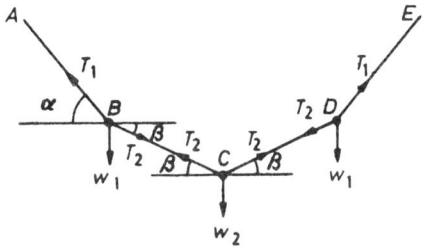

Figure 11.6

Example 1. Three particles of weights w_1, w_2, w_1, are attached at points B, C, D of a light inextensible string which hangs from two points A and E, at the same horizontal level. In equilibrium the system hangs symmetrically with AB, DE, each inclined at an angle α to the horizontal and BC, CD each inclined at an angle β to the horizontal. Show that

$$\frac{\cot \alpha}{\cot \beta} = \frac{w_2}{(2w_1 + w_2)}.$$

EQUILIBRIUM OF A PARTICLE

If ABCDE has the form of half a regular octagon, deduce that $w_1/w_2 = 1 + \sqrt{2}$ $[\cot 22\tfrac{1}{2}° = 1 + \sqrt{2}]$.

Let T_1, T_2 be the tension in AB and BC and hence by symmetry they are the tensions in ED and DC respectively.

Consider the equilibrium of particle w_2 at C. Resolving vertically (refer to *Figure 11.6*).

$$2T_2 \sin \beta - w_2 = 0 \qquad \ldots \text{(i)}$$

Consider the equilibrium of particle w_1 at B.
Resolving vertically

$$T_1 \sin \alpha - T_2 \sin \beta - w_1 = 0 \qquad \ldots \text{(ii)}$$

Resolving horizontally

$$T_1 \cos \alpha - T_2 \cos \beta = 0 \qquad \ldots \text{(iii)}$$

From equations (i) and (ii)

$$T_1 \sin \alpha - \tfrac{1}{2} w_2 = w_1$$

$$\therefore \quad T_1 \sin \alpha = w_1 + \tfrac{1}{2} w_2 \qquad \ldots \text{(iv)}$$

Substituting for T_1 and T_2 from equations (iv) and (i) in equation (iii)

$$\frac{(w_1 + \tfrac{1}{2} w_2)}{\sin \alpha} \cos \alpha = \frac{w_2}{2 \sin \beta} \cos \beta$$

$$\therefore \quad (w_1 + \tfrac{1}{2} w_2) \cot \alpha = \frac{w_2}{2} \cot \beta$$

$$\therefore \quad \frac{\cot \alpha}{\cot \beta} = \frac{w_2}{(2w_1 + w_2)}.$$

The internal angle of a regular octagon is 135 degrees. Therefore, by symmetry (refer to *Figure 11.6*).

$$2\beta + 135° = 180°$$

$$\therefore \quad \beta = 22 \cdot 5°$$

also

$$\alpha + 135° - \beta = 180°$$

$$\therefore \quad \alpha = 67 \cdot 5°$$

$$\therefore \quad \frac{w_2}{2w_1 + w_2} = \frac{\cot 67 \cdot 5°}{\cot 22 \cdot 5°}.$$

HARDER EXAMPLES

Now $\quad \cot\theta = \dfrac{1}{\tan\theta} = \dfrac{1}{\cot(90° - \theta)}$

$\therefore \quad \dfrac{w_2}{2w_1 + w_2} = \dfrac{1}{\cot^2 22\cdot5°}$

$\qquad\qquad\qquad = \dfrac{1}{(1 + \sqrt{2})^2}$

$\qquad\qquad\qquad = \dfrac{1}{3 + 2\sqrt{2}}$

$\therefore \quad (3 + 2\sqrt{2})w_2 = 2w_1 + w_2$

$\quad (2 + 2\sqrt{2})w_2 = 2w_1$

$\quad \dfrac{w_1}{w_2} = 1 + \sqrt{2}.$

Example 2. *A rough circular wire of radius r, centre O, is fixed in a vertical plane. Two rings A and B each of mass m can slide on the wire and are joined by a light elastic string of natural length r and modulus 0·5 mg. AB is horizontal and above O and the two rings are about to slip upwards. If angle $AOB = 2\theta$, prove that $2\sin\theta = 1 + 2\tan(\theta + \lambda)$ where $\mu \,(= \tan\lambda)$ is the coefficient of friction between a ring and the wire.*

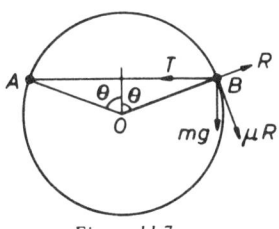

Figure 11.7

Let T be the tension in the elastic string,

$\therefore \quad T = 0\cdot5mg(2r\sin\theta - r)/r \quad$ (refer to *Figure 11.7*)

$\therefore \quad T = 0\cdot5mg(2\sin\theta - 1). \qquad\qquad \ldots\text{(i)}$

Let R be the reaction of the circular wire on the ring, its direction will be radially outwards. Since the rings are about to slip upwards,

EQUILIBRIUM OF A PARTICLE

the frictional force on the ring will be μR tangentially downwards (refer to *Figure 11.7*).

Resolving horizontally

$$T - R \sin \theta - \mu R \cos \theta = 0 \qquad \ldots \text{(ii)}$$

eliminating T from equations (i) and (ii)

$$mg \sin \theta - 0{\cdot}5mg = R \sin \theta + \mu R \cos \theta \qquad \ldots \text{(iii)}$$

Resolving vertically

$$mg = R \cos \theta - \mu R \sin \theta \qquad \ldots \text{(iv)}$$

dividing equation (iii) by equation (iv)

$$\sin \theta - 0{\cdot}5 = (\sin \theta + \mu \cos \theta)/(\cos \theta - \mu \sin \theta)$$

$$\therefore \qquad \sin \theta - 0{\cdot}5 = \frac{\tan \theta + \mu}{1 - \mu \tan \theta}$$

but $\qquad \mu = \tan \lambda$

$$\therefore \qquad \sin \theta - 0{\cdot}5 = \tan(\theta + \lambda)$$

$$2 \sin \theta = 1 + 2 \tan(\theta + \lambda).$$

Example 3. Find the least force that will support a particle of weight W on a rough plane inclined at an angle α to the horizontal. The angle of friction between the particle and plane is λ ($\lambda < \alpha$).

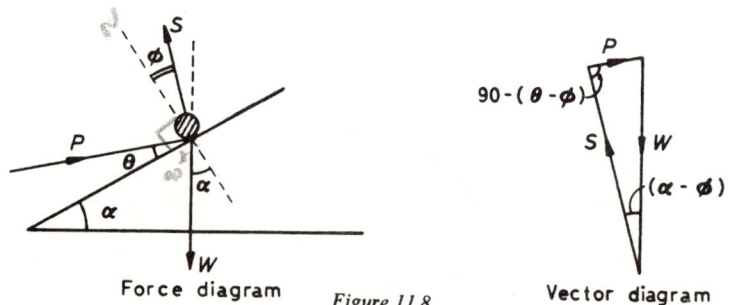

Force diagram   *Figure 11.8*   Vector diagram

Since $\lambda < \alpha$, the particle will tend to slip down the plane and must be supported by a force of magnitude P, inclined at an angle θ say, to the plane as shown in *Figure 11.8*.

P will be a minimum when the particle is about to slip (and $\phi = \lambda$), and θ has an optimum value. Thus putting $\phi = \lambda$, we have to find P as a function of θ, and choose θ to make P a minimum.

Applying the sine rule to the vector diagram

$$\frac{P}{\sin(\alpha - \phi)} = \frac{W}{\sin[90 - (\theta - \phi)]}$$

$$\therefore \quad P = \frac{W \sin(\alpha - \phi)}{\cos(\theta - \phi)}$$

or, since $\phi = \lambda$,
$$P = \frac{W \sin(\alpha - \lambda)}{\cos(\theta - \lambda)}.$$

As θ varies, $\cos(\theta - \lambda)$ has a maximum value of 1 when $(\theta - \lambda) = 0$. Hence the force P will have a minimum value of $W \sin(\alpha - \lambda)$ when $\theta = \lambda$.

*Exercises 11b**

1. Two small particles each of weight W are connected to a light inextensible string $ABCD$ at points B and C. The string hangs freely in a vertical plane with its ends A and D connected to two points on the same horizontal level. If in the equilibrium position BC is horizontal and $ABCD$ has the form of half a regular hexagon, find the tensions in the three strings AB, BC, CD.

2. A particle, weight W, rests on a rough plane which is inclined at an angle α to the horizontal. λ is the angle of friction between the plane and the particle. A force F acts on the particle in an upward direction making an angle θ with the plane, θ being measured in the same direction as α. Show that when the particle is on the point of moving up the plane, $F = W \sin(\alpha + \lambda)/\cos(\theta - \lambda)$.

Deduce the minimum value of F as θ varies.

3. Two rough rods OA, OB are fixed at right angles, OA being vertically downwards. Two small rings of weights $3W$ and $4W$ can slide on the two rods, the one of weight $3W$ on OA, the other on OB. The rings are joined by a light inextensible string and $\mu\ (= \frac{1}{2})$ is the coefficient of friction between each ring and the rod. Prove that in equilibrium the greatest inclination of the string to the vertical is $\tan^{-1}(\frac{7}{4})$ and find the tension in the string in this position.

4. A triangular wedge has two equally rough surfaces inclined in opposite directions at 60 degrees and 30 degrees to the horizontal. Two particles of weights 3 kg and 2 kg rest on the two surfaces and are joined by a light inextensible string passing over a smooth pulley at the vertex of the wedge. The string lies in a vertical plane which intersects the two surfaces in lines of greatest slope. If the heavier

* Exercises marked thus, †, have been metricized, *see* Preface.

particle rests on the steeper surface, and is just about to slip downwards, find μ the coefficient of friction between a particle and a surface of the wedge.

5. A and D are fixed points on a horizontal ceiling at a distance 52 cm apart. A light string $ABCD$ (of length greater than 52 cm) is attached to A and D at its ends and carries weights W g and 12 g at B and C respectively, where $AB = 24$ cm and $CD = 24$ cm. When hanging in equilibrium, AB is inclined at 45 degrees to the horizontal and CD is inclined at 60 degrees to the horizontal. Calculate (a) the inclination of BC to the horizontal, (b) the value of W. (London)†

EXERCISES 11

1. A particle is in equilibrium under the action of forces, F_1, F_2, F_3 and F_4. If $F_1 = 3i - 4j - k$, $F_2 = j + 2k$, $F_3 = -4i + 4j - k$, find F_4.

2. A particle of weight W hangs on the end of a light inelastic string. It is held in equilibrium with the string inclined at an angle θ to the vertical by a force F applied at right angles to the string. Find F and the tension in the string.

3. Find the least force required to move a particle of weight W along a rough horizontal surface if the angle of friction between the particle and the plane is λ.

4. A smooth circular wire is fixed with its plane vertical. A small bead of weight W is threaded on the wire and tied to its topmost point by an elastic thread, whose natural length is equal to the radius of the wire. If there is a position of equilibrium with the string stretched and inclined at an angle θ to the vertical, find the modulus of the string.

5. A mass of 100 g is suspended by two light inextensible strings of lengths 18 cm and 24 cm to two fixed points A and B on the same horizontal level. If $AB = 30$ cm, find the tensions in the strings.

6. A light inextensible string of length 4 m has its ends attached to two points, A and B, 2 m apart on the same horizontal level. A smooth ring of weight 8 kg is threaded on the string and held in equilibrium vertically below B by a horizontal force F. Find the magnitude of F and the tension in the string.

7. A particle of mass 5 kg rests in equilibrium on a rough plane which is inclined at 15 degrees to the horizontal. A force of 3 kg wt. is applied to the particle in a direction parallel to and down the plane. If the particle is just about to move, find μ the coefficient of friction between the particle and the plane.

EXERCISES

8. A light inextensible string has one of its ends attached to a fixed point A and the other end B to a small smooth ring of negligible weight. A similar string CBD has its end C attached to a fixed point on the same horizontal level as A, passes through the smooth ring at B and supports a freely hanging weight W at its other end D. In equilibrium the tension in AB is $0.75W$, find the inclinations of AB and BC to the vertical.

9. A rough circular wire of radius r, centre O is fixed in a vertical plane. Two rings each of mass m can slide on the wire and are joined by a light elastic string AB of natural length r and modulus mg. AB is horizontal and below O and the two rings are just about to slip downwards. If $\angle AOB = 2\theta$, prove that $2 \sin \theta = 1 - \tan(\theta + \lambda)$.

10. A long light string is tied at one end to a fixed point A. The string passes through a small smooth fixed ring B distant a from A and at the same level as A. A particle of weight $2W$ is attached to the other end of the string and hangs freely. A smooth bead C of weight $3W$ is free to slide on the string between A and B. Show that in the equilibrium position AC and BC are each inclined at an angle $\cos^{-1} \frac{3}{4}$ to the vertical.

The bead C is now fixed to the string at a distance a from A. Show that in the new equilibrium position (when C is not at B) the inclination of BC to the vertical is

$$\cos^{-1}\left(\frac{\sqrt{19}+1}{6}\right).$$

(London)

11. Show that, if a set of coplanar forces acting at a point is in equilibrium, the forces may be represented in magnitude and direction by the sides of a closed polygon taken in order. A light inextensible string of length greater than AB is attached at its ends to two fixed points A and B at the same horizontal level. A small smooth ring C of weight $5W$ is threaded on the string and hangs below AB. C is pulled aside by a horizontal force W in the vertical plane through AB. In equilibrium the angle $CAB = 30$ degrees and the vertical through C passes between A and B. Find graphically or otherwise (a) the angle CBA (b) the tension in the string.

(London)

12. A small ring A of weight W is constrained to slide on a rough horizontal rail, the coefficient of friction between the ring and the rail being μ. One end of a weightless inextensible string of length $2l$ is tied to the ring while the other end is secured to the rail at a point B such that $AB = 2x$. A weight $4W$ attached to the mid-point

C of the strings hangs in equilibrium. Show that
$$x \leqslant \frac{3\mu l}{\sqrt{\{4 + 9\mu^2\}}}.$$

The ring is just about to slide on the rail when an additional weight $2W$ is hung from C. If $\angle BAC = 60$ degrees, find μ. (WJEC)

13. A light elastic string is attached at its ends to two fixed points in a horizontal line, the distance between the points being equal to the natural length of the string. A particle of weight W, attached to the mid-point of the string, hangs freely in equilibrium when each half of the string makes an angle 30 degrees with the horizontal. Show that the modulus of elasticity of the string is $W(3 + 2\sqrt{3})$.

If the weight of the particle is increased by an amount w so that each half of the string makes an angle 60 degrees with the horizontal when in the new equilibrium position, find w in terms of W.

(London)

14. A particle of weight W is attached by two light inextensible strings each of length a to two fixed points distant a apart in a horizontal line. Write down the tension in either string.

One of the strings is now replaced by an elastic string of the same natural length, and it is found that in the new position of equilibrium this string has stretched to a length $5a/4$. Prove that the modulus of elasticity of this string is $7W/\sqrt{39}$, and show that the tension in the other string has been increased in the ratio $5:\sqrt{13}$. (London)

15. A light inextensible string $ABCD$ has two particles of 10 kg wt. and W kg wt. attached at B and C respectively. The ends of the string are attached to points A and D at the same horizontal level. In the equilibrium position, AB, BC and CD are inclined at angles of 30, 60 and 60 degrees respectively to the vertical. Find the value of W and the tensions in the strings.

16. A triangular wedge has two equally rough surfaces S_1, S_2 inclined in opposite directions at angles α and β to the horizontal respectively. A particle of mass M rests on S_1 and another of mass m on S_2. The particles are joined by a light inelastic string which passes over a smooth pulley at the vertex of the wedge. If M is about to slip downwards and $\tan \lambda$ is the coefficient of friction between a particle and a surface of the wedge show that

$$\frac{m}{M} = \frac{\sin(\alpha - \lambda)}{\sin(\beta + \lambda)}.$$

17. Two small rings P and Q, of weights w_1 and w_2, respectively ($w_1 > w_2$), can slide on a smooth circular wire, centre O, fixed in a

EXERCISES

vertical plane. They are connected by a light string which, in length, is less than the diameter of the circle. When the rings are in equilibrium with P higher than Q, and with the string taut and above O, the angle POQ is 2α. If the string makes an angle θ with the horizontal, prove that

$$(w_1 + w_2)\tan\theta = (w_1 - w_2)\tan\alpha.$$

Show also that the tension in the string is

$$\frac{2w_1 w_2 \sin\alpha}{\sqrt{(w_1^2 + w_2^2 + 2w_1 w_2 \cos 2\alpha)}}.$$ (WJEC)

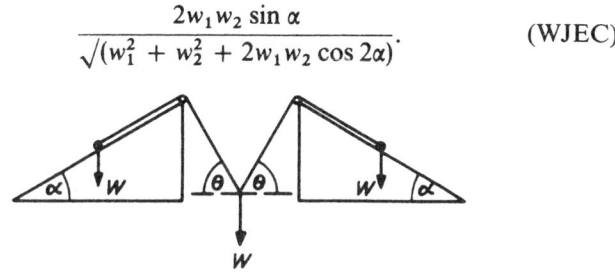

Figure 11.9

18. *Figure 11.9* shows two fixed planes each inclined at an angle α ($<30°$) to the horizontal. Two particles, each of weight W, rest one on each of the planes and are connected by a light inextensible string which passes over a small smooth pulley at the top of each plane and carries a smooth ring of weight W which hangs down between the planes. The coefficient of friction between the particles and the planes is μ. Prove that, if θ is the inclination of the string to the horizontal then, provided that the string is long enough and that $\mu > \frac{1}{2}\sec\alpha - \tan\alpha$, there will be equilibrium if

$$\operatorname{cosec}\theta < 2(\sin\alpha + \mu\cos\alpha).$$

What happens if $\mu < \frac{1}{2}\sec\alpha - \tan\alpha$? (Oxford)

19. Three light connected strings AB, BC and CD have a weight W attached at B and a weight $2W$ at C. AB is an inextensible string of length a. BC and CD are elastic strings each of natural length a and of the same modulus of elasticity. A and D are fixed to points at the same horizontal level. The system hangs freely with CD at 60 degrees to the horizontal and BC perpendicular to CD. Find the tensions in BC and CD in terms of W and the inclination of AB to the horizontal.

Show that the distance AD is $a(1 + \sqrt{3})$ and find the modulus of elasticity of the elastic strings. (London)

EQUILIBRIUM OF A PARTICLE

20. Two small rings A, B, each of weight W, are threaded on a fixed rough horizontal wire. The rings are connected by a light inelastic string, of length $2a$, to the mid-point C of which is attached a particle of weight $2W$. The system rests in equilibrium with $\angle ACB = 2\beta$.

(a) Find the tension in each part of the string.

(b) Find, in terms of W and β, the normal reactions and the frictional forces between the wire and the rings. (The directions of the frictional forces acting on the rings must be clearly shown in a diagram.)

If the coefficient of friction between each ring and the wire is $\frac{3}{8}$, show that $AB \leqslant 6a/5$. (JMB)

21. Two small rings of weights $3w$ and $5w$ are capable of sliding on a smooth circular wire of radius a fixed in a vertical plane. The rings are connected together by a light inextensible string of length $8a/3$ which passes over a smooth peg fixed at a height $a/3$ vertically above the highest point of the wire. The rings rest on opposite sides of the vertical through the peg. Find the reaction of the wire on each ring and show that the tension in the string is $15w/4$.

If the wire is uniform and of weight w, find the horizontal and vertical components of the external force required to keep the wire in position, indicating the directions clearly. (London)

22. A light inextensible string PQ is attached by its ends to two points at the same horizontal level. Five equal particles are attached to the string at points A, B, C, D, E. The system hangs in equilibrium with the horizontal projections of PA, AB, BC, CD, DE and EQ all equal to a. If the depth of C below PQ is $3a$, find the inclinations to the horizontal of PA, AB, and BC.

23. Two beads A and B, of mass m and $3m$ respectively, are threaded on a rough straight horizontal wire; the coefficient of friction between each bead and the wire is μ. A and B are joined by a smooth string on which a bead C, of mass $2m$, can slide freely. Show that, when the bead C hangs in equilibrium under the wire, the strings AC and BC are equally inclined to the horizontal. Show also, that when A and B are at the greatest distance apart consistent with equilibrium,

(a) the friction is limiting at A but not at B,

(b) angle $BAC = \cot^{-1} 2\mu$.

A horizontal force is now applied at B in the direction BA. Find the value of this force when the bead B is also on the point of slipping. (WJEC)

24. Points A, B, C and D on a horizontal ceiling lie at the corners of a square with sides 2 m. Four light, inextensible strings AO, BO,

204

EXERCISES

CO and *DO*, each of length 2 m, are attached to the ceiling at the points *A*, *B*, *C* and *D* and to a particle of mass 2 kg at the point *O*. Show that, in equilibrium, the point *O* is $\sqrt{2}$ m below the level of the ceiling. Find the tension in each of the four strings.

The strings *BO* and *CO* are now removed and replaced by a single, light, inextensible string which is attached to *O* and the point *E* which is the mid-point of the side *BC* of the square. If the point *O* remains at $\sqrt{2}$ m below the level of the ceiling, show that $EO = \sqrt{3}$ m and find the tensions in the strings *EO*, *AO* and *DO*.

(AEB)

25. Two rods *OA*, *OB* are fixed in a vertical plane with *O* uppermost, each rod making an acute angle α with the vertical. Two smooth rings of equal weight *W*, which can slide one on each rod, are connected by a light inextensible string, upon which slides a third smooth ring of weight 2*W*. Show that, in the symmetrical position of equilibrium, the angle between the two straight pieces of the string is 2β, where $\tan \beta = 2 \cot \alpha$.

Show that the tension in the string is $W\sqrt{(1 + 4\cot^2 \alpha)}$ and that the reaction between the rings and the rods is $2W \operatorname{cosec} \alpha$.

(WJEC)

26. A particle of mass $3m$ is tied to the end *C* and a particle of mass *m* is tied at the mid-point *B* of a light unstretched elastic string *ABC*. The end *A* of the string is fixed and a horizontal force of magnitude $4mg$ is applied to the particle at *C* so that the system hangs in equilibrium as shown in the diagram. Calculate

(i) the tensions in *BC* and *AB*,
(ii) the inclinations of *BC* and *AB* to the vertical.

Given that the modulus of elasticity of the string is $6mg$, show that, for this position of equilibrium, $AB:BC = (6 + 4\sqrt{2}):11$.

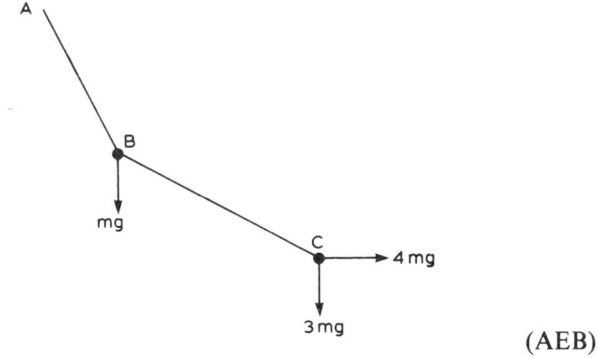

(AEB)

12
WORK, POWER AND ENERGY

12.1. WORK DONE BY A CONSTANT FORCE

Definition—When the point of application of a force F, *constant in magnitude and direction*, is given a displacement δr, then the work done by F is given by $W = F \cdot \delta r$.

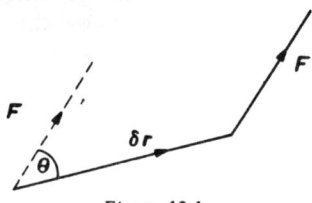

Figure 12.1

Hence, $W = F\delta r \cos\theta$ where θ is the angle between F and δr (refer to *Figure 12.1*). Thus the work done by F may be regarded either as the product of the force and the projection on it of the displacement, or as the product of the component of the force in the direction of the displacement and the distance moved.

In particular, if δr is perpendicular to F, $W = 0$, and if δr is parallel to F, and in the same sense, $W = F\delta r$ (if δr is parallel to F and in the opposite sense, $W = -F\delta r$).

SI unit of work ... N m = joule (J)

If a system of constant forces, $F_1, F_2, F_3, \ldots, F_n$, acts on a particle which undergoes a displacement δr, then

Work done $= F_1 \cdot \delta r + F_2 \cdot \delta r + \ldots F_n \cdot \delta r$

$\qquad = (\sum F) \cdot \delta r$ (by the distributive law for scalar products refer to Section 3.8)

$\qquad = P \cdot \delta r$

where P is the resultant of the system of forces, $F_1, \ldots, F_n$, and the

WORK DONE BY A CONSTANT FORCE

work done is the same as if the system of forces was replaced by its resultant.

If a single constant force F acts on a particle which undergoes several displacements, $\delta r_1, \delta r_2, \ldots \delta r_n$,

Work done $= F . \delta r_1 + F . \delta r_2 + \ldots + F . \delta r_n$

$\qquad = F . (\Sigma \delta r)$ (by the distributive law for scalar multiplication).

The work done is equal to the scalar product of F and the vector sum of the displacements.

Example 1. A particle moves from a point with position vector $2i - j$ to a point with position vector $3i + j + 2k$. Among the forces acting on it is a force $F = 2i - 4j - 3k$ which remains constant throughout the motion. Find the work done by F.

The displacement of the particle

$$\delta r = (3i + j + 2k) - (2i - j)$$
$$= i + 2j + 2k.$$

By definition, the work done by

$$F = F . \delta r$$
$$= (2i - 4j - 3k) . (i + 2j + 2k)$$
$$= -12 \text{ units.}$$

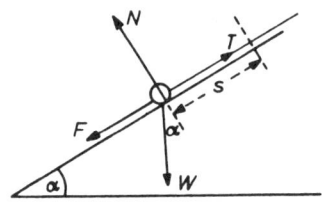

Figure 12.2

Example 2. A particle on a string is pulled a distance s up a plane inclined at an angle α to the horizontal. The forces acting on it are the tension T in the string, its weight W, the normal reaction N, with the plane and the frictional reaction F. Write down expressions for the work done by each of these forces.

WORK, POWER AND ENERGY

Using the formulae $W = F\delta r \cos\theta$, we have

work done by $T = Ts$,

work done by $W = Ws \cos(90° + \alpha) = -Ws \sin\alpha$

work done by $F = -Fs$

work done by $N = 0$.

When the work done by a force is negative ($= -x$ say), then x is called the work done *against* the force. For example, in the above problem, the work done against the weight is $Ws \sin\alpha$ and the work done against friction is Fs.

If a body can be said to exert a force F, then the work done by F is often called the work done *by the body*. For example, in order to raise a mass m through a height h, a man must exert a force at least equal and opposite to the weight mg of the particle. Then the work done by the man is mgh.

Example 3. One side of the roof of a bungalow is rectangular in shape and slopes at 45 degrees to the vertical. It is covered with 50 rows of tiles each containing 60 tiles. A tile measures 13 cm × 20 cm and weighs 6 kg. Find the work done, against gravity, in bringing the tiles from the eaves to their final positions on the roof.

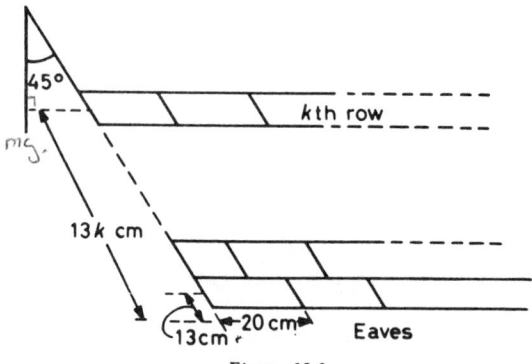

Figure 12.3

The weight F of a tile is 6g N acting vertically downwards and the displacement r_k of a tile of the kth row is of magnitude $13k$ cm or $0·13k$ metres in a direction at 45 degrees to the vertical.

Work done in bringing one tile into position $= F \cdot r_k$.

Each tile in a given row is displaced the same amount.

WORK DONE BY A CONSTANT FORCE

∴ Work done in bringing the kth row into position $= \sum_{1}^{60} F \cdot r_k$

$$= \left(\sum_{1}^{60} F\right) \cdot r_k$$

$$= 60F \cdot r_k$$

∴ Total work done in bringing all the rows into position $= \sum_{k=1}^{50} 60F \cdot r_k$

$$= 60F \cdot \sum_{k=1}^{50} r_k$$

$$= 360g \left(\sum_{k=1}^{50} 0 \cdot 13k\right) \cos 45°$$

$$= 46 \cdot 8g/\sqrt{2} \left(\sum_{k=1}^{50} k\right).$$

Now

$$\sum_{1}^{k=50} k = 1 + 2 + \ldots + 50 = \frac{50 \cdot 51}{2} \quad \text{(Sum of an A.P.)}$$

∴ Total work done $= \dfrac{46 \cdot 8 \times 9 \cdot 8}{\sqrt{2}} \cdot \dfrac{50 \cdot 51}{2}$

$$= 41\,340 \text{ J}.$$

The work done in bringing all the tiles from the eaves is 41·34 kJ.

Example 4. A pit shaft of rectangular cross-section 3 m × 4 m is to be sunk 70 m into the ground. The average density of material removed from the shaft is 900 kg/m³. Find the work done against gravity in excavating the shaft, assuming that the material moved is spread thinly on the ground.

Consider a layer of earth x m down of thickness δx.

Its volume $= 3 \times 4 \times \delta x$ m³

∴ its mass $= 3 \times 4 \times \delta x \times 900$

$$= 10\,800 \delta x \text{ kg}.$$

The force required to lift it against gravity is $10\,800 g\, \delta x$ N.

WORK, POWER AND ENERGY

$$\therefore \quad \text{Work done} = 10\,800g\,\delta x \times J.$$

$$\therefore \quad \text{Total work done} = \lim_{\delta x \to 0} \sum_{0}^{70} 10\,800gx\,\delta x$$

$$= \int_{0}^{70} 10\,800gx\,dx$$

$$= 10\,800g\left[\frac{x^2}{2}\right]_{0}^{70}$$

$$= 25\cdot83 \times 10^7\,J.$$

The work done in excavating the shaft is 258·3 MJ.

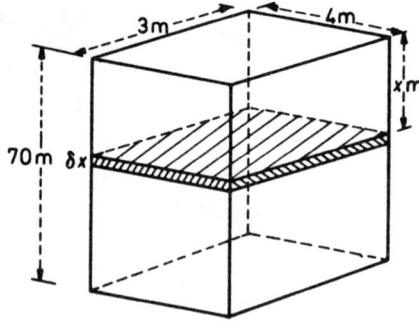

Figure 12.4

12.2. WORK DONE BY A VARIABLE FORCE

In the general case of a variable force **F** whose point of application moves along a curve from A to B, consider the curve to be split into n separate portions by the points $L, M \ldots$ (refer to *Figure 12.5*).

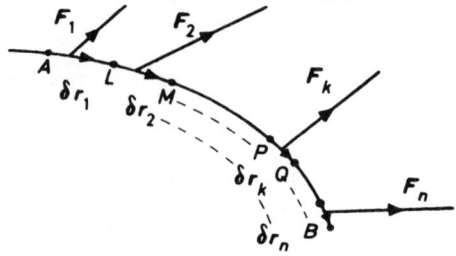

Figure 12.5

WORK DONE BY A VARIABLE FORCE

Let $\delta r_1, \delta r_2, \ldots$, represent the displacements $\overrightarrow{AL}, \overrightarrow{LM}, \ldots$. If $F_1, F_2, \ldots$, are the values of the variable force F at some points in $\delta r_1, \delta r_2, \ldots$, then in moving from A to B along the curve,

$$\text{Work done} \simeq F_1 \cdot \delta r_1 + F_2 \cdot \delta r_2 + \cdots$$

$$\simeq \sum_A^B (F \cdot \delta r).$$

In the limit as $n \to \infty$ and each displacement $\to 0$,

$$\text{work done} = \lim_{n \to \infty} \sum_A^B (F \cdot \delta r)$$

$$= \int_A^B F \cdot dr.$$

This integral is known as the *line integral* along the curve. We shall only consider cases in which the displacement $\overrightarrow{AB}$ is a straight line.

Example. A body of mass 50 kg is pulled in a straight line along a smooth horizontal surface from A to B by a force F. F acts in the direction from A to B and is of magnitude $2(2 - x/100)$ N, where x cm is the distance of the body from A. If $AB = 100$ m, find the work done by F.

$$\text{Work done} = \int_A^B F \cdot dx.$$

Since F is the direction of A to B, i.e. of dx,

$$\text{work done} = \int_A^B F \, dx$$

$$= \int_0^{100} 2\left(2 - \frac{x}{100}\right) dx$$

$$= 2\left[2x - \frac{x^2}{200}\right]_0^{100}$$

$$= 2\left[\left(200 - \frac{100^2}{200}\right) - 0\right]$$

$$= 300 \text{ J}.$$

WORK, POWER AND ENERGY

THEOREM

The work done by the tension in a spring or elastic thread, when its extension is reduced from x_2 to x_1, is $\lambda(x_2^2 - x_1^2)/2l$.

Consider the spring when its extension is x and a free end is at P, the other end being fixed at Q (refer to *Figure 12.6*).

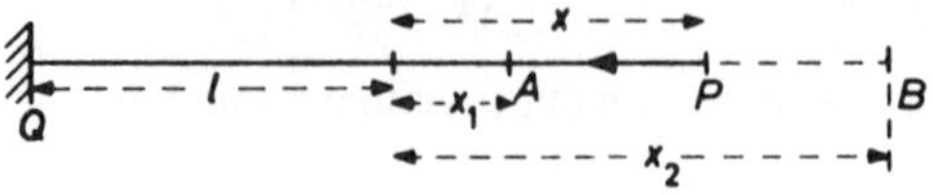

Figure 12.6

By Hooke's law the tension $T = -\lambda x/l$. Hence the work done as P moves from B to A is given by

$$W = \int_{x_2}^{x_1} T . dx = \int_{x_2}^{x_1} \left(-\frac{\lambda x}{l}\right) dx$$

$$= -\frac{\lambda}{l}\left[\frac{x^2}{2}\right]_{x_2}^{x_1}.$$

∴ Work done by $T = \dfrac{\lambda(x_2^2 - x_1^2)}{2l}$.

Exercises 12a

1. P, Q, R are three points whose position vectors are a, $a + b$ and $2a - b$ respectively. A particle moves from P to Q and then to R. Among the forces acting on the particle is a force $F = k(a + 2b)$. Find the work done by F as the particle moves (a) from P to Q, (b) from Q to R and (c) from R to P again.

2. Two forces $F_1 = 2i + j$ N and $F_2 = i - 3j$ N act on a particle which is given a displacement $15i - 10j$ metres. Find the work done by each force and the work done by their resultant.

3. A cart is pulled a distance 2 m up an inclined plane by a force F whose magnitude is 50 N and whose direction is inclined at an angle of 20 degrees to the plane. Find the work done by F on the cart.

4. A man and his equipment weigh 80 kg. He climbs to the top of a mountain 1 000 m high, how much work does he do against gravity?

5. A boy cycles 20 km. If the resistances to motion average 25 N, how much work does he do against the resistances?

6. A man and his cycle weigh 100 kg. He travels 0·5 km up an incline of $\sin^{-1}(\frac{1}{20})$. Find the work done against gravity. If, in addition to gravity, the other resistances to motion total 20 N, find the total work done by the cyclist.

POWER

7. Find the total work done against friction and gravity when a mass of 200 g is pulled a distance of 35 cm up a rough plane. The plane is inclined at 30 degrees to the horizontal and the coefficient of friction between the mass and the plane is $2/\sqrt{3}$.

8. A man and his cycle weigh 100 kg. To travel at a constant speed, a distance of 400 m down an incline of $\sin^{-1}(\frac{1}{150})$ the man has to do 1 200 J of work. Find the resistance to his motion.

9. A Venetian blind consists of a top fixed bar and 31 movable bars each of mass 100 g and of negligible thickness. When the blind is fully extended, the distance between each pair of consecutive bars is 8 cm. Find the work done in pulling up the blind.

10. A rectangular section of roof is covered by 20 horizontal rows of tiles each row containing 30 tiles. The section of the roof slopes at 40 degrees to the vertical and each tile weighs 5 kg and measures 45 cm × 30 cm. Find the work done against gravity in bringing the tiles from the eaves to their final position on the roof.

11. A bucket of mass 10 kg is attached to the end of a chain whose mass per unit length is 1 kg/m. The bucket is at the bottom of a well 20 m deep, find the work done in hauling the bucket and chain to the surface.

12. A railway truck is pulled a distance of 100 m along a straight track, starting at a point A, by a varying force F. The magnitude of F is $(6 - x/20)$ N, where x m is the distance moved from A. If the direction of F is inclined at 30 degrees to the track, find the work done by the force.

13. A light elastic thread of modulus 20 N and natural length 1 m, hangs unstretched. A man pulls one end until its length is 1·5 m. Find (a) the work done by the tension in the string (b) the work done by the man.

If he now extends the string by a further 0·5 m how much extra work does he do?

14. An elastic string of natural length 50 cm has a modulus of elasticity of 600 mN. Find the work done in stretching the string from an initial stretched length of 60 cm to 100 cm.

12.3. POWER

Definition—The *power* (or activity) of a force is the rate at which it does work, i.e.
$$\text{power } H = \frac{dW}{dt}.$$

SI unit of power ... J/S = watt (W)

WORK, POWER AND ENERGY

In the particular case when work is done at a constant rate, $H = W/t$.

Example. An engine working a conveyor belt raises gravel, through a height of 30 m, *at the rate of* 400 kg/min. *Neglecting work done in overcoming frictional and other resistances, calculate the rate at which the engine is working.*

In 1 min the work done in raising 400 kg through 30 m is 400g . 30 J.
Now
$$H = W/t$$

∴
$$H = \frac{400g \cdot 30}{60}$$

$$= 200 \times 9{\cdot}8.$$

∴ Power of the engine = 1 960 W

12.4. POWER AND VELOCITY

Consider a body working *instantaneously* with power H and exerting a force $\mathbf{F}$. By definition the work done by a force $\mathbf{F}$ is

$$W = \int_a^b \mathbf{F} \cdot d\mathbf{r} = \int_a^b \mathbf{F} \cdot \frac{d\mathbf{r}}{dt} dt.$$

$$W = \int_a^b (\mathbf{F} \cdot \mathbf{v}) \, dt$$

$$dW/dt = \mathbf{F} \cdot \mathbf{v}.$$

∴
$$\boxed{H = dW/dt = \mathbf{F} \cdot \mathbf{v}}$$

Thus at any instant the power of an engine can be measured by the scalar product of the force it exerts and the velocity with which the force moves. In the particular case when force and velocity are in the same direction $H = Fv$.

We shall first consider examples in which the working body moves at constant velocity, so that the acceleration is zero and the resultant force on it is zero.

POWER AND VELOCITY

Example 1. An engine draws a train, whose total mass is 300 tonne, along a straight horizontal track against resistances of 150 N/t. If its maximum speed is 15 m/s, find the maximum power output of the engine.

If the resistances remain constant, find the maximum speed of the train up an incline of 1 in 70.

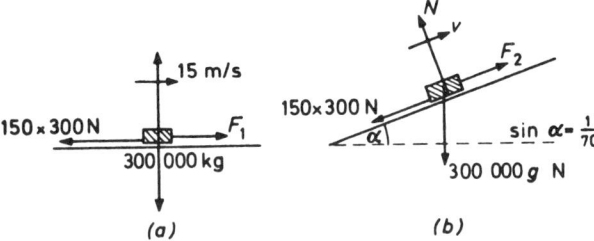

Figure 12.7

At *maximum* speed the train has zero acceleration. Hence, on the horizontal (refer to *Figure 12.7a*),

$$F_1 - 150 \times 300 = 0$$
$$\therefore F_1 = 150 \times 300 \text{ N},$$

F_1 being the tractive force exerted by the engine.

Now

$$H = \mathbf{F} \cdot \mathbf{v}$$
$$= Fv \quad \text{(since force and velocity are in the same direction).}$$
$$\therefore \text{Power of the engine} = (150 \times 300) \cdot 15$$
$$= 675\,000 \text{ W}.$$

Again when travelling at maximum speed (*v* say) up the incline the train has zero acceleration. Hence, referring to *Figure 12.7b*,

$$F_2 - 150 \times 300 - 300\,000 g \sin \alpha = 0$$
$$\therefore F_2 = 300(150 + 9\,800/70)$$
$$= 300 \times 290 \text{ N}.$$

At maximum power $H = Fv$ gives $675\,000 = 300 \times 290 v$.

$$\therefore v = 7 \cdot 759 \text{ m/s}.$$

215

WORK, POWER AND ENERGY

The maximum power output of the engine is 675 kW and it can pull the train up the incline at 7·76 m/s.

Example 2. *A cyclist and his machine together weigh 84 kg. Without pedalling he travels at 35 km/h down a straight road inclined at $\sin^{-1}(\frac{1}{12})$ to the horizontal. Assuming that the resistance to motion is proportional to the speed, what is his power output when travelling at a uniform speed of 27 km/h along a straight horizontal road?*

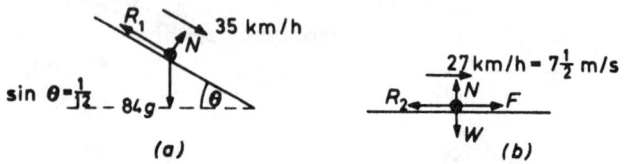

Figure 12.8

When the cyclist is travelling down the slope there is no power output and the speed is constant, therefore, the magnitude of the resistance R_1 to motion down the plane is equal to the component of his weight. (Refer to *Figure 12.8*.)

$$\therefore \qquad R_1 = 84g \sin \theta$$
$$= 84g \cdot \tfrac{1}{12}$$
$$= 7g \text{ N}.$$

The resistance to motion is proportional to the speed therefore, because

$$R = 7g \text{ N} \quad \text{at } 35 \text{ km/h}$$

$$\therefore \qquad R = 7g \cdot \frac{27}{35} \text{ N} \quad \text{at } 27 \text{ km/h}$$

$$\therefore \qquad R_2 = \frac{27g}{5} \text{ N}.$$

His speed along the horizontal being uniform the magnitude F of the force he exerts is equal to the magnitude of the resistance R_2.

$$\therefore \qquad F = \frac{27g}{5} \text{ N}$$

But $\qquad H = Fv$

$$\therefore \quad H = \frac{27g}{5} \cdot \frac{15}{2}$$

$$= 396 \cdot 9 \text{ W.}$$

The cyclist's power when travelling at 27 km/h is 397 W.

Exercises 12b

1. A small steamer is travelling at 7·5 m/s. If the effective power output of her engines is 1 500 kW, what is the resistance to motion?

2. A motor car whose weight is 1 500 kg is travelling at 10 m/s up an incline of $\sin^{-1}(\frac{1}{7})$ to the horizontal. If air and road resistances are neglected, what is the power output of the engine?

3. A locomotive pulls a train of mass 200 tonne up an incline of $\sin^{-1}(\frac{1}{140})$ to the horizontal at a uniform speed of 10 m/s. If the air and track resistances are 35 N/t, find the power output of the locomotive.

4. A cable car weighs 650 kg and carries a load of 850 kg. At one part of its journey the car is travelling upwards at a constant speed of 7·2 km/h, its direction of motion being inclined at an angle of $\sin^{-1}(\frac{1}{6})$ to the horizontal. The air and track resistances total $50g$ N. Find the power output of the engine hauling the car.

5. A train of mass M tonne is travelling at a steady speed of v km/h on the level ground. If the air and track resistances total R N/t, find the power output of the engine. The train now travels up an incline of $\sin^{-1}(\frac{1}{100})$ the air and track resistances remaining the same. If V km/h is its maximum speed on the level ground and U km/h its maximum speed up the incline, show that

$$\frac{V}{U} = 1 + \frac{10g}{R}.$$

6. The engine of a car works at a constant rate. The maximum speed of the car on the level is v_1 and the maximum speed up a given slope is v_2. Find the maximum speed of the car down the same incline, assuming that the air and track resistances remain the same in all three cases.

7. A locomotive pulls a train of mass 30 000 kg up a rack railway; at its steepest part the track is inclined at $\sin^{-1}(\frac{48}{100})$. If, at this point, the train is travelling at 1 m/s and the air and track resistances total 7 840 N, find the power output of the engine in Force de Cheval. [1 Force de Cheval = 735 W.]

8. The combined mass of a man and his bicycle is 90 kg. He cycles up a hill at 5 m/s. If his power output is 250 W and the resistances to motion, other than gravity, are 5 N, find the slope of the hill.

WORK, POWER AND ENERGY

9. A cyclist and his machine together weigh 100 kg. Without pedalling he goes at a constant speed of 27 km/h down a road inclined at $\sin^{-1}(\frac{1}{20})$ to the horizontal. At what rate must he work in order to go at the same constant speed (a) on the level (b) up an incline of $\sin^{-1}(\frac{1}{30})$, the resistances to motion, other than gravity, remaining constant.

In the following examples the resultant force is not necessarily zero and the mass can accelerate (or decelerate).

Example 3. *A car of mass* 1 500 kg *is capable of working at a maximum rate of* 20 kW. *It is being driven up a hill, inclined at* $\sin^{-1}(\frac{1}{12})$ *to the horizontal, at* 8 m/s. *If the resistances to motion, other than gravity, are* 150 N, *find the greatest acceleration it can have at that speed.*

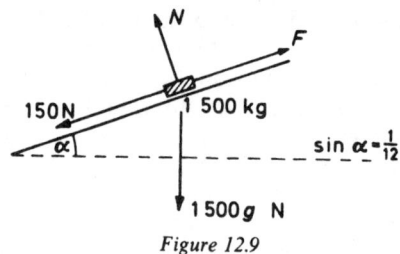

Figure 12.9

The maximum power of the car = 20 000 W. But if F is the force exerted by the engine,

$$H = Fv$$

∴ $\qquad 20\,000 = F \times 8$

∴ $\qquad F = 2\,500 \text{ N}.$

Applying $P = mf$ up the plane (refer to *Figure 12.9*),

$$2\,500 - 150 - 1\,500g \sin \alpha = 1\,500f$$

∴ $\qquad 2\,350 - \dfrac{1\,500 \times 9\cdot 8}{12} = 1\,500f$

∴ $\qquad 1\,125 = 1\,500f$

$$f = 0\cdot 75 \text{ m/s}^2.$$

When working at maximum power and travelling at 8 m/s up the slope, its acceleration is 0·75 m/s².

POWER AND VELOCITY

Example 4. A train of mass 300 tonne, begins to ascend an incline of 1 in 490 moving at 5 m/s and working at 450 kW. If it continues to work at this rate and the resistances to motion are 80 N/t, find the maximum speed it can reach. Find also the time taken to reach two-thirds of this maximum speed.

Let V be the maximum speed of the train. At speed v ($<V$) the resultant force P on the train is:

$$P = \frac{450\,000}{v} - 80 \times 300 - 300\,000g \times \frac{1}{490}$$

$$= \frac{450\,000}{v} - 24\,000 - \frac{300\,000 \times 9 \cdot 8}{490}$$

$$= \frac{450\,000}{v} - 30\,000.$$

Now $$P = mf$$

$\therefore$ $$\left(\frac{450\,000}{v} - 30\,000\right) = 300\,000 f$$

$$f = \frac{1}{10}\left(\frac{15}{v} - 1\right). \qquad \ldots \text{(i)}$$

The maximum speed is reached when $f = 0$. Therefore from (i), $V = 15$ m/s. At speeds less than V, equation (i) can be written

$$\frac{dv}{dt} = \frac{1}{10}\left(\frac{15}{v} - 1\right)$$

$\therefore$ $$\int \frac{v\,dv}{15 - v} = \frac{1}{10}\int dt$$

$$-\int\left(1 - \frac{15}{15 - v}\right)dv = \frac{1}{10}\int dt$$

$\therefore$ $$v + 15\log_e(15 - v) = C - \frac{t}{10}$$

when $t = 0$, $v = 5$,

$\therefore$ $$C = 5 + 15\log_e 10$$

$\therefore$ $$v + 15\log_e\left(\frac{15 - v}{10}\right) = 5 - \frac{t}{10}$$

219

WORK, POWER AND ENERGY

when $v = 10$

$$10 + 15 \log_e \left(\frac{1}{2}\right) = 5 - \frac{t}{10}$$

$$t = 10[15 \log_e 2 - 5]$$
$$= 50[3 \log_e 2 - 1]$$
$$= 53{\cdot}97 \text{ s}.$$

The maximum speed is 15 m/s and the time taken to reach two-thirds of this speed (from 5 m/s) is 54·0 s.

Exercises 12c

1. A car, of mass 1 200 kg, is travelling at 15 m/s on level ground. The maximum rate of working of the car is 15 kW. If the resistance to motion is 100 N, find the greatest possible acceleration it can have at this speed.

2. A car of mass 1 400 kg is travelling at 20 m/s on level ground. Its greatest possible acceleration at this speed is found to be 0·5 m/s². If the resistance to motion is 100 N, find the maximum rate of working of the car.

3. A train whose total mass is 240 tonne is travelling up an incline of $\sin^{-1}\left(\frac{1}{120}\right)$, the air and track resistances being 12 kN. The train is travelling at 16 m/s and the maximum rate of working of its engine is 640 kW. What is the maximum possible acceleration it can have at this speed?

4. A train whose total mass is M is travelling on level ground against resistances which total kv where v is its speed. The engine can either work at a maximum rate H, or exert a maximum tractive force P. Show that the maximum speeds attainable in each case are equal to $\sqrt{H/k}$ or to P/k.

5. A cyclist whose maximum rate of working is 75 W cycles down a slope of inclination $\sin^{-1}\left(\frac{1}{98}\right)$ at a constant, maximum, speed of 45 km/h. If the total mass of the rider and his machine is 80 kg, find the resistances opposing motion.

6. The cyclist in Question 5 finds that the resistances opposing motion, other than gravity, vary directly as his speed, and at 22·5 km/h they total 7 N. Find his maximum acceleration when travelling on a level road at 11·25 km/h.

7. The combined mass of a cyclist and his machine is 90 kg. While working at a steady rate of 150 W, he commences to accelerate up a hill. If the resistance to motion, other than gravity, is 10 N and, at the moment his speed is 3 m/s his acceleration is $\frac{2}{45}$ m/s², find the inclination of the hill.

WORK AND KINETIC ENERGY

8. A train of total mass 160 tonne is ascending a hill inclined at $\sin^{-1}(\frac{1}{280})$. With the engine working at half its maximum rate of 420 kW the train moves steadily at 63 km/h. Find the resistances to motion in N/t. If the engine is now made to work at maximum rate, find the immediate acceleration up the hill.

9. A hovercraft of mass 500 tonne is moving across smooth water against negligible resistances. If the power being used to propel it forward is constant at 1 000 kW, find the time taken to increase its speed from 30 m/s to 36 m/s.

12.5. WORK AND KINETIC ENERGY

Referring back to Section 12.1, we have that, for a particle subject to a resultant force P, the net work done in a displacement r is given by,

$$\text{net work done} = \int_A^B P \cdot dr$$

$$= \int_A^B mf \cdot dr \quad \text{(see Section 6.3)}$$

$$= m \int_A^B \frac{dv}{dt} \cdot dr \quad \text{(since } m \text{ is constant)}$$

$$= m \int_A^B \frac{dv}{dt} \cdot \frac{dr}{dt} dt$$

$$= m \int_A^B \left(\frac{dv}{dt} \cdot v\right) dt$$

$$= m \int_A^B \tfrac{1}{2} \frac{d(v \cdot v)}{dt} dt \quad \text{(refer to Exercise 3e No. 15)}$$

$$= \tfrac{1}{2} m [v \cdot v]_A^B$$

$$\text{net work done} = \tfrac{1}{2} m v_B^2 - \tfrac{1}{2} m v_A^2.$$

The quantity $\tfrac{1}{2}mv^2$ is defined as the *kinetic energy* (K.E.) of the particle (its units being the same as those of work done, i.e. joules).

WORK, POWER AND ENERGY

Hence equation (iii) states that the increase of kinetic energy of a particle during a displacement is equal to the net work done by the resultant force acting on the particle or more shortly:

> Final K.E. − initial K.E. = net work done
>
> This result is often called the *Principle of Work* and can be conveniently used to solve some problems which involve speed and distance (but not time and acceleration).

Example 1. A smooth circular wire is fixed with its plane vertical and has a small bead threaded on it. If the bead is projected from the lowest point of the wire at 4 m/s, find the vertical height through which it has risen when its speed is 2 m/s.

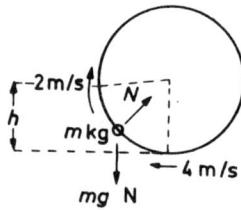

Figure 12.10

Let the particle have mass m kilogrammes. Then the forces acting on it are shown in *Figure 12.10*. Since N is always at right angles to the motion of the particle it does no work. By the principle of work,

$$\text{Final K.E.} - \text{initial K.E.} = \text{work done}$$

$$\therefore \quad \tfrac{1}{2}m2^2 - \tfrac{1}{2}m4^2 = -mgh + 0$$

$$\therefore \quad 2^2 - 4^2 = -2 \times 9{\cdot}8h$$

$$\therefore \quad h = 1{\cdot}663 \text{ m.}$$

The particle has risen through a height of 1·63 m when its speed is 2 m/s.

222

WORK AND KINETIC ENERGY

The principle of work can be extended to several particles considered together. Thus, for a system of particles, the total increase in kinetic energy of the system is equal to the total work done on the system.

Example 2. A particle of mass m_1 lies at rest on a rough horizontal table. The coefficient of friction between the particle and the table is μ. It is connected by a light inelastic string, passing over a smooth pulley, to a freely hanging mass m_2. If the system is released from rest, find the speed u of each mass after m_1 has moved a distance d along the table.

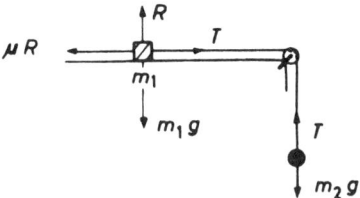

Figure 12.11

The reaction R is equal to $m_1 g$ and hence $\mu R = \mu m_1 g$. Since the string is inextensible m_2 falls a distance d, and hence

$$\text{work done by gravity} = m_2 g d,$$
$$\text{work done against friction} = \mu m_1 g d,$$
$$\text{work done by tensions} = -Td + Td = 0.$$

By the principle of work,

$$\text{final K.E.} - \text{initial K.E.} = \text{work done}$$

so
$$(\tfrac{1}{2}m_1 u^2 + \tfrac{1}{2}m_2 u^2) - 0 = m_2 g d - \mu m_1 g d$$
$$\tfrac{1}{2}u^2(m_1 + m_2) = (m_2 - \mu m_1)gd$$
$$u^2 = 2gd(m_2 - \mu m_1)/(m_1 + m_2).$$

Example 3. A particle of mass m, is attached to the mid-point of a light elastic string AB, of natural length 15a and modulus 105mg/16. The ends A and B of the string are attached to two points on the same horizontal level, distance 15a apart. The particle is released from rest on the same horizontal level as AB with the string just taut. Find its speed after it has fallen a distance 4a to a point Q.

WORK, POWER AND ENERGY

Since the string is just taut, the particle P must start from a point midway between A and B. Referring to *Figure 12.12*, we have

$$BQ^2 = BP^2 + PQ^2$$
$$= (15a/2)^2 + (4a)^2 = 289a^2/4$$
$$\therefore BQ = 17a/2.$$

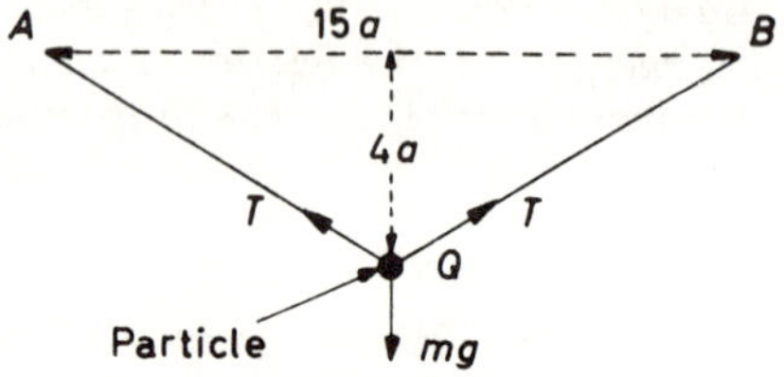

Figure 12.12

The work done *by* gravity is $mgh = mg4a$. The work done *by* the tension in the string $= -2 \times \frac{1}{2}\lambda e^2/l$ (refer to Section 12.2).

$$= -2 \times \frac{1}{2} \times \frac{105}{16}mg \frac{(17a/2 - 15a/2)^2}{15a/2}$$

$$= -\frac{105}{16}mga \bigg/ \frac{15}{2}$$

$$= -7mga/8.$$

By the principle of work, final K.E. − initial K.E. = net work done

$$\therefore \quad \tfrac{1}{2}mv^2 - 0 = 4mga - 7mga/8$$

$$\therefore \quad v^2 = 6\tfrac{1}{4}ga$$

$$v = 5\sqrt{ga}/2.$$

Exercises 12d

1. A particle of mass 5 kilograms executes simple harmonic motion with amplitude 2 metres and period 12 seconds. Find the maximum kinetic energy of the particle, leaving your answer in terms of π.

Initially the particle is moving with its maximum kinetic energy. Find the time that elapses until the kinetic energy is reduced to one quarter of the maximum value, and show that the distance moved in this time is $\sqrt{3}$ metres.

(Standard formulae relating to simple harmonic motion may be quoted without proof.) (JMB)

2. A particle is projected directly up a rough plane inclined at an angle $\sin^{-1}\frac{3}{5}$ to the horizontal with a speed of 7 m/s. Find the distance it travels before coming to rest, if the coefficient of friction between particle and plane is $\frac{1}{2}$.

3. A light elastic string, of modulus $3g$ N and natural length 1 m, has one end fixed at O and to the other is attached a particle P of mass 3 kg. P is held at O and then released. If P first comes to instantaneous rest at A, what is the net change in kinetic energy between O and A? Find the distance OA.

4. A particle is free to move in a smooth circular tube which is fixed with its plane vertical. The radius of the tube is a and the particle is projected from the lowest point with speed v. If the particle just performs complete revolutions, find v in terms of g and a.

5. A particle of mass m is attached to one end of a light elastic string of modulus $2mg$ and natural length l. The other end of the string is attached to a fixed point A and the string hangs vertically with the particle held a distance l below A. If the particle is allowed to fall, find the distance travelled before it comes to rest.

6. A cricket ball is thrown in an upward direction with a speed of 21 m/s. At what speed is it travelling when it reaches a height of 10 m?

7. A curling stone is projected across the level surface of a frozen pond at 7 m/s. If the coefficient of friction is 0·05, how far will it slide?

8. Two particles of masses 80 g and 60 g are connected by a light inextensible string passing over a smooth pulley. They are released from rest with the string taut. Find their speeds after the heavier particle has fallen a distance of 20 cm.

9. If, in Question 8, the heavier particle had been travelling upwards at 70 cm/s, how far would it travel before coming to rest?

10. A wedge has two smooth faces inclined in opposite directions at angles α, β to the horizontal. Particles A and B of masses m_1 and m_2 rest on the faces and are joined by a light inextensible string passing over a smooth pulley at the vertex of the wedge. If m_1 rests on the face inclined at angle α and the system is released from rest, find the speed of the particles after they have each travelled a distance d.

11. In Question 10, if the faces are rough and μ is the coefficient of friction in each case, what would be the speed?

12. Two particles, each of mass 30 g, rest on a smooth horizontal table. They are connected by a light elastic string of modulus $0·06g$ mN and natural length 24 cm. The particles are held apart by

WORK, POWER AND ENERGY

two forces of $0{\cdot}09g$ mN and then released. Find their speeds when the distance apart is reduced to 36 cm.

13. A particle is moving with speed V directly up a rough plane inclined at an angle α to the horizontal. If λ is the angle of friction, show that it will travel a distance $V^2 \cos \lambda / 2g \sin(\alpha + \lambda)$ before coming to rest.

14. A light inextensible string has one of its ends attached to a fixed point. Its other end is threaded around a smooth pulley A of mass m_1, over a fixed smooth pulley B and attached to a mass m_2, which hangs freely. A mass M is attached to A $(M + m_1 > m_2)$ and the system released from rest. Find the speed of M after it has moved a distance a, assuming all the portions of the string are vertical and m_2 does not reach the fixed pulley B.

15. A light elastic string has a natural length of 20 cm and modulus $0{\cdot}4g$ mN. The string is attached to two points A and B 30 cm apart on a smooth table and a particle of mass 10 g is attached to its mid-point. The particle is displaced a distance 4 cm from its equilibrium position to a point in the line AB, and released. Find its speed v on passing through its equilibrium position. Could the principle of work be used to find v, if the displacement had been greater than 5 cm?

12.6. CONSERVATIVE FORCES AND POTENTIAL ENERGY

We have seen in Section 12.1, that when a force F has its point of application displaced from A to B

$$\text{work done} = \int_A^B \mathbf{F} \cdot d\mathbf{r}.$$

For some forces the work done depends on the path taken. For a block of wood resting on a rough horizontal table the work done against friction in moving the block from A to B depends on the length of the path taken. Sometimes the work done by F depends only on the positions of A and B and is independent of the path taken, such forces are called *Conservative Forces*.

For such a force it follows that the work done in moving from A to B is equal and opposite to the work done in moving from B to A. Therefore, the work done by a conservative force, when its point of application follows *any* closed path, is zero. The reverse is also true if the work done by a force, when its point of application follows *any* closed path is zero, the force is a conservative force.

Any constant force is conservative because, referring to Section 12.1, the work done by a *constant* force F is given by $\mathbf{F} \cdot (\Sigma \, \delta \mathbf{r})$ and

CONSERVATIVE FORCES AND POTENTIAL ENERGY

for a closed circuit $\Sigma \, \delta r = 0$, therefore the work done is zero, therefore F is a conservative force. Thus the weight of a particle (where the work done depends only on the change of vertical height) is conservative.

Also it can be shown that the tension in a spring or elastic thread is a conservative force.

Frictional forces are not, however, conservative since, as indicated earlier, the work done depends on the path taken.

Example 1. Show that the force F is conservative where $F = x^2 a i + bj + ck$ and x is distance measured parallel to unit vector i.

From Section 12.2 the work done by a variable force is

$$\int_A^B F \cdot dr = \int_A^B (x^2 ai + bj + ck) \cdot dr.$$

Now
$$r = xi + yj + zk$$

∴
$$dr = i\,dx + j\,dy + k\,dz.$$

∴ Work done from A to B

$$= \int_A^B (x^2 ai + bj + ck) \cdot (dx\,i + dy\,j + dz\,k)$$

$$= \int_A^B (x^2 a\,dx + b\,dy + c\,dz)$$

$$= a\int_A^B x^2\,dx + b\int_A^B dy + c\int_A^B dz$$

$$= a\left[\frac{x^3}{3}\right]_A^B + b[y]_A^B + c[z]_A^B$$

$$= a\left[\frac{x_B^3 - x_A^3}{3}\right] + b[y_B - y_A] + c[z_B - z_A]$$

and the value of this depends only on the initial and final positions of A and B and is therefore independent of the path taken. It follows that F is conservative.

WORK, POWER AND ENERGY

We can now define *Potential Energy* (P.E.) at a point A for any conservative force F as minus the work done by the force [i.e., the work done against the force] in moving from an arbitrary* base point Q to the point A

$$\text{Potential energy at } A = -\int_Q^A \mathbf{F} \cdot d\mathbf{r} = \int_A^Q \mathbf{F} \cdot d\mathbf{r}$$

Thus at a height h, mgh is the gravitational P.E. of a particle of mass m; and $\lambda x^2/2l$ may be regarded as the P.E. of a stretched string.

We also have that:

$$\text{work done} = \int_A^B \mathbf{F} \cdot d\mathbf{r}$$

$$= \int_A^Q \mathbf{F} \cdot d\mathbf{r} + \int_Q^B \mathbf{F} \cdot d\mathbf{r}$$

$$= \int_A^Q \mathbf{F} \cdot d\mathbf{r} - \int_B^Q \mathbf{F} \cdot d\mathbf{r} \quad \text{[by the usual properties of definite integrals]}$$

$$= \text{P.E. at } A - \text{P.E. at } B.$$

From Section 12.5, work done $= \tfrac{1}{2}mv_B^2 - \tfrac{1}{2}mv_A^2$. ....(ii)

Therefore from equations (i) and (ii)

$$\text{P.E. at } A - \text{P.E. at } B = \tfrac{1}{2}mv_B^2 - \tfrac{1}{2}mv_A^2$$

$$\therefore \quad \text{P.E. at } A + \tfrac{1}{2}mv_A^2 = \text{P.E. at } B + \tfrac{1}{2}mv_B^2$$

or the sum of K.E. and P.E. is constant for a conservative force. This is the *Principle of Conservation of Energy* for a particle in motion under conservative forces. It may be used in place of the Principle of Work in solving problems in which there are no non-conservative forces.

Example 2. A mass m slides a distance d from A to B down a plane inclined at an angle α to the horizontal. If it was initially at rest at A, find its speed at B. Consider two cases (a), when the plane is smooth (b), when the plane is rough, coefficient of friction μ.

Case (a)

Since the plane is smooth, the only force acting is gravity—a conservative force.

* In practice, we choose the most convenient point as the base point.

EXERCISES

∴　　　　K.E. + P.E. at A = K.E. + P.E. at B.

Taking B as the base point for the P.E.

$$0 + mgd \sin \alpha = \tfrac{1}{2}mv^2 + 0$$

∴ $$v = \sqrt{(2gd \sin \alpha)}$$

Case (b)

A non-conservative frictional force μR, where $R = mg \cos \alpha$ is now acting. Therefore we use the Principle of Work. (Work is done *by* gravity and *against* friction.)

Final K.E. − initial K.E. = work done

so $$\tfrac{1}{2}mv^2 - 0 = mgd \sin \alpha - \mu Rd$$

∴ $$\tfrac{1}{2}mv^2 = mgd \sin \alpha - \mu dmg \cos \alpha$$

∴ $$v = \sqrt{2gd(\sin \alpha - \mu \cos \alpha)}.$$

EXERCISES 12*

1. A particle of unit mass is acted upon at time t by a force defined by $\mathbf{F} = 2\mathbf{i} - 4t\mathbf{j} + 2t\mathbf{k}$. When $t = 0$ the velocity of the particle is $-4\mathbf{i} + 3\mathbf{k}$. Determine its velocity at time t and hence find an expression for the power of $\mathbf{F}$ at time t.　　　　(JMB)

2. A particle of mass 100 kg moving freely in a straight line is acted on by a force which changes its speed from 10 m/s to 24 m/s. Find the work done by the force on the particle.

The particle is now acted on by another force which brings it to rest, find the work done by this second force.

3. A locomotive has a maximum power output of 200 Force de Cheval. It is pulling a train of mass 20 000 kg up a rack railway inclined at $\sin^{-1}(\tfrac{1}{4})$. If the air and track resistances total 1 kN, find the maximum steady speed which can be maintained (1 Force de Cheval = 735 W).

4. A particle of mass 56 g lies on a rough plane inclined at $\sin^{-1}(\tfrac{1}{7})$. It is attached by a light inextensible string passing over a smooth pulley to a freely hanging weight of 14 g. The masses are initially moving with a speed of 5 m/s. If the speed decreases to 2 m/s, find the net work done. The frictional force F remains constant and the distance travelled during the decrease of speed is 50 m. Find the magnitude of F (neglect air resistance).

* Exercises marked thus, †, have been metricized, *see* Preface.

WORK, POWER AND ENERGY

5. The combined mass of a cyclist and his bicycle is 100 kg. When he ascends a slope of inclination $\sin^{-1}(\frac{1}{10})$, his constant speed is 3 m/s. If his rate of working is 330 W, find the resistance to motion other than gravity. If he continues to work at the same rate, find his initial acceleration when the road becomes level.

6. A sledge of mass 9 kg is dragged at constant speed a distance of 1 m up an icy slope of inclination 30 degrees to the horizontal by a rope. The coefficient of friction is $\frac{1}{30}$. Find the work done by (a) the force exerted by the rope (b) the friction (c) gravity.

7. An engine of mass 120t can maintain a constant speed of 17 m/s along a straight horizontal track when it is working at 170 kW. At what power would it have to work when climbing an incline of $\sin^{-1}(\frac{1}{147})$ at the same speed, the frictional resistances remaining the same?

8. A particle of mass m lies on a rough plane which is inclined at an angle α to the horizontal. It is attached to a fixed point on the plane by a light elastic string of modulus $3mg/2$ and natural length l. Initially, the particle is below the point of suspension, with the string just taut and in the line of greatest slope of the plane. It is released from rest, show that it will travel a distance

$$4l \sin(\alpha - \lambda)/3 \cos \lambda$$

before coming to rest again where λ is the angle of friction.

9. The resistance to a train of mass 240 tonne travelling at a steady speed of v m/s on a level track is $(a + bv)$ N/t. When travelling at a steady speed of 20 m/s the power output of the engine is 160 kW and at a steady speed of 10 m/s it is 60 kW, find a and b.

10. A motor car, of weight W kN, can travel along a straight level road at a uniform speed of v m/s and can climb a hill of inclination α to the horizontal at $9v/15$ m/s. Assuming that the resistances to motion other than gravity are R N, and the power output of the car H kW in each case, prove that (a) $3R = 500W \sin \alpha$ (b) $2H = 3Wv \sin \alpha$.

11. A number n of uniform wooden planks, each of weight W, thickness t and width a, lie in a pile of height nt on horizontal ground. The planks are taken from the pile and placed one on top of another to form a vertical wall of height na. Find the work done against gravity. (JMB, part)

12. Explain the terms *work* and *power*.

Find, to the nearest whole number, the power in kilowatts required to raise 3 t of water per minute through 30 m vertically and deliver it through a horizontal pipe of which 1 m contains 10 kg of water. (London)†

13. When a cyclist is travelling at a speed v, the resistances to

motion are $k(v^2 - 24v + 180)$ (k a positive constant). Sketch a graph of his power output H against his speed v for values of v from 0 to 20. Show that the graph has a local maximum at $v = 6$. If H_0 is his power output at $v = 6$, find for what value of v his power output is again H_0.

14. A car of mass 1 600 kg is travelling on a level road. The resistance to motion is kv N, where v is its speed in m/s. The engine works at a constant rate of 16 kW and the car's maximum speed is 20 m/s. Find the distance in which a speed of 10 m/s can be reached from rest.

15. An electric locomotive is travelling on a straight level track against negligible resistances. The speed of the locomotive increases from 9 m/s to 18 m/s while it travels 600 m. Find the time taken if the locomotive was (a) exerting a constant pull (b) working at constant power.

16. A cyclist is working at a constant rate of 120 W as he rides along a level road at v m/s. The total mass of the rider and his machine is 96 kg and the resistances to motion are $3v/2$ N. Show that his acceleration is $(80 - v^2)/64v$ m/s^2, and find the time taken for his speed to increase from 4 m/s to 6 m/s.

17. A particle starts from rest and moves in a straight line under the action of a force of magnitude $k(10t - t^2)$, where t seconds is the time since starting. Find an expression in terms of t for the power output by the force. Hence, show that this power is a maximum, when $t = 10 - \sqrt{10}$.

18. A uniform chain of length 3 m weighs 10 kg, it is flexible with very small links. Two metres of the chain rests on a rough horizontal table top with the other 1 m hanging vertically. The chain is perpendicular to the smooth edge of the table. If μ is the coefficient of friction and the chain is released from rest, show that, when the length of the hanging portion is x, the work done *by* friction during a small displacement δx is approximately $-\tfrac{10}{3}\mu g(3 - x)\,\delta x$ joule, and the work done *by* gravity is $\tfrac{10}{3}gx\,\delta x$ joule. If $\mu = 0.3$, find the total work done on the body from the moment of release until the last link leaves the table. Deduce the speed with which the last link leaves the table.

19. A bead of mass m is threaded on a smooth circular wire centre O, radius a, which is fixed in a vertical plane. A light elastic string of natural length a and modulus $3mg$ connects the bead to the lowest point A of the wire. The bead is projected from A with a speed v, show that if $v > \sqrt{7ga}$, the bead will make complete revolutions of the wire.

20. The engine of a train is working at a constant rate. The maximum speed of the train up a certain incline is v_1 and the maximum speed down the same incline is v_2. Prove that the maximum speed on the level is $2v_1v_2/(v_1 + v_2)$, assuming that the resistance to the motion is constant and the same in all three cases.
(London, part)

21. A train of total mass 150 t is travelling on a horizontal track at a steady speed of 72 km/h, the power developed by the engine being 220 kW. Frictional resistance is 10 kN. Calculate the air resistance.

If the air resistance varies as the square of the speed and the engine is drawing the same train up an incline of angle $\sin^{-1}\frac{1}{280}$ to the horizontal, at a steady speed of 36 km/h, calculate the power developed. (It may be assumed that frictional resistance is unchanged.) (London)†

22. The frictional resistance to the motion of a train is always k times the total weight of the engine and coaches. An engine of weight $2W$ works at constant horse-power throughout and attains a maximum speed U when pulling 6 coaches each of weight W up an incline. Down the same incline with only 4 coaches, the engine can attain a maximum speed $2U$. Find the maximum speed of the engine when pulling 2 coaches on the level and also find its acceleration in this case when it is travelling with a speed U on the level.
(London)

23. An engine working at 300 kW pulls a train of mass 150 t along a level track. Determine the frictional resistance in N/t if the acceleration is 0·16 m/s² when the speed is 30 km/h.

If the power and the frictional resistances remain constant, find the gradient which the train can ascend at a steady speed of 30 km/h.
(WJEC, part)†

24. A particle of unit mass moves in a straight line under the action of a tractive force and a resistance kv^2, where v is the speed and k is constant.

(a) If the tractive force is P (constant) and the particle increases its speed from u to $2u$ over a distance a, show that

$$P = ku^2\left(\frac{4e^{2ka} - 1}{e^{2ka} - 1}\right)$$

and find the corresponding time taken.

(b) If the tractive force works at constant power and is equal to P when $v = u$, find the distance over which the speed increases from u to $2u$.
(London)

EXERCISES

25. An engine, of mass 50 t, moves from rest on a horizontal track against constant resistances and for the first 30 s of its motion its acceleration is $\frac{3}{100}(30 - t)$ m/s^2, where t seconds is the time from the start. If the force exerted by the engine is initially 50 kN, find the magnitude of the resistances and an expression for the force exerted by the engine while t is less than 30.

Calculate the speed of the engine after 20 s and the power at which it is then working. (WJEC)†

26. A train of mass 200 tonne ascends a hill of inclination α to the horizontal, where $\sin \alpha = \frac{1}{100}$. The frictional resistance is 75 N/t. If the locomotive works at a rate of 220 kW, what is the acceleration in m/s^2 when the speed is 18 km/h?

[Take g as 9·8 m/s^2.]

Assuming that the locomotive continues to work at the same rate and that the resistance to motion is unchanged, find the maximum speed which the train is capable of attaining whilst ascending the hill.

(London, part)†

27. A car of mass M moves on a level road against a resisting force which is proportional to the speed. The tractive force works at a constant rate H. If the maximum speed attainable under these conditions is V, prove that a speed $u \, (< V)$ can be reached from rest after moving a distance x where

$$\frac{Hx}{MV^3} = \tfrac{1}{2} \log_e \left(\frac{V + u}{V - u} \right) - \frac{u}{V}.$$

Find also the time required to attain a speed $\tfrac{1}{2} V$ from rest.

(London)

28. A train of mass M moves on a straight horizontal track. At speeds less than V the resultant force on the train is constant and equal to P; at speeds not less than V the rate of working of the force is constant and equal to PV. Show that a speed $v \, (> V)$ is attained from rest in the time

$$\frac{M(V^2 + v^2)}{2PV},$$

and find the corresponding distance travelled. (JMB)

29. A train of total weight W kilonewtons is drawn on the level against a resistance Rv^2 kilonewtons, where v is the speed in m/s. The greatest pull which the engine can exert is P kilonewtons and the greatest power which it can develop is H kilowatts. The train starts from rest and the engine exerts its greatest pull until the greatest power is developed. Show that, at this instant, the speed

WORK, POWER AND ENERGY

of the train is V m/s and that it has been acquired in a distance s metres where

$$V = \frac{H}{P} \quad \text{and} \quad s = \left(\frac{W}{2gR}\right) \log_e \left(\frac{P}{P - RV^2}\right).$$

What is the greatest steady speed at which the train can travel on the level? (WJEC)†

30. A light elastic string of modulus $6mg$ and unstretched length l, is extended by an amount x. Derive an expression for the energy stored in the string. One end of the string is tied to the highest point of a smooth circular wire, radius l, which is fixed in a vertical plane. The other end is attached to a small ring of mass m, which is free to move along the wire. If initially, the ring is gently displaced from rest at the lowest point of the wire, find the speed of the ring (a) at the moment when the string becomes slack (b) when the ring reaches its highest point on the wire. (JMB)

31. A train of weight 150 tonne is being pulled on a level track by a tractive force of 40 kN against a resistance of $(8\,000 + 20v^2)$ N, where the speed is v m/s. Show that the distance in m taken to accelerate from rest to 72 km/h is given by

$$7\,500 \int_0^{20} \frac{v\,dv}{40^2 - v^2}.$$

At the moment when the train has reached this speed the power is shut off and the brakes are applied. The resistance now totals $(18\,000 + 20v^2)$ N. Show that the train travels from rest to rest in a distance of 2 500 m approximately. (London)†

32. A train has a total mass of 275 tonne. It starts from rest and travels up a straight incline of $\sin^{-1}\left(\frac{1}{385}\right)$. The engine works at a constant rate of 450 kW and the resistance to motion other than gravity is 80 N/t. Find the acceleration of the train when its speed is v m/s. Hence, find the greatest possible speed of the train and find the time during which the train acquires half of this speed from rest.

33. A train of mass M moves on a straight horizontal track. At speeds less than V the resultant force on the train is constant and equal to P; at speeds not less than V the rate of working of the force is constant and equal to PV. Show that a speed v $(> V)$ is attained from rest in a time

$$\frac{M(V^2 + v^2)}{2PV}$$

EXERCISES

and that the distance travelled in this time is
$$\frac{M(V^3 + 2v^3)}{6PV}.$$
(WJEC)

34. A particle of mass 4 units moves so that its position vector at time t is
$$r = i \sin 2t + j \cos 2t + k(t^2 - 2t).$$

Find
  (i) the kinetic energy of the particle at time t,
  (ii) the work done by the resultant force in the time interval $t = 0$ to $t = 2$,
  (iii) the resultant force acting on the particle,
  (iv) the time when the acceleration of the particle is perpendicular to its direction of motion. (AEB)

13

IMPULSE AND IMPACT

13.1. IMPULSE OF A CONSTANT FORCE

A constant force F, acting for the interval of time from t_0 to t_1, is said to exert an *impulse* I where

$$I = F(t_1 - t_0).$$

Impulse is a vector quantity and in the particular case of a constant force has the direction of the force.

SI unit of impulse ... N s

Example. A particle is acted on by a force of $2i - 2j + k$ N for 0·2 seconds, what is the impulse of the force?

$$\begin{aligned} I &= F(t_1 - t_0) \\ &= (2i - 2j + k)\,0·2 \\ &= 0·4i - 0·4j + 0·2k \text{ N s}. \end{aligned}$$

13.2. IMPULSE OF A VARIABLE FORCE

When F is a variable force acting for an interval of time t_0 to t_1, we divide the range $t_0 \to t_1$ into n sub-intervals, $\delta t_1, \delta t_2, \ldots, \delta t_n$.

Let
  F_1 be some value of F in the first sub-interval δt_1

  F_2 be some value of F in the second sub-interval δt_2

  . . . .

  F_n be some value F in the nth sub-interval δt_n

Then, as $F_1, F_2, \ldots, F_n$ can each be taken as approximately constant in their respective sub-intervals,

$$\begin{aligned} I &\simeq F_1\,\delta t_1 + F_2\,\delta t_2 + \cdots + F_n\,\delta t_n \\ &\simeq \sum_{t_0}^{t_1} F\,\delta t. \end{aligned}$$

236

IMPULSE AND MOMENTUM

In the limit, as the width of each sub-interval tends to zero, we have

$$I = \lim_{\delta t \to 0} \sum_{t_0}^{t_1} F \, \delta t = \int_{t_0}^{t_1} F \, dt \quad \text{(refer to Section 3.6).}$$

(If F varies in magnitude but has a constant direction, I also has this constant direction.)

Example. *A particle is acted on by a force of $t^2 i - 2tj$ newton where t is the time in seconds. If the force acts for the interval of time from $t = 0$ to $t = 2$ seconds, what is the impulse of the force?*

$$\begin{aligned} I &= \int_{t_0}^{t_1} F \, dt \\ &= \int_0^2 (t^2 i - 2tj) \, dt \\ &= [\tfrac{1}{3} t^3 i - t^2 j]_0^2 \\ &= \tfrac{8}{3} i - 4j \text{ N s.} \end{aligned}$$

It follows immediately from the additive properties of integrals, that

$$\int_{t_0}^{t_1} \left(\sum F \right) dt = \sum \int_{t_0}^{t_1} F \, dt,$$

i.e. the impulse of a finite number of concurrent forces is equal to the vector sum of the impulses of the separate forces.

13.3. IMPULSE AND MOMENTUM

In practical problems concerning blows and impacts, the forces are very large and act for short times. In general, we cannot measure the varying magnitudes. In such cases the impulse is measured by the effect it produces. Thus,

$$\begin{aligned} I = \int_{t_0}^{t_1} F \, dt &= \int_{t_0}^{t_1} \frac{d(mv)}{dt} \, dt \quad \text{(refer to Section 6.2)} \\ &= \int_{(mv)_0}^{(mv)_1} d(mv) \\ &= (mv)_1 - (mv)_0. \end{aligned}$$

If the mass m is constant,

$$I = mv_1 - mv_0.$$

IMPULSE AND IMPACT

The quantity mv is known as the *Momentum* (refer to Section 6.1) and is a vector quantity with the same units as impulse.

$$\text{Impulse} = \text{Change of Momentum}$$

Example 1. *A particle of mass 10 kg moving in a straight line, is acted on by a force which changes its speed from 36 km/h to 54 km/h. Find the impulse exerted.*

$$36 \text{ km/h} = 10 \text{ m/s}, \qquad 54 \text{ km/h} = 15 \text{ m/s}$$

$$\therefore \quad \text{Magnitude of impulse} = mv_1 - mv_0$$
$$= 10 \times 15 - 10 \times 10$$
$$= 50 \text{ N s}.$$

Example 2. *A racing car of mass 1 200 kg travelling on a horizontal track at 144 km/h, strikes a vertical crash barrier at an angle of 30 degrees and rebounds with a speed of 90 km/h. If the direction of the rebound makes an angle of 140 degrees with the original direction of motion, find the impulse given to the car by the barrier.*

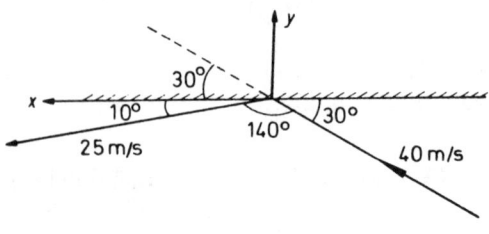

Figure 13.1

If v_0 and v_1 are the initial and final velocities, $v_0 = 144 \text{ km/h} = 40 \text{ m/s}$, and $v_1 = 90 \text{ km/h} = 25 \text{ m/s}$.

Taking axes Ox and Oy along and perpendicular to the crash barrier (refer to *Figure 13.1*),

$$v_0 = 40[i \cos 30° + j \sin 30°]$$
$$= 40(i\sqrt{3}/2 + j/2)$$
$$= 34 \cdot 64 i + 20 j \text{ m/s}$$

IMPULSE AND MOMENTUM

Also
$$v_1 = 25(i\cos 10° - j\sin 10°)$$
$$= 25(0{\cdot}9848i - 0{\cdot}1736j)$$
$$= 24{\cdot}62i - 4{\cdot}34j \text{ m/s.}$$

Now
$$I = m(v_1 - v_0).$$

Hence impulse received by car
$$= 1\,200[(24{\cdot}62 - 34{\cdot}64)i + (-4{\cdot}34 - 20)j]$$
$$= 1\,200[-10.02i - 24{\cdot}34j]$$
$$= -12\,024i - 29\,210j \text{ N s.}$$

Example 3. A cricket ball weighing 160 g is moving horizontally directly towards a batsman. Its speed just before it hits the bat is 30 m/s. It leaves his bat at 40 m/s at 90 degrees to its original direction, and at an angle of 45 degrees with the horizontal. Find the magnitude of the impulse imparted by the batsman to the ball.

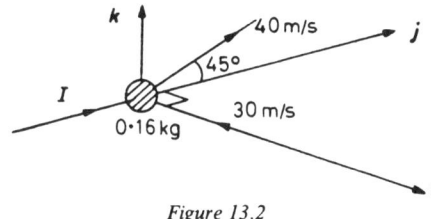

Figure 13.2

Take the batsman as origin, axes Ox, Oy horizontally along the pitch and perpendicular to it and Oz vertical. Let i, j and k be unit vectors parallel to Ox, Oy and Oz respectively (refer to *Figure 13.2*).

The initial velocity is $-30i$ m/s, the final velocity is in a direction given by $j + k$. The unit vector in this direction is $(1/\sqrt{2})(j + k)$, therefore, the final velocity is $(40/\sqrt{2})(j + k)$ m/s.

Now
$$I = mv_1 - mv_0$$

∴ Impulse received by ball, $I = 0{\cdot}16 \times \dfrac{40}{\sqrt{2}}(j + k) + 0{\cdot}16 \times 30i$

∴
$$|I| = 0{\cdot}16\sqrt{30^2 + \left(\frac{40}{\sqrt{2}}\right)^2 + \left(\frac{40}{\sqrt{2}}\right)^2}$$
$$= 0{\cdot}16 \times 50$$
$$= 8 \text{ N s.}$$

239

Exercises 13a

1. A particle is acted on by a force of $3i - 4j$ newton for 3 s. What is the magnitude of the impulse of the force?

2. A particle is acted on by a force of $(3t^2 - 1)i + 5t^3 j$ newton, where t is the time in seconds. If the force acts for the interval of time $t = 2$ to $t = 4$ s, what is the impulse?

3. A particle of mass 500 kg moving at 30 m/s, alters its speed to 45 m/s in the same direction when acted on by an impulse I. Find the magnitude of I.

4. A particle of mass 25 kg originally moving in a straight line at 10 m/s is acted on by an impulse I. The magnitude of I is 550 N s and it acts directly against the motion of the particle. Find the new speed of the particle.

5. A ball of mass 200 g travelling horizontally at 6 m/s strikes a vertical wall at right angles and rebounds with a speed of 4 m/s. Find the magnitude of the impulse given to the ball.

6. In a game of football, the ball, of mass 400 g travelling horizontally at 20 m/s, hits the cross-bar of the goal posts at right angles. The impulse imparted by the cross-bar to the ball is 12 N s, find the speed with which the ball rebounds.

7. A football weighing 400 g is passed along the ground to a player, with a speed of 6 m/s. He returns it along the ground at a speed of 8 m/s. If the direction of motion of the ball is altered by 90 degrees find the magnitude of the impulse given to the ball by the player.

8. A racing car of mass 1 t (tonne) is travelling on a horizontal track at 198 km/h. It strikes a vertical crash barrier and rebounds at a speed of 126 km/h at an angle of 120 degrees to its original direction of motion. Find the magnitude and direction of the impulse transmitted by the barrier to the car.

9. A particle weighing 1 kg is travelling at 50 m/s, what is the magnitude of the impulse which will change its direction by 60 degrees leaving its speed unchanged?

10. A football weighing 1 kg is moving horizontally towards a player. Its speed just before it reaches him is 6 m/s. He passes it at a speed of 12 m/s at 60 degrees to its original direction and at an angle of 60 degrees with the horizontal. Find the magnitude of the impulse given by the player to the ball.

13.4. INELASTIC IMPACTS

Example 1. A pile of mass m_1 is being driven into the ground by a piledriver of mass m_2. At the moment of impact the piledriver is

INELASTIC IMPACTS

moving vertically downwards with speed u and the pile is stationary. Immediately after the impact they both move with a common speed V. Find V in terms of m_1, m_2 and u, and the change in the total kinetic energy of the system.

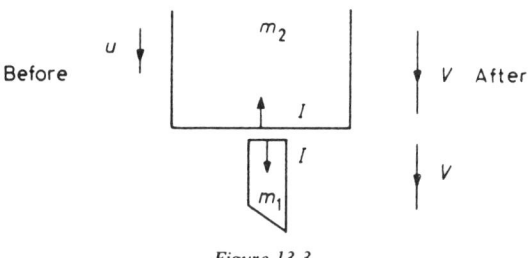

Figure 13.3

Referring to *Figure 13.3*, since the impulse is equal to the change of momentum

for the piledriver $\quad -I = m_2(V - u) \quad \quad \ldots (i)$

for the pile $\quad \quad I = m_1(V - 0) \quad \quad \ldots (ii)$

adding equations (i) and (ii)

$$0 = m_2(V - u) + m_1 V \quad \quad \ldots (iii)$$

∴ $\quad m_2 u = (m_1 + m_2)V$

$$V = \frac{m_2 u}{(m_1 + m_2)}.$$

From equation (iii) we see that when considering the two particles together, provided there is no outside impulse, the change of momentum is zero. This is an example of the principle of conservation of momentum (refer to Section 16·2), where

momentum before impact = momentum after impact

∴ $\quad m_2 u = (m_1 + m_2)V$

$$V = \frac{m_2 u}{m_1 + m_2} \quad \text{as before.}$$

Kinetic energy before impact $= \tfrac{1}{2} m_2 u^2 + 0$

kinetic energy after impact $= \tfrac{1}{2}(m_2 + m_1)V^2$

IMPULSE AND IMPACT

$$\therefore \text{ change in K.E.} = \tfrac{1}{2}m_2 u^2 - \tfrac{1}{2}(m_2 + m_1)V^2$$

$$= \tfrac{1}{2}m_2 u^2 - \tfrac{1}{2}(m_2 + m_1)\frac{m_2 u^2}{(m_2 + m_1)^2}$$

from equation (i)

$$= \tfrac{1}{2}m_2 u^2 \left(1 - \frac{m_2}{m_2 + m_1}\right)$$

$$= \tfrac{1}{2}m_2 u^2 \left(\frac{m_1}{m_2 + m_1}\right)$$

$$= \tfrac{1}{2}\frac{m_2 m_1 u^2}{m_2 + m_1}.$$

This quantity is essentially positive. Thus there is a *loss* of kinetic energy. This is true in nearly all cases of impact as part of the energy is converted into other forms such as heat, sound, light etc.

Example 2. A gun of mass M is free to move horizontally but not vertically. The angle of elevation of its barrel is θ and it fires a shell of mass m. Show that initially the direction of motion of the shell is at an angle $\tan^{-1}[(M + m)\tan\theta/M]$ to the horizontal.

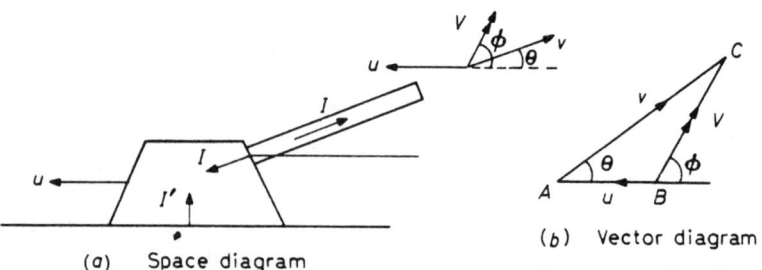

(a) Space diagram   (b) Vector diagram

Figure 13.4

Referring to *Figure 13.4*, the shell is reacted on by the gases and there is an equal and opposite reaction on the gun. The gun is free to move horizontally and the horizontal component of the reaction imparts a backward velocity of magnitude u to the gun. Thus, as the shell leaves the barrel of the gun it has a velocity v relative to the barrel of the gun, and the backward velocity u of the gun. The resultant V of these two velocities is shown in *Figure 13.4*.

The gun is only free to move horizontally and the vertical component of the reaction due to the explosion is equal to the reaction of

INELASTIC IMPACTS

the ground. The horizontal momentum of the gun is equal to the horizontal momentum of the shell.

$$\therefore \qquad Mu = mV\cos\phi \qquad \ldots\text{(i)}$$

From the sine rule applied to the vector diagram in *Figure 13.4(b)*,

$$\frac{\sin\angle ACB}{u} = \frac{\sin\theta}{V}$$

i.e.
$$\frac{\sin(\phi - \theta)}{u} = \frac{\sin\theta}{V}. \qquad \ldots\text{(ii)}$$

From equations (i) and (ii),

$$\frac{u}{V} = \frac{m\cos\phi}{M} = \frac{\sin(\phi - \theta)}{\sin\theta}$$

$$\therefore \quad m\cos\phi\sin\theta = M\sin\phi\cos\theta - M\cos\phi\sin\theta$$

$$\therefore \quad M\sin\phi\cos\theta = (m + M)\cos\phi\sin\theta$$

$$\therefore \quad M\tan\phi = (m + M)\tan\theta$$

$$\therefore \quad \phi = \tan^{-1}\left[\frac{(m + M)\tan\theta}{M}\right].$$

Exercises 13b

1. A truck, moving at 3 m/s, bumps into and is automatically coupled to another truck of half its mass, which is initially at rest. What is their common speed after impact?

2. A pile-driver, of mass 2 tonne, falls through a height of 2·5 m onto a pile of mass 0·5 t. What is the momentum of the pile-driver just before the impact? Immediately after the impact the driver and pile move together. Find their common speed.

3. A rocket, of mass M and velocity V, is free to move in space. It emits a blast of gas of mass m with a velocity, relative to the rocket, of v. Show that its new velocity is $V - mv/(M - m)$.

4. A body, of mass 50 g, is moving with a speed of 7 m/s. It meets a body of mass 120 g moving in the exactly opposite direction with a speed of 10 m/s. If they coalesce into one body, find its new speed and the loss of K.E. during the impact.

5. A pile-driver, of mass 4 tonne, falls through a height of 2·5 m onto a pile of mass 1 t. The pile-driver does not rebound after

impact. If the pile is driven 20 cm into the ground, find the resistance of the ground (assumed uniform).

6. A gun mass M can move on a horizontal plane. It fires horizontally a shell of mass m. If u is the velocity of the shell relative to the gun, find the velocities of the gun and the shell relative to the plane.

7. A block of wood of mass M is suspended by a light inextensible string. A bullet of mass m is fired horizontally into the wood with a speed u. The wood and the embedded bullet swing until the block of wood is at a height h above its original position. Show that $mu = (M + m)\sqrt{2gh}$.

8. Two particles of masses m_1 and m_2 and velocities u_1 and u_2 respectively, collide and stick together. Show that the energy lost in the collision can be expressed as

$$\frac{1}{2}\frac{m_1 m_2}{m_1 + m_2}(u_2 - u_1).(u_2 - u_1).$$

13.5. IMPULSIVE TENSIONS IN STRINGS

If two particles are connected by an inextensible string, any relative movement of the particles, which causes the string to tighten, will produce an impulsive tension in the string. For a light string this impulse acts in the direction of the string and has an equal and opposite effect on both particles.

Example 1. *A particle of mass M lying on the ground is connected, by means of a light inextensible string passing over a smooth pulley to a mass m. After the mass m has fallen through a height h, the string tightens and the mass M begins to rise. Find the impulse applied to M when the string tightens and the initial speed.*

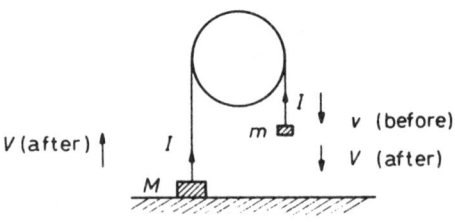

Figure 13.5

IMPULSIVE TENSIONS IN STRINGS

Referring to *Figure 13.5*. Immediately before the string tightens the speed of *m* is given by

$$v^2 = u^2 + 2fs$$

i.e.
$$v^2 = 0 + 2gh$$

∴
$$v = \sqrt{2gh}. \qquad \ldots (i)$$

If *I* is the impulse and *V* the common speed of the two particles immediately after the string tightens, then, considering the two masses separately,

$$mV - mv = -I$$

and
$$MV - 0 = I. \qquad \ldots (ii)$$

Adding $mV + MV - mv = 0$.

∴
$$V = \frac{mv}{(m + M)}.$$

From equation (i)
$$V = \frac{m\sqrt{2gh}}{(m + M)}. \qquad \ldots (iii)$$

To find *I* we substitute from equation (iii) in equation (ii) and

$$I = \frac{mM\sqrt{2gh}}{(m + M)}.$$

Example 2. Two balls A and B of masses 200 g and 300 g respectively, lie on a smooth horizontal table connected by a taut, light, inextensible string. The mass B is given an impulse I, of magnitude 0·03 N s in a direction inclined at 30 degrees to the string AB and away from A. Find the velocities of A and B immediately after the blow.

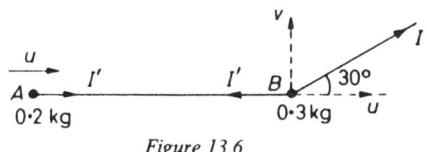

Figure 13.6

Considering the system as a whole, the internal impulsive tensions in the string (*I'*) acting on *A* and *B* cancel. Therefore the total momentum of the system is due to the impulse *I*.

245

IMPULSE AND IMPACT

Let B move initially with speeds u and v in the direction of AB and perpendicular to it respectively (refer to *Figure 13.6*). AB is a taut string, therefore, intially A has the speed u in the direction of AB.

In a direction of AB, $I\cos 30°$ is equal to the total change in momentum of A and B.

$\therefore \qquad\qquad 0\cdot 03 \cos 30° = 0\cdot 3u + 0\cdot 2u$

$\therefore \qquad\qquad 0\cdot 025\,98 = 0\cdot 5u$

$\therefore \qquad\qquad u = 0\cdot 051\,96 \text{ m/s}.$

In the direction perpendicular to AB,

$$0\cdot 03 \sin 30° = 0\cdot 3v + 0$$

$\therefore \qquad\qquad 0\cdot 015 = 0\cdot 3v$

$\therefore \qquad\qquad v = 0\cdot 05 \text{ m/s}.$

If V is the initial velocity of B,

$$|V| = \sqrt{0\cdot 051\,96^2 + 0\cdot 05^2}$$
$$= 0\cdot 072\,11 \text{ m/s} = 7\cdot 2 \text{ cm/s},$$

and the direction of V makes an angle $\tan^{-1}(0\cdot 962\,3) = 43\cdot 9°$ with AB produced.

The initial velocity of A is $0\cdot 051\,96$ m/s $= 5\cdot 20$ cm/s in the direction of AB.

Example 3. Three particles A, B and C, of masses 4, 6 and 8 kg, respectively, lie at rest on a smooth horizontal table. They are connected by taut light inextensible strings AB and BC and $\angle ABC = 120$ degrees. An impulse I is applied to C. If the magnitude of I is 88 N s and it acts in the direction BC, find the initial speeds of A, B and C.

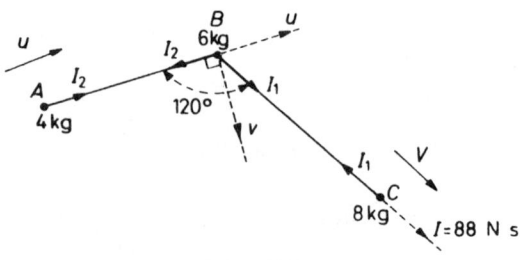

Figure 13.7

INPULSIVE TENSIONS IN STRINGS

Referring to *Figure 13.7*, let I_1 and I_2 be the impulsive tensions in the strings BC and AB respectively. Since A is acted on only by I_2, its initial speed u will be in the direction of AB. Since the string AB is taut, B must have a speed u in the direction AB. Also, let B have a speed v in a direction perpendicular to AB [this is necessary because I_1 acts in a different direction to I_2]. Finally, since both I and I_1, the impulses acting on C, have the same direction BC, let C have an initial speed V in the direction BC.

Since BC is taut, the speeds of B and C in the direction BC are equal.

$\therefore \qquad u \cos 60° + v \cos 30° = V$

$\therefore \qquad u/2 + v\sqrt{3}/2 = V.$ .... (i)

Considering the motion of A,

$$4u = I_2.$$ .... (ii)

Considering the motion of B along and perpendicular to AB,

$$6u = I_1 \cos 60° - I_2.$$

$\therefore \qquad 6u = I_1/2 - I_2$ .... (iii)

$$6v = I_1 \cos 30°$$

$\therefore \qquad 6v = I_1\sqrt{3}/2.$ .... (iv)

Considering the motion of C,

$$8V = 88 - I_1.$$ .... (v)

Eliminating I_2 from equations (ii) and (iii)

$$6u = I_1/2 - 4u$$

$\therefore \qquad I_1 = 20u.$ .... (a)

Substituting from equation (a) in equation (iv),

$$6v = 20u\sqrt{3}/2$$

$\therefore \qquad v = \dfrac{5}{\sqrt{3}} u.$ .... (b)

Substituting from equation (b) in equation (i),

$$\dfrac{u}{2} + \dfrac{5}{\sqrt{3}} \cdot \dfrac{\sqrt{3}}{2} u = V$$

$\therefore \qquad V = 3u.$ .... (c)

IMPULSE AND IMPACT

Substituting for V and I_1 [equations (a) and (c)] in equation (v),
$$8.3u = 88 - 20u$$
$$\therefore \qquad 44u = 88$$
$$\therefore \qquad u = 2.$$

From (b) $\qquad v = 10/\sqrt{3}.$

From (c) $\qquad V = 6.$

$\therefore \qquad$ speed of A is 2 m/s,

speed of B is $\sqrt{u^2 + v^2} = \sqrt{4 + \frac{100}{3}}$
$$= 4\sqrt{\tfrac{7}{3}} \text{ m/s}$$

speed of C is 6 m/s.

Exercises 13c*

1. A particle, of mass M, is connected by means of a light inextensible string passing over a smooth pulley to a mass m. ($M > m$). The system is moving with speed v when the mass M hits the floor. If M does not rebound from the floor, find the impulsive tension in the string when it is jerked off the ground by the mass m.

2. Two particles A and B, each of mass 5 kg, lie on a smooth horizontal table at a distance 5 cm apart. They are connected by a light inextensible string of length 13 cm. A is given an impulse 65 N s in a direction perpendicular to AB. Find the impulsive tension in the string when it becomes taut and the initial velocity of B.

3. Two balls A and B, of masses 50 kg and 75 kg respectively, lie at rest on a smooth horizontal table, distance 5 m apart. They are connected by a light inextensible string of length 10 m. B is given an impulse I of magnitude 300 N s in the direction AB. Find the speed of A when it is first jerked into motion, and the impulsive tension in the string at this moment. Would the initial speed of A be altered if the string AB had been taut at the moment the impulse was applied?

4. Two equal particles A and B, each of mass 2 kg, lying close together at the edge of a table, are joined by a light inelastic string length 2·5 m. If B is gently pushed over the edge of the table, find the speed of the particles when A begins to move, and the impulsive tension in the string. [Assume that the height of the table is greater than 2·5 m.]

* Exercises marked thus, †, have been metricized, *see* Preface.

5. A particle A of mass M lies on a smooth horizontal table. It is connected by a taut light inextensible string, passing over a smooth pulley at the edge of the table, to another particle of mass m which is hanging freely. An impulse I is applied to A in the direction of the string but away from the edge of the table. Find the initial speed of the two particles.

6. Two particles each of mass M are connected by a taut light inextensible string passing over a smooth pulley. The particles are hanging freely at rest. A third particle of mass $3M$ falls freely through a height h, strikes one of the particles and coalesces with it. Find the initial speed of the particles.

7. Three particles A, B and C, each of mass 2 kg, lie at rest on a smooth horizontal table. They are connected by two taut, light, inextensible strings so that $\angle ABC = 135$ degrees. An impulse I, of magnitude 70 N s in a direction parallel to AB, is given to C. Find the initial speeds of A, B and C and the impulsive tension in AB.

8. Three particles, each of mass m, lie in order in a straight line on a smooth horizontal table. They are connected by two taut light inelastic strings AB and BC. C is given an impulse I in a direction 45 degrees with the line ABC and away from A and B. Find the impulsive tensions in the two strings.

9. A particle A mass m lies at rest on a smooth horizontal table. It is connected by two taut inelastic strings to two other particles B, C mass, $3m$ and $2m$, respectively. The angle BAC is 2θ. An impulse is given to A in a direction bisecting the angle BAC and away from B and C. Find the angle at which the particle A begins to move.

10. A string 80 cm long connects two masses m kg and $2m$ kg which rest on a smooth horizontal table. The mass m is near the edge, and the mass $2m$ is 40 cm from m in a direction perpendicular to the edge, the string between the two masses being slack. The mass m is gently pushed over the edge of the table. Prove that the velocity of the other mass when it is jerked into motion is 14/15 m/s, and find the time that elapses after the jerk takes place until the mass $2m$ reaches the edge of the table. (London)†

13.6. DIRECT IMPACT OF ELASTIC BODIES

In considering impacts of two or more elastic bodies we shall assume that they are smooth and therefore, that the mutual reaction acts only along the common normal at the point of impact. We say that the two bodies impinge *directly* if the direction of motion of each is along this common normal.

IMPULSE AND IMPACT

In Example 1 of Section 13.4, it was pointed out that provided there are no outside impulses the total momentum is unaltered and this will apply in the following problems. The bodies are generally taken as equal spheres.

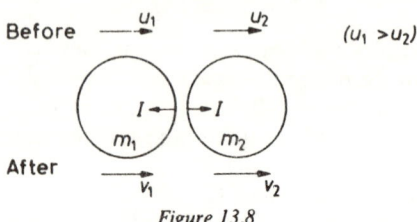

Figure 13.8

Consider two bodies of masses m_1 and m_2 which impinge directly, let u_1, u_2 be their speeds before impact and v_1, v_2 their speeds after impact. Since the bodies are supposed to be smooth and they impinge directly, u_1, u_2, v_1, v_2 are all in the direction of the common normal. Since the total momentum is conserved, we have that

$$m_1 u_1 + m_2 u_2 = m_1 v_1 + m_2 v_2 \qquad \ldots\text{(i)}$$

This one equation is not sufficient to calculate v_1 and v_2 and we have recourse to Newton's experimental law. If the velocities both before and after impact are taken relative to the same body, then, for two bodies impinging directly, their relative velocity after impact is equal to a constant (e) times their relative velocity before impact and *in the opposite direction*. e is known as the *coefficient of restitution*. For the two bodies depicted in *Figure 13.8*

$$v_1 - v_2 = -e(u_1 - u_2). \qquad \ldots\text{(ii)}$$

In the case of oblique impact, the result holds for the components of the velocities in the direction of the common normal at impact. The value of e has to be found by experiment and varies from 0 for completely inelastic bodies to practically 1 for nearly perfectly elastic bodies.

The two results (i) and (ii) enable us to find v_1 and v_2 but it is extremely important to note that these formulae will only be true if u_1, u_2, v_1, v_2 stand for speeds in the same direction.

Example 1. *A sphere of mass* 3 kg *moving at* 6 m/s, *impinges directly on another sphere of mass* 5 kg, *moving in the same direction at* 3 m/s. *If* $e = \frac{2}{3}$, *find their speeds after impact.*

DIRECT IMPACT OF ELASTIC BODIES

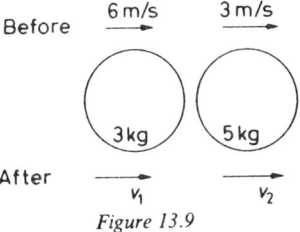

Figure 13.9

Let v_1, v_2 be their speeds after impact (refer to *Figure 13.9*).

By Newton's Law $\quad v_1 - v_2 = -\tfrac{2}{3}(6 - 3)$

$\therefore \qquad\qquad\qquad v_1 - v_2 = -2.$ ....(i)

Since the momentum is conserved,

$\qquad\qquad$ momentum before = momentum after impact.

$\therefore \qquad\qquad 6 \times 3 + 3 \times 5 = 3v_1 + 5v_2$

i.e. $\qquad\qquad\qquad 33 = 3v_1 + 5v_2.$ ....(ii)

Multiplying equation (i) by 5 and adding to equation (ii),

$\qquad\qquad 33 - 10 = (3v_1 + 5v_2) + (5v_1 - 5v_2)$

$\therefore \qquad\qquad\qquad 23 = 8v_1$

$\therefore \qquad\qquad\qquad v_1 = 2\tfrac{7}{8}$ m/s.

Substituting for v_1 in equation (i),

$$v_2 = 4\tfrac{7}{8} \text{ m/s.}$$

Example 2. A sphere, of mass 100 kg moving at 16 m/s, impinges directly on another sphere of mass 500 kg, moving in the opposite direction at 5 m/s. If $e = \tfrac{2}{3}$, find their speeds after impact and the magnitude of the impulse given to each sphere.

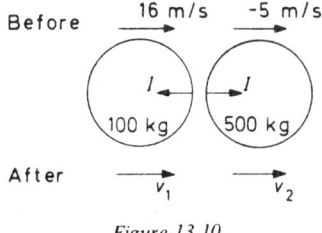

Figure 13.10

251

IMPULSE AND IMPACT

Let v_1, v_2 be their speeds after impact. Referring to *Figure 13.10*, we note that the second sphere has a speed of -5 m/s along the line of centres.

By Newton's law $\quad v_1 - v_2 = -\frac{2}{3}[16 - (-5)]$,

or $\qquad\qquad\quad v_1 - v_2 = -14.$ .... (i)

Since the momentum is conserved,

$$100 \times 16 + 500 \times (-5) = 100v_1 + 500v_2.$$

$\therefore \qquad\qquad\qquad -9 = v_1 + 5v_2.$ .... (ii)

Subtracting equation (i) from equation (ii),

$$6v_2 = 5$$

$\therefore \qquad\qquad\qquad v_2 = \frac{5}{6}$ m/s.

Substituting for v_2 in equation (i),

$$v_1 = -13\frac{1}{6} \text{ m/s.}$$

The magnitude of the impulse is equal to the change of momentum of either of the balls.

$\therefore \qquad$ magnitude of the impulse $= 500[\frac{5}{6} - (-5)]$

$$= 2916\frac{2}{3} \text{ N s.}$$

The speed of the first sphere changes from $+16$ m/s to $-13\frac{1}{6}$ m/s, so that it rebounds after impact. Similarly, the speed of the second sphere changes from -5 m/s to $+\frac{5}{6}$ m/s so that it also rebounds after impact.

Example 3. A ball moving with speed u impinges directly on another ball of the same mass. If the second ball was stationary before the impact and $e = \frac{1}{2}$, find the loss of kinetic energy during impact.

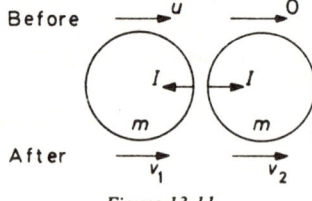

Figure 13.11

Let m be the mass of each ball (refer to *Figure 13.11*).

By Newton's law $\quad v_1 - v_2 = -\frac{1}{2}(u - 0),$

or $\qquad\qquad\quad v_1 - v_2 = -\frac{1}{2}u.$ .... (i)

DIRECT IMPACT OF ELASTIC BODIES

The momentum is conserved.

$$\therefore \qquad mu + 0 = mv_1 + mv_2$$
$$\therefore \qquad u = v_1 + v_2. \qquad \ldots (ii)$$

Adding equations (i) and (ii),

$$2v_1 = \tfrac{1}{2}u$$
$$\therefore \qquad v_1 = \tfrac{1}{4}u. \qquad \ldots (iii)$$

Substituting for v_1 in equation (ii),

$$v_2 = \tfrac{3}{4}u. \qquad \ldots (iv)$$

Now the loss of K.E. is given by

$$\tfrac{1}{2}mu^2 + 0 - \tfrac{1}{2}mv_1^2 - \tfrac{1}{2}mv_2^2$$
$$= \tfrac{1}{2}mu^2 - \tfrac{1}{2}m(\tfrac{1}{4}u)^2 - \tfrac{1}{2}m(\tfrac{3}{4}u)^2 \quad \text{by (iii) and (iv)}$$
$$= \tfrac{1}{2}mu^2 - \tfrac{1}{32}mu^2 - \tfrac{9}{32}mu^2$$
$$= \tfrac{3}{16}mu^2.$$

Exercises 13d

1. A ball, of mass 3 kg moving at 8 m/s, impinges directly on another ball whose mass is 5 kg. If the second ball is moving at 3 m/s in the same direction and $e = \tfrac{2}{5}$, find their velocities after the impact.

2. A ball, of mass 4 kg moving at 5 m/s, impinges directly on another ball of mass 7 kg moving at 10 m/s in the opposite direction. If $e = \tfrac{2}{3}$, find their velocities after the impact.

3. A sphere, of mass 600 g moving at 12 cm/s, impinges directly on another sphere of mass 1800 g, which is stationary. If $e = \tfrac{1}{3}$, find their velocities after impact and the fraction of the initial kinetic energy lost during impact.

4. A railway wagon of mass 10 tonne is moving at 1 m/s. It collides with another wagon of mass 40 tonne moving at 0·2 m/s in the same direction. The first wagon remains at rest after the impact. Find the coefficient of restitution between the two wagons.

5. A sphere, of mass m moving with speed V, impinges directly on an equal sphere moving with speed v in the same direction. Show that the magnitude of the impulse on either of the spheres is $\tfrac{1}{2}m(1 + e)(V - v)$, where e is the coefficient of restitution.

6. Three smooth equal spheres A, B and C each of mass m lie, in order, on a smooth horizontal table with their centres in a straight line. B and C are initially at rest and A is moving with speed u along the line of centres towards B. After the impact, B collides with C. If e

is the coefficient of restitution between each pair of the balls, find the velocities of the three spheres after the second impact.

7. A railway wagon, of mass m moving with a speed u, impinges directly on a similar wagon which is stationary. Show that the loss of kinetic energy during the impact is $\frac{1}{4}mu^2(1 - e^2)$, where e is the coefficient of restitution.

8. Three smooth spheres A, B and C of equal radii and masses m, $2m$, $4m$ respectively, lie at rest on a smooth horizontal table. Their centres are in a straight line and B is between A and C. A is projected towards B with a speed u. After the impact, B continues along the line of centres and strikes C. If $e = \frac{1}{2}$, show that after the second impact both A and B are at rest, and find the loss of kinetic energy after the two impacts.

9. Two balls A and B of equal radii, and masses 4 kg and 5 kg respectively, lie on a smooth horizontal floor. A is given a velocity u and impinges directly on B, which then hits a smooth vertical wall normally. After rebounding from the wall, B hits A a second time. If B is brought to rest after the second impact with A, show that $2e^3 - 3e^2 - 3e + 2 = 0$ where e is the coefficient of restitution between the two balls and between B and the wall. Verify that $e = \frac{1}{2}$ is the only practical solution of the equation.

10. A sphere, of mass m, impinges directly on a stationary sphere of mass $3m$. If two-thirds of the original kinetic energy is lost, find the coefficient of restitution between the two spheres.

13.7. OBLIQUE IMPACT OF ELASTIC BODIES

Consider two smooth bodies, masses m_1 and m_2, which impinge obliquely. Let their speeds be u_1 and u_2 in directions making angles α and β respectively, with the common normal (refer to *Figure 13.12*).

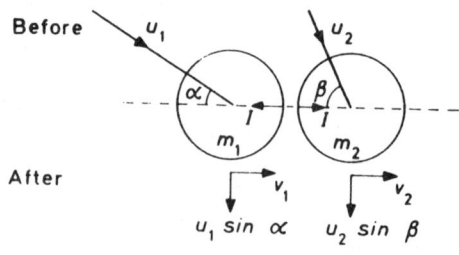

Figure 13.12

OBLIQUE IMPACT OF ELASTIC BODIES

Since the bodies are assumed to be smooth and the impulse has no component perpendicular to the common normal, the components of their velocities perpendicular to the common normal $u_1 \sin \alpha$, $u_2 \sin \beta$ are unaltered.

The components of their velocities along the common normal can be found in the same manner as for direct impact. Let v_1, v_2 be the components along the common normal of the velocities after impact. As we remarked at the beginning of Section 13.6, for oblique impact Newton's law holds for the component velocities along the common normal

$$\therefore \quad v_1 - v_2 = -e(u_1 \cos \alpha - u_2 \cos \beta). \quad \ldots (i)$$

Since the momentum is conserved

$$m_1 u_1 \cos \alpha + m_2 u_2 \cos \beta = m_1 v_1 + m_2 v_2. \quad \ldots (ii)$$

Equations (i) and (ii) enable us to find v_1 and v_2 and since the components perpendicular to the common normal are known, we can find the velocities of the two spheres after impact.

Example 1. Two smooth elastic spheres A and B of equal radii and masses 80 kg and 72 kg respectively, lie at rest on a smooth horizontal surface. A is projected towards B with a speed of 60 m/s and strikes B obliquely at an angle of 30 degrees with the line of centres. If $e = \frac{1}{2}$, find the velocities of the two spheres after impact.

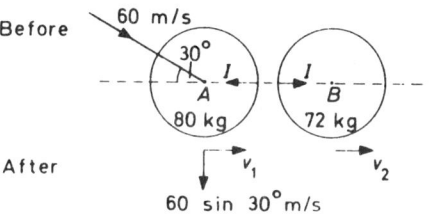

Figure 13.13

Let AB be the line of centres. The components of the velocities perpendicular to AB, $60 \sin 30°$ and 0 are unaltered by the impact. Along AB (refer to *Figure 13.13*). By Newton's law

$$v_1 - v_2 = -\tfrac{1}{2}(60 \cos 30° - 0)$$

$$\therefore \quad v_1 - v_2 = -25\cdot 98. \quad \ldots (i)$$

IMPULSE AND IMPACT

Since the momentum is conserved,

$$80.60 \cos 30° + 0 = 80v_1 + 72v_2$$

$$\therefore \qquad 4157 = 80v_1 + 72v_2$$

or $\qquad 519\cdot 6 = 10v_1 + 9v_2.$ ....(ii)

Adding $9 \times$ equation (i) to equation (ii)

$$19v_1 = 285\cdot 8$$

$$\therefore \qquad v_1 = 15\cdot 04.$$

By substitution in (ii) $\qquad v_2 = 41\cdot 02.$

Hence the velocity of B after impact is $41\cdot 0$ m/s in the direction of AB.

The velocity of A after impact has components $15\cdot 04$ m/s along AB and 30 cm/s perpendicular to AB.

$$\therefore \qquad |v_A| = \sqrt{15\cdot 04^2 + 30^2} = 33\cdot 6 \text{ m/s}$$

and the direction of v_A makes an angle of

$$\tan^{-1}(30/15\cdot 04) = 63\cdot 4°$$

with AB in the direction AB.

Example 2. A smooth sphere, of mass 10 kg *moving horizontally with a speed of* 9 m/s, *impinges on another smooth sphere of mass* 8 kg *moving horizontally with a speed of* 6 m/s. *If their directions of motion at impact are inclined at* 45 *and* 60 *degrees respectively to the line of the common normal,* $e = \frac{2}{3}$, *and their radii are equal, find their speeds after impact.*

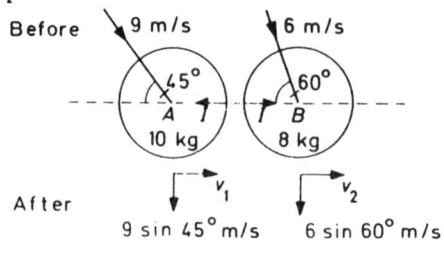

Figure 13.14

OBLIQUE IMPACT OF ELASTIC BODIES

Let A and B be the centres of the two spheres. The components $9 \sin 45°$ and $6 \sin 60°$ perpendicular to AB are unaltered by the impact (refer to *Figure 13.14*). Along AB, by Newton's experimental law

$$v_1 - v_2 = -\tfrac{2}{3}(9 \cos 45° - 6 \cos 60°)$$

$$= -\tfrac{2}{3}(9/\sqrt{2} - 3).$$

$$\therefore \quad v_1 - v_2 = -3\sqrt{2} + 2. \qquad \ldots (i)$$

By conservation of momentum, $10 \times 9 \cos 45° + 8 \times 6 \cos 60° = 10v_1 + 8v_2$, where v_1, v_2 are the components of their velocities along AB after impact.

$$\therefore \quad 45\sqrt{2} + 24 = 10v_1 + 8v_2. \qquad \ldots (ii)$$

Adding $8 \times$ equation (i) to equation (ii)

$$18v_1 = 40 + 21\sqrt{2}$$

$$\therefore \quad v_1 = \frac{40 + 21\sqrt{2}}{18} = 3 \cdot 872.$$

Substituting in (i) $\quad v_2 = \dfrac{4 + 75\sqrt{2}}{18} = 6 \cdot 115.$

The speed of the 10 kg sphere is

$$\sqrt{v_1^2 + (9 \sin 45°)^2} = \sqrt{3 \cdot 872^2 + 6 \cdot 364^2}$$

$$= 7 \cdot 45 \text{ m/s}.$$

The speed of the 8 kg sphere is

$$\sqrt{v_2^2 + (6 \sin 60°)^2} = \sqrt{6 \cdot 115^2 + 5 \cdot 196^2}$$

$$= 8 \cdot 02 \text{ m/s}.$$

Example 3. *A small smooth sphere S is suspended by a light inextensible string of length 2a from a fixed point P, whose distance from a smooth*

IMPULSE AND IMPACT

vertical wall is $a\sqrt{3}$. *The sphere is projected horizontally towards the wall with a speed of* $\sqrt{(37ga/5)}$. *Find the velocity with which it strikes the wall. If the coefficient of restitution between the sphere and the wall is* $\frac{1}{3}$, *find the loss of kinetic energy on impact.*

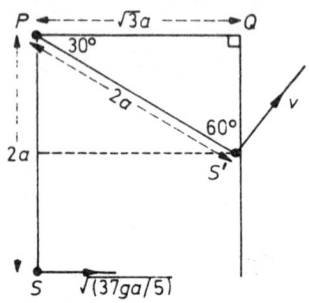

Figure 13.15

Let S' be the point at which the sphere strikes the wall and PQ the perpendicular from P to the wall. By trigonometry the angles in $\triangle PQS'$ are $30°, 60°, 90°$ (refer to *Figure 13.15*), and the length QS' is a.
Let v be the speed of the sphere at S' and m its mass:
By the Principle of Conservation of Energy (refer to Section 12.6)

P.E. at S + K.E. at S = P.E. at S' + K.E. at S'

Taking PQ as the level of zero potential

$$-mgPS + \tfrac{1}{2}m(37ga/5) = -mgQS' + \tfrac{1}{2}mv^2$$

$\therefore \quad \tfrac{1}{2}v^2 = \tfrac{1}{2}(37ga/5) - mg(PS - QS')$

$\quad\quad\quad\; = \tfrac{1}{2}(37ga/5) - mg(2a - a)$

$\therefore \quad v^2 = 37ga/5 - 2ga$

$\therefore \quad v = 3\sqrt{3ga/5}.$

The velocity on impact is of magnitude $3\sqrt{3ga/5}$ and is in a direction perpendicular to PS', that is, at an angle of 30 degrees to the upward vertical.

In this case the sphere on impact is only free to move vertically and hence we apply the principle of conservation of momentum vertically. If v_1 and v_2 are the vertical and horizontal components of the velocity of the sphere after impact,

$$mv \cos \theta = mv_1$$

$$\therefore \quad m3\sqrt{3ga/5} \cos 30° = mv_1$$

$$\therefore \quad v_1 = \tfrac{9}{2}\sqrt{ga/5}.$$

Applying Newton's law along the common normal, that is, perpendicular to the wall

$$v_2 - 0 = -\tfrac{1}{3}(v \sin 30° - 0)$$

$$\therefore \quad v_2 = -\tfrac{1}{3} 3\sqrt{3ga/5}\tfrac{1}{2} - 0$$

$$v_2 = -\tfrac{1}{2}\sqrt{3ga/5}.$$

Loss of kinetic energy
$$= \tfrac{1}{2}mv^2 - \tfrac{1}{2}mv_1^2 - \tfrac{1}{2}mv_2^2$$
$$= \tfrac{1}{2}m9 \times 3ga/5 - \tfrac{1}{2}m\tfrac{81}{4}ga/5 - \tfrac{1}{2}m\tfrac{1}{4}3ga/5$$
$$= \tfrac{3}{8}mga.$$

Exercises 13e

1. Two smooth spheres A, B, of equal radii and masses m and $2m$ kg respectively, rest on a smooth horizontal table. A is projected towards B with a speed of 16 m/s and strikes B obliquely, so that on impact the line of centres is inclined at an angle of 60 degrees with the direction of motion of A. If $e = \tfrac{1}{2}$, find the speeds of A and B after impact.

2. Two equal smooth spheres of mass m rest on a horizontal table. A is projected towards B so that on impact the line of centres is inclined at 45 degrees to the direction of motion of A. If $e = \tfrac{1}{3}$, find the angle through which the direction of motion of A is turned through by the impact.

3. A smooth sphere is moving on a smooth horizontal plane. It strikes a fixed smooth vertical wall at an angle α to the wall and rebounds in a direction making an angle β with the wall. Show that $\tan \beta = e \tan \alpha$, where e is the coefficient of restitution.

4. A smooth sphere A is moving on a smooth horizontal table. It strikes obliquely another smooth sphere of equal size and mass,

which is stationary. After the impact, the directions of motion of the two spheres make equal angles ϕ with the original direction of movement of A. Prove that $e = \tan^2 \phi$, where e is the coefficient of restitution.

5. A smooth sphere of mass 200 g, moving horizontally with a speed of 26 cm/s, impinges on another smooth sphere of equal radius and mass 300 g. The second sphere is moving horizontally with a speed of 20 cm/s. If the directions of motion at impact are inclined at $\tan^{-1} \frac{5}{12}$ and $\tan^{-1} \frac{3}{4}$ to the line of centres, and $e = \frac{1}{2}$, find their speeds after impact.

6. A small metal ball falls vertically and strikes a fixed smooth plane inclined at an angle $\theta (<45°)$ to the horizontal. If the coefficient of restitution is $\frac{2}{3}$ and the ball rebounds horizontally, find what fraction of the kinetic energy is lost during the impact.

7. Two smooth spheres A and B of the same radii but of masses $2m$ and m respectively rest on a horizontal table. A is projected towards B with a speed u and strikes B obliquely so that the line of centres is inclined at an angle of 60 degrees with the direction of motion of A. The coefficient of restitution is $\frac{1}{2}$. Find the loss of kinetic energy due to the impact.

8. Two smooth spheres A and B, each of mass 10 kg and the same radius, are moving towards one another on a smooth horizontal table. They impinge on one another. At the moment of impact their speeds are 6 m/s and 8 m/s respectively, and the direction of motion of A is inclined at 30 degrees to AB and the direction of motion of B is inclined at 45 degrees to BA. If $e = \frac{1}{2}$, find the components of their velocities after impact in the direction AB. Hence find the loss of kinetic energy during impact.

9. Two elastic particles P and Q of masses $3m$ and m respectively are at rest and in contact, each being freely suspended from a fixed point O by a light inelastic string of length a. The particle P is drawn aside to a point A where the string is taut and makes an acute angle α with the downward vertical. P is then released from rest and impinges directly with Q, the coefficient of restitution being $\frac{1}{2}$. The particle Q reaches a point B on AO produced at the instant when the string OB becomes slack. Show that $\cos \alpha = \frac{17}{177}$ and find the speed of Q at this instant. (London)

10. Two smooth spheres A and B of equal radii but of masses 20 kg and 10 kg respectively, are on a smooth horizontal table. B is at rest and A moving at 4 m/s strikes it obliquely at an angle of 60 degrees with the line of centres. After the impact, find A's speed, B's speed, the angle between A's and B's directions of motion and the loss of kinetic energy, given $e = \frac{1}{2}$.

EXERCISES

EXERCISES 13*

1. A cricket ball weighing 160 g is travelling horizontally at 25 m/s directly towards a batsman just before it hits the bat. He returns the ball directly down the pitch. On leaving the bat it is travelling at 20 m/s horizontally. Find the impulse imparted to the ball by the bat.

2. The water from a fire hose issues from a circular nozzle of diameter 50 mm and strikes a vertical wall at right angles at 40 m/s. If the water does not rebound from the wall, find the thrust on the wall. (Assume that 1 m^3 of water weighs 1 000 kg.)

3. A gun of mass 5 tonne is free to move on a horizontal plane. It fires a shell of mass 15 kg with a speed of 600 m/s relative to the ground. If the angle of elevation of the gun is 45 degrees, find the initial direction of motion of the shell relative to the ground.

4. A bullet of mass 60 g is fired horizontally with a speed of 500 m/s into a block of wood of mass 30 kg. The block of wood is suspended by a light inextensible string of length 5 m. Find the angle through which the block and embedded bullet swing. (Give your answer to the nearest degree.)

5. A fire engine picks up water from a lake and delivers it at the same level through a circular nozzle of diameter 80 mm at a speed of 30 m/s. The jet strikes a vertical wall at right angles at this speed and the water does not rebound from the wall. Find (a) the effective power of the engine, (b) the steady force exerted on the wall.
[1 m^3 of water weighs 1 000 kg]   (London)†

6. Two particles, each of mass m, lying on a smooth horizontal table, are connected by a light inextensible string of length 5a. Initially, they are at points A and B, where $AB = 3a$. They are projected simultaneously at right angles to AB and in opposite directions, each with speed V. Find the impulse in the string when it becomes taut.

7. Two particles, of masses M and m, are connected by a light inextensible string. They are projected simultaneously from the same point on a smooth horizontal table, with speeds U and u respectively, in horizontal directions at right angles. Show that, after the string becomes taut, both particles move at the same angle θ to the direction of the string at the instant of tightening, where

$$\tan \theta = \frac{(M + m)Uu}{MU^2 - mu^2}.$$

Show also that the loss of kinetic energy due to the tightening

* Exercises marked thus, †, have been metricized, see Preface.

of the string is

$$\frac{1}{2}\frac{Mm}{M+m}(U^2+u^2).$$

8. Two particles A and B, of masses 10 kg and 15 kg, respectively, are connected by a light inextensible string which passes over a smooth pulley. Initially, A is moving upwards and B downwards with speeds of 12 m/s. B is instantaneously stopped and released. Show that the string becomes taut again when the first particle is at rest, and find the impulsive tension in the string as it tightens.

9. Two small equal masses are tied to the ends of an inelastic string of length 16 m. They lie very close together on the ground and one mass is projected away from the other with a velocity whose horizontal component is 6 m/s and vertical component 16 m/s. Show that the string becomes taut when this mass is at the highest point of its path. Find the speeds of the masses immediately after the string becomes taut. (Take $g = 10$ m/s^2.) (London)†

10. Two particles A and B, of masses $3m$ and $2m$ respectively, lie at rest on a smooth horizontal table. A is projected directly towards B with a speed u. If the coefficient of restitution is $\frac{2}{3}$, find the speed with which B begins to move. After moving a distance d, B strikes a vertical plane at right angles and rebounds. If the coefficient of restitution between B and the plane is also $\frac{2}{3}$, how far does B travel from the barrier before it next collides with A? Find the impulse between the particles on their second impact.

11. Two spheres A and B, of masses 4 kg and 2 kg, respectively, are travelling in the same direction on a smooth horizontal table at speeds of 3 m/s and 2 m/s. A impinges directly on B, find their speeds after impact, the impulsive reaction between the spheres and the fractional loss of kinetic energy during impact. The coefficient of restitution is $\frac{1}{3}$.

12. An elastic ball falls vertically onto a horizontal plane with speed u. It continues to bounce. If the coefficient of restitution is e, find the total distance the ball travels after the first impact and the time taken to travel that distance.

13. Two particles, of equal mass, are connected by an inextensible string of length $2a$. They are placed at points A and B on a smooth floor, such that AB is perpendicular to a smooth wall, A being distant $3a$ from the wall and B being distant a from the wall. The particle at A is projected towards the wall with velocity u at an angle $\tan^{-1} 6$ with the wall. If the string next becomes taut when the moving particle has reached a point C on the floor such that angle $ABC = \tan^{-1}(\frac{4}{3})$, find the coefficient of restitution between the

EXERCISES

particle and the wall. Find the speed with which the particle at *B* begins to move. (London)

14. Three small smooth particles *A*, *B* and *C*, of equal mass, are connected, *A* to *B* and *B* to *C* by two inextensible strings of equal length. The particles lie in a straight line on a rough horizontal table with *B* in between and touching *A* and *C*. *A* is projected horizontally away from *B* in the direction of *CBA* with a speed of 12 m/s. If the coefficient of friction μ is $\frac{1}{4}$, find the speeds with which *B* and *C* begin to move.

15. A sphere *A*, of mass 4 kg, and a sphere *B*, of mass 2 kg are travelling in opposite directions along a straight line on a smooth horizontal table at speeds of 10 m/s and 5 m/s respectively. If the coefficient of restitution is $\frac{1}{3}$, find their velocities after impact. If the impulsive reaction between the spheres is equivalent to a constant force *F* acting for 0·02 seconds, show that $F = 1\,000/3$ N.

If the impulse between the spheres were increased by 20 per cent, find the new velocities after impact and show that the coefficient of restitution would be $\frac{3}{5}$. (WJEC)†

16. Two smooth spheres of masses 30 g and 40 g, but of equal radii, move with their centres on the line $r = \lambda i$. They impinge upon one another moving with velocities $18i$ and $-30i$ cm/s respectively. If $e = \frac{3}{4}$, find their velocities after impact and the kinetic energy which is lost in the collision.

17. Four equal particles, each of mass *m*, lie on a smooth horizontal table. They are joined by four equal, taut, inextensible strings which form the sides of a square *ABCD*. An impulse *I* is given to one particle *A* in the direction *CA* diagonally outwards from the square. Find the initial speeds of the four particles.

18. Two particles moving with speeds u_1, u_2 in the same straight line ($u_1 > u_2$), impinge directly. If m_1, m_2 are their respective masses and *e* the coefficient of restitution, show that the loss of kinetic energy on impact is

$$\frac{1}{2} \frac{m_1 m_2}{m_1 + m_2}(u_1 - u_2)^2(1 - e^2).$$

19. A particle of mass $3m$, moving with speed $4v$, impinges directly on a particle of mass $2m$, moving in the opposite direction with a speed $3v$, and is brought to rest. Prove that the velocity of the second particle is reversed in direction, but unchanged in magnitude, by the impact, and that the coefficient of restitution between the particles is $\frac{3}{7}$.

What is the loss of kinetic energy at the impact? (Oxford)

20. Two parallel walls are 12 m apart. A point P on the floor is 3 m from one wall and a point Q is 4 m from the other wall, the line PQ being perpendicular to the walls. A particle is projected with speed 1 m/s from P directly towards the nearer wall and simultaneously a second particle is projected from Q with speed 2 m/s directly towards the wall nearer to it. The floor is smooth and the coefficients of restitution between the particles and the walls, and between the particles themselves, are in each case $\frac{3}{4}$. The particles meet at a point X and the first particle is brought to rest by the impact. Find the distance of X from the nearer wall and the ratio of the masses of the particles.

Find also the time that will elapse before a second collision between the particles occurs. (London)†

21. A smooth sphere impinges directly on a stationary smooth sphere of double the mass. If $\frac{10}{27}$ of the original kinetic energy is lost during the impact, find the coefficient of restitution between the spheres.

Verify that the velocity of the centre of mass of the two spheres is unchanged by the impact. (London)

22. Two smooth spheres, A and B, of equal radii and masses, m and $2m$, respectively, are lying at rest on a smooth horizontal table. A is given a velocity $v = ai + aj$ and strikes B. At the moment of impact the equation of the line of centres is $r = \lambda i$, where λ is a parameter. If the velocities of A and B after the impact are $u_1 j$ and $u_2 i$ respectively, show that, the coefficient of restitution is 0·5 and find u_1 and u_2 in terms of a.

23. An elastic particle of mass m is projected horizontally with velocity u from the centre of a circular ring, of mass nm and radius a, lying at rest on a smooth horizontal table. If e is the coefficient of restitution between the particle and the ring, find the velocities of the ring and the particle after each of the first two impacts.

Show that at the moment of the second impact the ring has moved through a distance $2a(1 + e)/(n + 1)e$. (JMB)

24. Two smooth vertical walls stand on a smooth horizontal floor and intersect at an acute angle θ. A particle on the floor is projected horizontally at right angles to one wall and away from it. After one impact with each wall the particle is moving parallel to the first wall struck. If the coefficient of restitution between the particle and each wall is e, show that $(1 + 2e)\tan^2 \theta = e^2$.

Show that if the particle leaves the second wall in a direction parallel to the first wall struck, θ cannot exceed 30 degrees.

(London)

25. Three equal spheres A, B and C rest in that order in a straight

EXERCISES

line on a smooth horizontal plane. If A is set moving towards B, and if the coefficient of restitution between any two spheres is $\frac{1}{2}$, show that there are altogether three collisions and that the final velocities of the spheres are in the ratios

$$13:15:36. \qquad \text{(London)}$$

26. Two particles A and B moving on a straight line always have the same acceleration a. At time $t = 0$ their velocities are u_1 and u_2 respectively. What is the velocity of B relative to A at time t?

A particle P of mass m is dropped from rest at a fixed point O on to a horizontal plane, its speed immediately before striking the plane being U. At the instant at which P rebounds from the plane, a second particle Q of mass $2m$ is dropped from rest at O. If e is the coefficient of restitution between P and the plane, show that P will strike the plane a second time before striking Q if $e < \frac{1}{2}$.

If $e = \frac{3}{4}$, show that P and Q collide at a time $2U/(3g)$ after Q has been dropped, and find the speed of P immediately after the collision if the coefficient of restitution between P and Q is $\frac{2}{3}$.

[The impacts may be assumed to take place instantaneously.]

(WJEC)

27. Three equal spheres, centres A, B, C, lie at rest in a straight line on a smooth horizontal table. The coefficient of restitution between any two spheres is e. A is projected with speed U to strike B directly which then strikes C. Find the speeds of the spheres after the two collisions and show that A strikes B again whatever the value of e.

28. If in Question 27, the spheres had had masses m, λm and $\lambda^2 m$ respectively, and $e \geqslant \lambda$, show that there would only be two collisions.

29. If in Question 27, the spheres had had masses $2m$, $7m$ and $14m$ respectively, and $e = \frac{1}{2}$, find their velocities after two collisions.

30. Two equal spheres B and C, each of mass $4m$, lie at rest on a smooth horizontal table. A third sphere A, of the same radius as B and C but of mass m, moves with velocity V along the line of centres of B and C. The sphere A collides with B, which then collides with C. If A is brought to rest by the first collision, show that the coefficient of restitution between A and B is $\frac{1}{4}$.

If the coefficient of restitution between B and C is $\frac{1}{2}$, find the velocities of B and C after the second collision. Show that the total loss of kinetic energy due to the two collisions is $27mV^2/64$.

(JMB)

31. Two smooth spheres A and B of equal radii and masses m and $m/2$, are lying at rest on a smooth horizontal table. A is given a velocity $u\mathbf{i}$ and strikes B. At the moment of impact the equation of the line of centres is $\mathbf{r} = \lambda\mathbf{i} - \sqrt{3}\lambda\mathbf{j}$, where λ is a parameter. If the

coefficient of restitution is 0·5, show that the velocity of B after impact is $(u\mathbf{i} - \sqrt{3}u\mathbf{j})/4$ and find the velocity of A.

32. Prove that the work done in stretching a string, of elastic modulus λ, from its natural length l to a length $l + x$ is $\frac{1}{2}(\lambda x^2/l)$.

The ends of a light elastic string, of modulus Mg and natural length $2a$, are fixed to two points P and Q of a smooth horizontal table, the length of PQ being $2a$. A particle of mass M is at rest on the table, attached to the mid-point of the string. A second particle of mass m is projected along the table at right-angles to PQ and strikes the first particle, being thereby reduced to rest. The string is stretched to a maximum length of $4a$ before bringing the first particle instantaneously to rest. This particle then recoils and strikes the second particle. Prove that the coefficient of restitution between the particles is m/M, that the initial velocity of the second particle is $M\sqrt{(2ag)}/m$ and that the final velocity of this particle is $\sqrt{(2ag)}$.

(Oxford)

33. A particle is projected from a point on the floor and hits the ceiling, which is smooth. The height of the ceiling is λ times the height which the particle would have reached in the absence of the ceiling, and the coefficient of restitution is e. Prove that the distance from the starting point to the point where the particle hits the floor is

$$\tfrac{1}{2}[1 - (1 + e)(1 - \lambda)^{1/2} + \{e^2 + \lambda(1 - e^2)\}^{1/2}]$$

times the distance in the absence of the ceiling. (Oxford)

34. A smooth circular tray is fixed horizontally and an elastic particle is projected along the surface from a point A just inside the rim. The first two impacts with the rim occur at points B and C, such that the arc AC subtends a right angle at the centre of the tray and B is on this arc. Assuming that impulses with the rim are horizontal and that the coefficient of restitution is $\tfrac{2}{3}$, show that the direction of projection makes an angle $\tan^{-1} 2$ with the diameter through A and find the ratio of the times taken to describe AB and BC.

(London)

35. A passageway has a smooth horizontal floor and two long smooth parallel vertical walls which meet the floor along lines AB and DC respectively, distant a apart. A small sphere is projected from A in a direction making an acute angle θ with AB so as to strike a point of DC. At the $2n$th impact the sphere strikes AB at B. The coefficient of restitution between the ball and each wall is e. Prove that the length of AB is

$$\frac{a(1 - e^{2n}) \cot \theta}{e^{2n-1}(1 - e)}.$$

Determine the time taken to complete the journey from A to B if the speed of projection is u. (London)

36. A smooth circular horizontal table is surrounded by a smooth rim whose interior surface is vertical. Two equal particles are projected simultaneously with speed V along the table from a point A of the rim in different directions each making 30 degrees with the diameter AB through A. If the coefficient of restitution e of each particle with the rim of the table is greater than $\frac{1}{3}$, prove that, after one impact at the rim for each particle, the particles meet at a point of the line AB. If when they meet they coalesce, prove that their common velocity subsequently is

$$\frac{\sqrt{3}}{4}(1-e)V. \qquad \text{(JMB)}$$

37. Two smooth uniform spheres A and B, of the same radius but of masses m and km respectively, rest on a horizontal table. A is projected along the table towards B with speed u so as to strike B obliquely at an angle 60 degrees with the line of centres. The coefficient of restitution is $\frac{1}{2}$. Show that B's speed after impact is $3u/[4(k+1)]$ and find A's speed after impact.

If, after impact, A moves in a direction making an acute angle $\tan^{-1}(2\sqrt{3})$ with B's direction of motion, find the value of k and show that the loss of kinetic energy on impact is $\frac{1}{32}mu^2$. (London)

38. A smooth inelastic sphere of mass M lies on a smooth horizontal table, and a second smooth completely inelastic sphere of mass m falls on it. At the moment of impact the line of centres makes an angle α with the vertical, and the speed of the falling sphere is u. Prove that the speed of the first sphere after impact is

$$\frac{mu \sin \alpha \cos \alpha}{M + m \sin^2 \alpha}.$$

Prove that the loss of kinetic energy is

$$\frac{Mmu^2 \cos^2 \alpha}{2(M + m \sin^2 \alpha)}.$$

39. Two smooth elastic spheres of masses m_1 and m_2 and of equal radii lie at rest on a smooth horizontal floor. The mass m_1 is projected along the floor in a direction parallel to a smooth vertical elastic wall and strikes m_2 obliquely. Subsequently m_2 strikes the wall at an angle α with the wall. The coefficients of restitution between m_1 and m_2 and between m_2 and the wall are each e. If the

IMPULSE AND IMPACT

final velocities of the spheres are parallel, show that

$$m_2(1 + e)^2 \cos^2 \alpha = (m_1 + m_2)e.$$ (London)

40. A smooth sphere A of mass m, moving with speed u, strikes a stationary sphere B of mass $3m$ obliquely. If the direction of motion of A is turned through a right angle, find the impulsive reaction between the spheres. [Let e be the coefficient of restitution.]

41. Two smooth spheres of equal radii, masses m and $2m$, are at rest on a smooth horizontal table. The first ball A is projected along the table and strikes the second one B obliquely. After the impact the directions of motion of A and B make equal angles θ with the original direction of A. If the coefficient of restitution is $\frac{1}{2}$, show that $\theta = 45$ degrees.

42. A ball is dropped from the roof of a lift ascending with an acceleration f m/s^2. If the height of the lift is h m and the coefficient of restitution between the ball and the floor of the lift is e, show that the time which elapses before the ball stops bouncing is

$$\frac{1+e}{1-e}\sqrt{\left(\frac{2h}{f+g}\right)} \text{ s.}$$ (WJEC)†

43. Two identical smooth spheres A and B moving on a horizontal table with velocity vectors $2V\boldsymbol{i}$ and $V\boldsymbol{j}$ respectively, collide and the equation of the line of centres at that instant is $\boldsymbol{r} = \lambda \sin \alpha \boldsymbol{i} - \lambda \cos \alpha \boldsymbol{j}$, where λ is a parameter. If the velocity vector of the sphere B after the collision is $u\boldsymbol{i}$, and the coefficient of restitution between the spheres is $\frac{1}{3}$, prove that $\tan \alpha$ equals either 1 or $\frac{1}{3}$, and in each case find u and the velocity vector of the sphere A after the collision.

(London)

44. Two smooth spheres A and B have masses $2m$ and m respectively, and velocity vectors $3u\boldsymbol{i} + 4u\boldsymbol{j}$ and $-4u\boldsymbol{i} + 3u\boldsymbol{j}$ respectively, when they collide with their line of centres parallel to the unit vector $\boldsymbol{i}$. If the impact causes a loss of energy equal to the original kinetic energy of the sphere B, prove that the coefficient of restitution between the spheres is $\sqrt{(23/98)}$. (London)

45. When a body of mass $3m$ is moving in a straight line with speed u, it explodes. As a result it splits into two bodies A and B, of respective masses m and $2m$, which move in the same straight line as before, but in opposite directions. Given that the extra energy created by the explosion is $3mu^2$, find the speeds of A and B.

The body B then immediately strikes a body of mass M which is at rest and also free to move in the same straight line. If the impact is perfectly elastic, prove that B will subsequently strike A if $M > 6m$. (Oxford)

EXERCISES

46. State the law of conservation of momentum for two interacting particles which move in the same straight line, and derive the law from Newton's laws of motion.

Two particles, P of mass $4m$ and Q of mass $5m$, are subject to a mutual interaction which is variable and of unknown magnitude, and no other force acts on them. Initially Q is at rest and P is moving with speed u directly away from Q. Find the velocity of Q when P has velocity v.

At a certain instant P comes to rest. Find the total net work which has been done against the forces of the interaction since the initial instant.

When P comes to rest the force of interaction is zero, and it remains zero until Q collides with P. Immediately after the collision the velocity of P is $\frac{1}{2}u$. Find the coefficient of restitution between the particles. (JMB)

47. A particle of mass 4 units moves on a smooth horizontal table which is in the i–j plane. The particle, which is at rest at the point $r = 0$ at time $t = 0$, is acted upon by a force F, where $F = 24(i + tj)$ at time t.

Find
(i) the cartesian equation of the path of the particle,
(ii) the velocity of the particle at time $t = 2$ and hence, or otherwise, the vector equation of the tangent to the path at this instant.

When $t = 2$ an impulse $-8(i + j)$ is applied to the particle. As a result the particle splits up into two fragments, of equal mass, which move in the directions i and $-i + j$. Calculate the speeds of the fragments immediately after the impulse. (AEB)

48. A water pump raises 50 kg of water a second through a height of 20 m. The water emerges as a jet with speed 50 m/s. Find the kinetic energy and the potential energy given to the water each second and hence find the effective power developed by the pump.

Given that the jet is directed at 30° above the horizontal, find the further height attained by the water.

The jet of water impinges at its highest point directly against a vertical wall. Show that the force exerted on the wall is at least $1250\sqrt{3}$ N.

[Take $g = 10 \text{ m/s}^2$.] (London)

14
PROJECTILES

14.1. BASIC THEORY

WE shall now consider the motion in *two dimensions* of a particle of mass m moving under gravity. Suppose that it is projected with a velocity whose magnitude is u making an angle θ with the horizontal (refer to *Figure 14.1*).

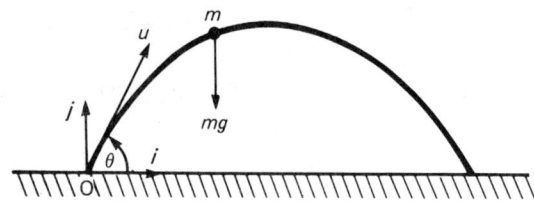

Figure 14.1

Two simplifying assumptions will be made:
(i) the air resistance is negligible,
(ii) the motion is near the Earth's surface so that the acceleration g due to gravity is uniform.

Then since at all times the only force acting on the particle will be its weight mg, we have, applying $\mathbf{P} = m\mathbf{f}$,

$$-mg\mathbf{j} = m\mathbf{f}.$$

This gives the equation governing the motion as

$$\frac{d\mathbf{v}}{dt} = -g\mathbf{j} \qquad \ldots(A)$$

Initially (i.e. when $t = 0$), the velocity of the particle has components $u \cos \theta$ horizontally and $u \sin \theta$ vertically upwards. Thus this initial velocity is

$$\mathbf{v}_0 = u \cos \theta \mathbf{i} + u \sin \theta \mathbf{j}$$

(where $\mathbf{i}$ and $\mathbf{j}$ are unit vectors in the horizontal and vertical directions respectively).

BASIC THEORY

Now integrating equation (A) above, we have

$$v = -gtj + c.$$

The constant vector c can be found by putting $v = v_0$ when $t = 0$. This gives $c = u \cos \theta i + u \sin \theta j$ so that

$$v = -gtj + u \cos \theta i + u \sin \theta j$$

or
$$v = u \cos \theta i + (u \sin \theta - gt)j. \qquad \ldots(B)$$

Similarly, since $v = dr/dt$, the equation (B) can be integrated to give the displacement

$$r = r_0 + u \cos \theta t i + (u \sin \theta t - \tfrac{1}{2}gt^2)j. \qquad \ldots(C)$$

Here r_0, the integration constant, is the initial displacement.

The two equations (B) and (C) will (together with energy considerations) form the basis of our approach to solving projectile problems. (The reader will note after studying our vector equations that the particle is moving horizontally at uniform speed and vertically with uniform acceleration. So problems can be solved by considering these two motions separately and using the formulae established in previous chapters. This is an alternative approach that is widely used, and indeed used in earlier editions of this book.)

Example 1. A particle is projected from a point O with a speed of 30 m/s in a direction making an angle of 30° with the horizontal. After 2 seconds it passes through a point Q. Find the position vector of the point Q, the magnitude of the velocity at Q and the vector equation of the tangent to the path at Q.

Refer to *Figure 14.1* with $u = 30$ m/s, $\theta = 30°$ and O as origin. Then from equations (B) and (C) we have

$$v = 30 \cos 30° i + (30 \sin 30° - gt)j$$
$$r = 30 \cos 30° t i + (30 \sin 30° t - \tfrac{1}{2}gt^2)j.$$

Hence at Q (where $t = 2$) we have

$$v_Q = 15\sqrt{3}i + (15 - 9\cdot 8 \times 2)j$$

giving $\qquad v_Q = 25\cdot 98i - 4\cdot 6j \quad$ and $\quad |v_Q| = 26\cdot 38.$

Also at $t = 2$

$$r_Q = 15\sqrt{3} \times 2i + (15 \times 2 - 4\cdot 9 \times 2^2)j$$

giving $\qquad r_Q = 51\cdot 96i + 10\cdot 4j.$

PROJECTILES

The velocity vector v_Q gives the direction of the tangent to the path at Q and r_Q is the position vector of Q. So from Section 3.12, the vector equation of the tangent to the path at Q is

$$r = r_Q + \lambda v_Q$$

or $$r = (51{\cdot}96i + 10{\cdot}4j) + \lambda(25{\cdot}98i - 4{\cdot}6j).$$

Thus we have that after 2 seconds the particle is moving with a speed of 26·4 m/s and is at a point whose position vector is $52{\cdot}0i + 10{\cdot}4j$. Just at this moment the particle is moving along the line whose vector equation is

$$r = (52{\cdot}0i + 10{\cdot}4j) + \lambda(26{\cdot}0i - 4{\cdot}6j).$$

Example 2. A particle is fired, from the top of a cliff of height 49 m, *with a speed of* 14 m/s *at an angle of* 45 *degrees above the horizontal. Find the maximum height reached and the point where the particle enters the sea.*

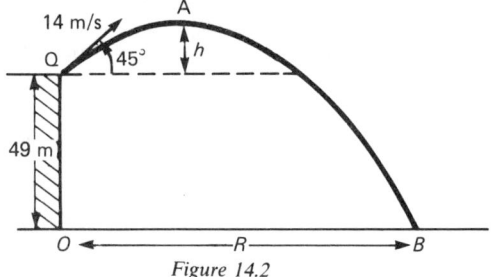

Figure 14.2

Refer to *Figure 14.2*.
For this motion the velocity vector is

$$v = 14 \cos 45°i + (14 \sin 45° - gt)j$$
$$= 7\sqrt{2}i + (7\sqrt{2} - 9{\cdot}8t)j.$$

Then, taking the foot of the cliff, O, as origin, the displacement vector is

$$r = 49j + 14 \cos 45° \, ti + (14 \sin 45° \, t - \tfrac{1}{2}gt^2)j$$
$$= 7\sqrt{2}ti + (49 + 7\sqrt{2}t - 4{\cdot}9t^2)j.$$

At the maximum height, point A, the j component of v will be zero so that $v_A = 7\sqrt{2}i$ and $|v_A| = 7\sqrt{2}$. Then, since we already have $|v_Q| = 14$, applying the principle of work between Q and A gives

final K.E. − initial K.E. = work done
$$\tfrac{1}{2}m(7\sqrt{2})^2 - \tfrac{1}{2}m(14)^2 = -mgh$$

BASIC THEORY

giving $$49 - 98 = -9 \cdot 8h$$
so that $$h = 5.$$

At B the j component of r is zero so that
$$4 \cdot 9 + 7\sqrt{2}t - 4 \cdot 9t^2 = 0$$
$$\therefore \quad 4 \cdot 9t^2 - 9 \cdot 899t - 49 = 0$$
and $$t = 4 \cdot 329 \quad \text{or} \quad t = -2 \cdot 309.$$

Taking the positive root, $r_B = (7\sqrt{2} \times 4 \cdot 329)i = 42 \cdot 85i$, so that the range R is $42 \cdot 85$ m.

Thus the maximum height reached is $49 + 5 = 54$ m, and the particle enters the sea at a point distant 42.9 m from the foot of the cliff.

Example 3. A particle is fired from a point on a horizontal plane with speed u, at an angle of θ with the horizontal. Find the time of its flight and its range. Deduce the maximum range.

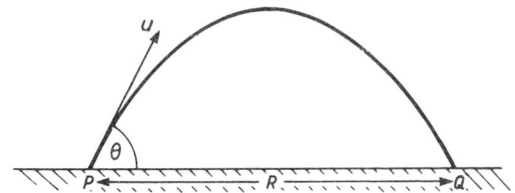

Figure 14.3

Refer to *Figure 14.3*.
For this particle we have at time t

velocity vector $\quad v = u \cos \theta i + (u \sin \theta - gt)j \quad \ldots$(i)

displacement vector $\quad r = u \cos \theta t i + (u \sin \theta t - \tfrac{1}{2}gt^2)j. \quad \ldots$(ii)

Let Q be the point at which the particle meets the plane and T its time of flight. Then since, at Q, the j component of r is zero,
$$u \sin \theta t - \tfrac{1}{2}gt^2 = 0$$
giving $$t = 0 \quad \text{or} \quad t = (2u \sin \theta)/g.$$

The value $t = 0$ refers to the point P and so the time of flight to Q is $T = (2u \sin \theta)/g$. Substituting this value for t in (ii) we have
$$r_Q = u \cos \theta \times [(2u \sin \theta)/g]i.$$

273

PROJECTILES

Hence range $\quad |r_Q| = (2u^2 \sin\theta \cos\theta)/g$

i.e. $\quad\quad\quad\quad R = (u^2 \sin 2\theta)/g.$

For a given speed of projection but varying angle, R will be greatest when $\sin 2\theta = 1$,

so $\quad\quad\quad\quad R_{max} = u^2/g.$

It is interesting to note that $\sin 2\theta = 1$ when $2\theta = 2n\pi + \pi/2$. The only case of practical interest is when $n = 0$ and $\theta = \pi/4$. This shows that for maximum range the projectile should be fired at 45° to the horizontal.

Exercises 14a

1. A particle is projected with a speed of 32 m/s at an angle of 40 degrees to the horizontal. Find the horizontal and vertical components of its velocity after 2 seconds.

2. A particle is projected from a point O with a speed of 7 m/s at an angle of 30 degrees to the horizontal. Find its horizontal and vertical displacements from O after $\frac{5}{7}$ths of a second.

3. A particle is projected from a point O with a speed of 28 m/s at an angle of $\tan^{-1} 2$ to the horizontal. Find its vertical displacement at the moment when its horizontal displacement is 40 m.

4. A stone is thrown with a speed of 14 m/s at an angle of projection of 60 degrees. Find (a) the greatest height reached (b) the time of flight (c) the range on a horizontal plane.

5. The greatest range of a projectile on a horizontal plane is 20 000 metres. Find its range when the angle of projection is (a) 30 degrees (b) $22\frac{1}{2}$ degrees (c) 75 degrees.

6. A ball is thrown with a speed of 28 m/s at an angle of projection of $\tan^{-1} 2$. Find (a) the greatest height reached (b) its vertical height when the direction of motion is inclined at an angle of 45 degrees to the horizontal.

7. A particle is projected from a point O at an angle of projection of 45 degrees. It passes through a point P whose horizontal and vertical displacements from O are $3a$ and a respectively. Find the speed of the particle when it passes through P.

8. A bullet is projected with a speed of 350 m/s at an angle of projection of $\tan^{-1}\frac{3}{4}$. Find (a) its range on a horizontal plane through the point of projection (b) the greatest height reached (c) its speed and direction when at a height of 1 250 m.

9. A particle is projected from a point O. After 5 seconds its horizontal and vertical displacements from O are 60 m and 57·5 m respectively. If the particle is still rising, find its initial velocity.

274

ADVANCED EXAMPLES

10. The furthest distance a man can throw a stone is 45 m. What is the time of flight and how high does the stone rise?

14.2. ADVANCED EXAMPLES

Example 1. A particle is projected under gravity from a point O with speed u at an angle of projection θ. Obtain expressions for the horizontal and vertical displacements x and y at time t after projection, and deduce an equation for the path of the particle.

A particle can be projected at 28 m/s and must just clear a wall which is 5 m high and 20 m from the pont of projection. Find the two possible angles of projection.

As in previous examples, with O as origin the displacement vector at time t will be given by

$$r = u \cos \theta t \mathbf{i} + (u \sin \theta t - \tfrac{1}{2}gt^2)\mathbf{j}.$$

So the horizontal and vertical displacements at time t are

$$x = u \cos \theta t \qquad \ldots(i)$$

and
$$y = u \sin \theta t - \tfrac{1}{2}gt^2. \qquad \ldots(ii)$$

From (i) $t = x/(u \cos \theta)$ and substituting this value in (ii)

$$y = u \sin \theta [x/(u \cos \theta)] - \tfrac{1}{2}g[x/(u \cos \theta)]^2$$

or
$$y = x \tan \theta - \frac{gx^2}{2u^2 \cos^2 \theta} \qquad \ldots(iii)$$

which is the equation of the path of the particle. The path is a parabola with axis vertically downwards.

In the second part of the question, since the particle just clears the wall, the top of the wall will be a point on the path. Thus (20, 5) is a point whose coordinates satisfy equation (iii). So substituting and noting that $u = 28$ m/s, we have

$$5 = 20 \tan \theta - (9 \cdot 8 \times 20^2)/(2 \times 28^2 \cos^2 \theta)$$

giving
$$5 = 20 \tan \theta - 2 \cdot 5 \sec^2 \theta$$

or
$$5 = 20 \tan \theta - 2 \cdot 5(1 + \tan^2 \theta),$$

i.e.
$$2 \cdot 5 \tan^2 \theta - 20 \tan \theta + 7 \cdot 5 = 0.$$

Solving this quadratic equation gives

$$\tan \theta = 0 \cdot 3944 \quad \text{or} \quad \tan \theta = 7 \cdot 6$$

PROJECTILES

so that $\theta = 21\cdot5°$ or $82\cdot5°$.

These are required angles of projection to just clear the wall.

Example 2. A bird is 10 m vertically above a man who throws a stone at an angle of projection of θ. The bird is flying with a uniform speed of 14 m/s in a direction making 60 degrees with the horizontal. Show, that, for the stone to hit the bird, $\theta \geq 75°$.

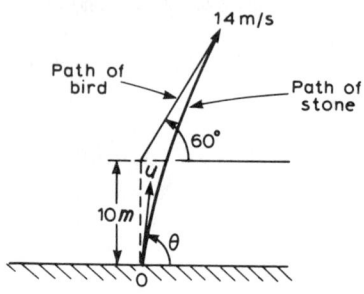

Figure 14.4

For the stone to hit the bird, the bird and stone must be in the same place at the same time t. That is, their displacement vectors must be equal (refer to *Figure 14.4*).

Then taking horizontal and vertical axes through O where the man is standing

For the stone $\quad r = u\cos\theta t\,i + (u\sin\theta t - \tfrac{1}{2}gt^2)j \quad\quad \ldots(i)$

For the bird $\quad r = 10j + 14\cos 60° t\,i + 14\sin 60° t\,j$

$\quad\quad\quad\quad\quad r = 7t\,i + (7\sqrt{3}t + 10)j. \quad\quad \ldots(ii)$

Equating components of (i) and (ii) gives

$$u\cos\theta t = 7t \quad\quad \ldots(a)$$

and $\quad u\sin\theta t - \tfrac{1}{2}gt^2 = 7\sqrt{3}t + 10. \quad\quad \ldots(b)$

From (a) $u = 7/\cos\theta$, and substituting this in (b), with $g = 9\cdot8$, gives

$$(7/\cos\theta)\sin\theta t - 4\cdot9t^2 = 7\sqrt{3}t + 10$$

or $\quad 4\cdot9t^2 - (7\tan\theta - 7\sqrt{3})t + 10 = 0.$

ADVANCED EXAMPLES

This is a quadratic equation giving two values of t. For the equation to have real roots

$$(7\tan\theta - 7\sqrt{3})^2 \geqslant 4 \times 4{\cdot}9 \times 10$$
$$(\tan\theta - \sqrt{3})^2 \geqslant 4,$$

so that $\quad \tan\theta - \sqrt{3} \leqslant -2 \quad$ or $\quad \tan\theta - \sqrt{3} \geqslant 2.$

The first inequality gives negative values of t, so we have

$$\tan\theta \geqslant 2 + \sqrt{3}$$

or $\quad\quad\quad\quad\quad\quad\quad \theta \geqslant 75°.$

In problems concerning projectiles landing on inclined planes, it is sometimes convenient to express our vectors in terms of $\hat{a}, \hat{b}$ unit vectors along and perpendicular to the plane respectively (see *Figure 14.5*).

Example 3. A bullet is projected from a point 0 on a plane inclined at 45 degrees to the horizontal. The velocity of projection is u at an angle of $\tan^{-1}(\frac{1}{2})$ to the line of greatest slope of the plane. Find the range of the bullet on the plane and show that it meets the plane at right angles.

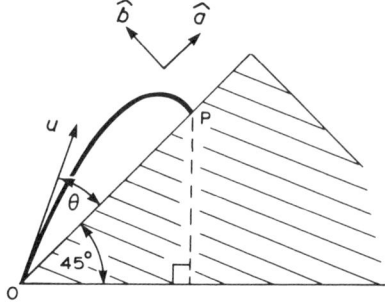

Figure 14.5

Let P be the point at which the bullet strikes the plane, so that OP is the range (refer to *Figure 14.5*). Also note that if θ is the angle the line of projection makes with the plane then $\tan\theta = \frac{1}{2}$ and, since θ is acute, $\sin\theta = 1/\sqrt{5}$ and $\cos\theta = 2/\sqrt{5}$.

277

PROJECTILES

The resultant acceleration $(-g\boldsymbol{j})$ of the particle can be resolved up the plane and perpendicular to the plane giving components $-g \sin 45°$ and $-g \cos 45°$ respectively;

i.e. $$\boldsymbol{f} = -g \sin 45° \hat{\boldsymbol{a}} - g \cos 45° \hat{\boldsymbol{b}}.$$

Then, as in Section 14.1, since $\boldsymbol{f} = d\boldsymbol{v}/dt$, integration of the above gives
$$\boldsymbol{v} = -g \sin 45° t \hat{\boldsymbol{a}} - g \cos 45° t \hat{\boldsymbol{b}} + \boldsymbol{v}_0.$$

$\boldsymbol{v}_0$, the initial velocity, has the value $u \cos \theta \hat{\boldsymbol{a}} + u \sin \theta \hat{\boldsymbol{b}}$ so that
$$\boldsymbol{v} = (u \cos \theta - g \sin 45° t) \hat{\boldsymbol{a}} + (u \sin \theta - g \cos 45° t) \hat{\boldsymbol{b}}. \quad \ldots (i)$$

Now again, since $\boldsymbol{v} = d\boldsymbol{r}/dt$ and $\boldsymbol{r} = \boldsymbol{0}$ when $t = 0$, we obtain
$$\boldsymbol{r} = (u \cos \theta t - \tfrac{1}{2} g \sin 45° t^2) \hat{\boldsymbol{a}} + (u \sin \theta t - \tfrac{1}{2} g \cos 45° t^2) \hat{\boldsymbol{b}}. \quad \ldots (ii)$$

Now at P the $\hat{\boldsymbol{b}}$ component of $\boldsymbol{r}$ is zero, so that
$$u \sin \theta t - \tfrac{1}{2} g \cos 45° t^2 = 0,$$

giving $t = 0$ or $t = (2u \sin \theta)/(g \cos 45°)$. The first of these refers to the point O, while the second gives us the time of flight from O to P.

Substituting this time of flight back in (ii)

$$\boldsymbol{r}_P = \left[u \cos \theta \left(\frac{2u \sin \theta}{g \cos 45°} \right) - \tfrac{1}{2} g \sin 45° \left(\frac{2u \sin \theta}{g \cos 45°} \right)^2 \right] \hat{\boldsymbol{a}} + 0 \hat{\boldsymbol{b}}.$$

So $$OP = |\boldsymbol{r}_P| = u \cos \theta \left(\frac{2u \sin \theta}{g \cos 45°} \right) - \tfrac{1}{2} g \sin 45° \left(\frac{2u \sin \theta}{g \cos 45°} \right)^2$$
$$= \frac{2u^2 \sin \theta}{g \cos 45°} (\cos \theta - \sin \theta) \quad [\text{note that } \sin 45° = \cos 45°]$$
$$= \frac{2u^2 (1/\sqrt{5})}{g(1/\sqrt{2})} \left(\frac{2}{\sqrt{5}} - \frac{1}{\sqrt{5}} \right)$$
$$OP = \frac{2\sqrt{2} u^2}{5g}.$$

Substituting the time of flight into (i) gives
$$\boldsymbol{v}_P = (u \cos \theta - 2u \sin \theta) \hat{\boldsymbol{a}} + (u \sin \theta - 2u \sin \theta) \hat{\boldsymbol{b}}$$
$$= [u(2/\sqrt{5}) - 2u(1/\sqrt{5})] \hat{\boldsymbol{a}} + [u(1/\sqrt{5}) - 2u(1/\sqrt{5})] \hat{\boldsymbol{b}}$$
$$= 0 \hat{\boldsymbol{a}} + [-u(1/\sqrt{5})] \hat{\boldsymbol{b}}$$

and since the $\hat{\boldsymbol{a}}$ component is zero the particle hits the plane at right angles.

ADVANCED EXAMPLES

Example 4. *A particle is projected from a point O on a plane inclined at an angle α to the horizontal. If the velocity of projection is u at an angle θ to the horizontal, find the range of the particle on the inclined plane. For a given value of u, deduce the maximum range up the plane and down the plane.*

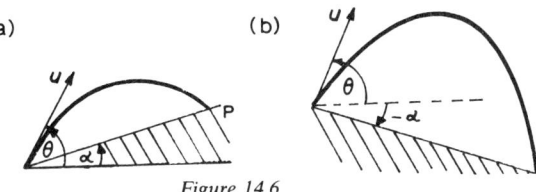

Figure 14.6

Refer to *Figure 14.6(a)*.

As in Example 3 we can resolve along (in an upwards direction) and perpendicular to the plane and here the velocity vector and displacement vector at time t are:

$$v = [u \cos(\theta - \alpha) - g \sin \alpha t]\hat{a}$$
$$+ [u \sin(\theta - \alpha) - g \cos \alpha t]\hat{b} \quad \ldots(i)$$
$$r = [u \cos(\theta - \alpha)t - \tfrac{1}{2}g \sin \alpha t^2]\hat{a}$$
$$+ [u \sin(\theta - \alpha)t - \tfrac{1}{2}g \cos \alpha t^2]\hat{b}. \quad \ldots(ii)$$

Then again at P the $\hat{b}$ component of r is zero so that

$$u \sin(\theta - \alpha)t - \tfrac{1}{2}g \cos \alpha t^2 = 0$$

giving $t = 2u \sin(\theta - \alpha)/(g \cos \alpha)$ as the time of flight.

Substituting this value back in (ii) then gives us

$$\text{Range } R = |r_P| = u \cos(\theta - \alpha)\frac{2u \sin(\theta - \alpha)}{g \cos \alpha}$$
$$- \tfrac{1}{2}g \sin \alpha \left[\frac{2u \sin(\theta - \alpha)}{g \cos \alpha}\right]^2$$
$$= \frac{2u^2 \sin(\theta - \alpha)}{g \cos^2 \alpha}[\cos(\theta - \alpha)\cos \alpha - \sin(\theta - \alpha)\sin \alpha]$$
$$= \frac{2u^2 \sin(\theta - \alpha)}{g \cos^2 \alpha} \cos \theta$$
$$= \frac{u^2}{g \cos^2 \alpha}[2 \sin(\theta - \alpha) \cos \theta]$$
$$= \frac{u^2}{g \cos^2 \alpha}[\sin(2\theta - \alpha) - \sin \alpha].$$

PROJECTILES

From this formula we see that if θ varies with u fixed the range R will be a maximum when $\sin(2\theta - \alpha) = 1$.

i.e.
$$R_{max} = \frac{u^2}{g\cos^2\alpha}(1 - \sin\alpha)$$

$$R_{max} = \frac{u^2}{g(1 - \sin^2\alpha)}(1 - \sin\alpha)$$

$$R_{max} = \frac{u^2}{g(1 + \sin\alpha)}.$$

To find the maximum range *down* the plane we can rework the problem with $(\theta - \alpha)$ replaced by $(\theta + \alpha)$, and the acceleration component $-g\sin\alpha$ replaced by $+g\sin\alpha$ down the plane (refer to *Figure 14.7(b)*). Alternatively we may regard firing down the plane as being equivalent to firing up a plane which makes a negative angle with the horizontal. So replacing α by $-\alpha$ in our formula we have

$$R_{max} = \frac{u^2}{g(1 + \sin(-\alpha))}$$

$$R_{max} = \frac{u^2}{g(1 - \sin\alpha)}.$$

Example 5. A shell is fired with speed u from a point on a cliff of height h above sea level. Find the greatest horizontal distance the shell can cover before landing in the sea.

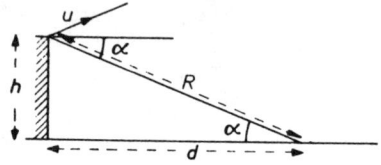

Figure 14.7

If d is the required maximum distance since h is constant, R (refer to *Figure 14.8*) is also a maximum, but R_{max} down an inclined plane is given by (refer to Example 4)

$$R_{max} = u^2/(g(1 - \sin\alpha)) \qquad \ldots \text{(i)}$$

Also
$$h/R_{max} = \sin\alpha \qquad \ldots \text{(ii)}$$

From equations (i) and (ii)
$$R_{max} = \frac{u^2}{g(1 - h/R_{max})}$$

$\therefore \quad R_{max}(1 - h/R_{max}) = u^2/g$

ADVANCED EXAMPLES

and
$$R_{max} = (u^2/g) + h$$

$$\therefore \quad d = \sqrt{(R_{max}^2 - h^2)} = \sqrt{\left[\left(\frac{u^2}{g} + h\right)^2 - h^2\right]}$$

$$= \sqrt{\left[\frac{u^4}{g^2} + 2\frac{hu^2}{g} + h^2 - h^2\right]}$$

$$= \frac{u}{g}\sqrt{(u^2 + 2gh)}.$$

Exercises 14b

1. Two particles A and B are projected simultaneously from a point O and both particles strike a horizontal plane through O at the same point P, A arriving first. If OP is 6 m and the speeds of projection of A and B are each 910 cm/s, verify that the angle of projection of A is $\sin^{-1}(\frac{5}{13})$.

2. A particle has a horizontal range of 15 m when projected at an angle θ to the horizontal with speed u. The greatest height reached is 5 m, find u and θ.

3. A cricket ball is thrown with a speed of 21 m/s. Find the greatest range on a horizontal plane. If the distance the ball is thrown is 22·5 m, find the two possible angles of projection.

4. A particle is projected from a point O at an angle of projection of 60 degrees with speed u. At the same time a second particle is projected from O with speed u but with a different angle of projection. If the two particles land at the same point, find the distance between them after t seconds.

5. A particle is projected with speed V. R_1, R_2, R are the maximum ranges up a plane, down the same plane and on the horizontal, respectively, at this speed of projection. Use the results of Example 4, Section 14.2, to show that
$$\frac{1}{R_1} + \frac{1}{R_2} = \frac{2}{R}.$$
Show also that the direction of projection in each case bisects the angle between the plane and the vertical.

6. A particle is projected, with a speed of 600 m/s at an angle of projection of 45 degrees, from the foot of a plane of inclination 30 degrees. Find the range on the plane and the time of flight.

7. A particle is projected from a point A with a speed of 40 m/s at an angle of projection of 30 degrees. Find the rate at which its distance from A is increasing 2 seconds after projection. [Hint, use $V \cdot \hat{r} = \dot{r}$.]

PROJECTILES

8. Show that the range of a projectile on an inclined plane is given by
$$R = (2u^2)[\sin\theta \cos(\theta + \alpha)]/(g\cos^2\alpha),$$
where u is the speed of projection, α the inclination of the plane and θ the angle of projection with respect to the line of greatest slope of the plane. Hence, show that the ratio of the maximum range down an inclined plane to the maximum range up the plane is
$$(1 + \sin\alpha):(1 - \sin\alpha).$$

9. A particle is projected at an angle θ to the horizontal. Show that it will strike an inclined plane through the point of projection at right angles if $\cot\alpha = 2\tan(\theta - \alpha)$, where α is the angle of inclination of the plane.

10. A man A stands on a cliff of height h and notices another man B on the seashore a distance k from the foot of the cliff. Simultaneously, A throws a stone with speed u and angle of projection α and B throws a stone with speed $2u$ and angle of projection β away from the cliff, the trajectories of the two stones being in the same plane. If the two stones collide, show that,
$$2\sin(\beta + \varepsilon) = \sin(\alpha + \varepsilon), \quad \text{where } \varepsilon = \tan^{-1}(h/k).$$

14.3. PROJECTILES AND IMPACT

When the motion of a projectile is interrupted by impacts, we can deal with each part of the motion separately and link them together. For the impacts we note that when a particle meets a smooth plane, only the component of velocity perpendicular to the plane is affected. Newton's law tells us that the speed away from the plane will be e times the speed of approach, where e is the coefficient of restitution.

Example 1. A ball is projected from a point A on smooth level ground. It strikes a vertical wall normally and returns to A after bouncing once on the ground. The wall is also smooth and the coefficient of restitution between the ball and the wall is $\tfrac{1}{2}$. Find the coefficient of restitution between the ball and the ground.

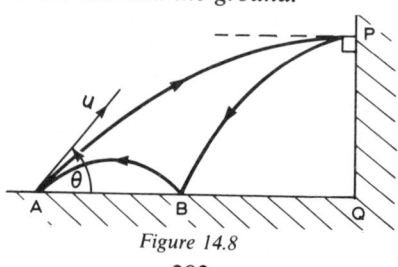

Figure 14.8

PROJECTILES AND IMPACT

Refer to *Figure 14.8*.

In the following analysis
the variable t_1 will refer to time measured from the point A,
the variable t_2 will refer to time measured from the point P,
the variable t_3 will refer to time measured from the point B.
Whichever part of the motion the particle is in, its displacement r will always be measured from the point A as origin, while the velocity will be given by $v = dr/dt$ except at the points of impact, where we apply Newton's law of restitution (refer to Section 13.6).

Motion from A to P

For this part of the motion our velocity and displacement vectors are as usual

$$v = u \cos \theta \, i + (u \sin \theta - gt_1)j$$
$$r = u \cos \theta \, t_1 i + (u \sin \theta \, t_1 - \tfrac{1}{2}gt_1^2)j.$$

At P

Here the particle is moving horizontally. This means that the j component of the velocity here is zero and t_1 has the value $u \sin \theta / g$. Substituting this value into the vectors above gives

$$r_P = u \cos \theta (u \sin \theta / g) i + [u \sin \theta (u \sin \theta / g) - \tfrac{1}{2}g(u \sin \theta / g)^2]j$$
$$= (u^2 \sin \theta \cos \theta / g) i + (\tfrac{1}{2}u^2 \sin^2 \theta / g)j. \qquad \ldots \text{(i)}$$

In addition, just before impact the velocity is $u \cos \theta \, i$ which becomes, because $e = \tfrac{1}{2}$, $-\tfrac{1}{2} u \cos \theta \, i$ immediately after impact.

Motion from P to B

Using the velocity just derived as our initial velocity here, the general velocity for this part of the motion is

$$v = -\tfrac{1}{2}u \cos \theta \, i - gt_2 j.$$

Integrating this equation,

$$r = r_P - \tfrac{1}{2}u \cos \theta \, t_2 i - \tfrac{1}{2}gt_2^2 j.$$

Hence from equation (i)

$$r = (u^2 \sin \theta \cos \theta / g) i + (\tfrac{1}{2}u^2 \sin^2 \theta / g)j - \tfrac{1}{2}u \cos \theta \, t_2 i - \tfrac{1}{2}gt_2^2 j$$
$$= (u^2 \sin \theta \cos \theta / g - \tfrac{1}{2}u \cos \theta \, t_2) i + [(\tfrac{1}{2}u^2 \sin^2 \theta / g) - \tfrac{1}{2}gt_2^2]j.$$

At B

Here the j component of r will be zero, which means that at B

$$\tfrac{1}{2}u^2 \sin^2 \theta / g - \tfrac{1}{2}gt_2^2 = 0$$

and hence the value of t_2 at B is $u \sin \theta / g$.

PROJECTILES

This gives $r_B = [u^2 \sin\theta \cos\theta/g - \frac{1}{2}u\cos\theta(u\sin\theta/g)]i$
$$= (\tfrac{1}{2}u^2 \sin\theta \cos\theta/g)i. \qquad\qquad ....(ii)$$

Again using this value of t_2 the velocity just before impact at B will be $-\frac{1}{2}u\cos\theta i - u\sin\theta j$, which becomes $-\frac{1}{2}u\cos\theta i + eu\sin\theta j$ immediately after impact. Here e is the coefficient of restitution at the ground.

Motion from B back to A

Again using the velocity just derived as initial velocity, the general velocity for this part of the motion is
$$v = -\tfrac{1}{2}u\cos\theta i + (eu\sin\theta - gt_3)j.$$
Integrating this equation,
$$r = r_B - \tfrac{1}{2}u\cos\theta t_3 i - (eu\sin\theta t_3 - \tfrac{1}{2}gt_3^2)j.$$
Hence from equation (ii)
$$r = (\tfrac{1}{2}u^2 \sin\theta \cos\theta/g)i - \tfrac{1}{2}u\cos\theta t_3 i - (eu\sin\theta t_3 - \tfrac{1}{2}gt_3^2)j.$$

At A

Here the vector r_A is identically zero so that both the i and j components are zero. This means that at A, t_3 satisfies the simultaneous equations
$$\tfrac{1}{2}u^2 \sin\theta \cos\theta/g - \tfrac{1}{2}u\cos\theta t_3 = 0$$
$$eu\sin\theta t_3 - \tfrac{1}{2}gt_3^2 = 0.$$
Eliminating t_3 from these two equations gives
$$\tfrac{1}{2}u^2 \sin\theta \cos\theta/g - \tfrac{1}{2}u\cos\theta(2eu\sin\theta/g) = 0$$
or
$$(\tfrac{1}{2}u^2 \sin\theta \cos\theta/g)(1 - 2e) = 0$$
giving
$$e = \tfrac{1}{2}.$$

Example 2. *A ball falls freely from a height of $2\tfrac{1}{2}$ m onto a point P of a smooth plane. The plane is inclined at an angle of 60 degrees to the horizontal. The ball rebounds to meet the plane again at Q. If the coefficient of restitution between the ball and the plane is $\tfrac{1}{2}$, find the length PQ.*

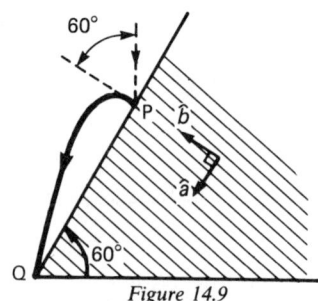

Figure 14.9

∴

PROJECTILES AND IMPACT

Consider first the motion before the ball hits the plane, and apply the principle of work:

final K.E. − initial K.E. = work done,

i.e. $\frac{1}{2}mv^2 - 0 = mg \times 2\frac{1}{2}$

giving $v = \sqrt{5g}$.

vertically downwards.

Let $\hat{a}$ and $\hat{b}$ be unit vectors parallel (in a *downwards** direction) and perpendicular to the plane (see *Figure 14.9*). Then $v = \sqrt{5g}$ can be resolved into $\sqrt{5g} \sin 60°$ parallel to the plane and $-\sqrt{5g} \cos 60°$ normal to the plane.

After impact only the normal component will be altered and, since $e = \frac{1}{2}$, it becomes $\frac{1}{2}\sqrt{5g} \cos 60°$. Hence, assuming $t = 0$ at the point P, which is taken as origin, then in

Motion from P to Q
the initial velocity is

$$v_P = \sqrt{5g} \sin 60° \hat{a} + \tfrac{1}{2}\sqrt{5g} \cos 60° \hat{b}.$$

The acceleration due to gravity can be resolved into $g \sin 60°$ and $-g \cos 60°$ in the directions of $\hat{a}$ and $\hat{b}$ respectively, so that in general

$$v = (\sqrt{5g} \sin 60° + g \sin 60° t)\hat{a} + (\tfrac{1}{2}\sqrt{5g} \cos 60° - g \cos 60° t)\hat{b}.$$

Integrating this equation to obtain the displacement vector r and remembering that, at the origin P, $t = 0$ and $r = 0$ we have that

$$r = (\sqrt{5g} \sin 60° t + \tfrac{1}{2}g \sin 60° t^2)\hat{a} + (\tfrac{1}{2}\sqrt{5g} \cos 60° t - \tfrac{1}{2}g \cos 60° t^2)\hat{b}.$$

At Q the b component of r will be zero; hence

$$\tfrac{1}{2}\sqrt{5g} \cos 60° t_Q - \tfrac{1}{2}g \cos 60° t_Q^2 = 0$$

$$t_Q = \sqrt{5/g}.$$

Substituting this value in r,

$$r_Q = [\sqrt{5g} \sin 60° \sqrt{5/g} + \tfrac{1}{2}g \sin 60°(5/g)]\hat{a} + 0\hat{b}.$$

Hence

$$PQ = |r_Q| = 5 \sin 60°(1 + \tfrac{1}{2})$$
$$= 6·495.$$

The distance bounced down the plane is 6·50 metres.

---

* Note that in this question it is more convenient to take the positive direction *down* the plane.

285

Exercises 14c
1. A particle falls freely from a height of 2·5 m onto a point *P* of a smooth plane. The angle of inclination of the plane is 30 degrees. The particle rebounds and meets the plane again at *Q*. If the distance *PQ* is 1·2 m, find the coefficient of restitution between the ball and the plane.

2. A ball is projected from a point *O* and strikes a smooth vertical wall returning, without bouncing, to the point of projection. The time of flight is 2 seconds, the coefficient of restitution between the ball and the wall is $\frac{1}{3}$ and the horizontal distance from the point of projection is 490 cm. Find the angle of projection.

3. A ball is thrown from a point *A* and strikes a smooth vertical wall, rebounds and lands at *A*. The coefficient of restitution between the ball and the wall is *e*, and *R* is the horizontal range from *A* in the absence of the wall. Prove that the horizontal distance from *A* to the wall is $(eR)/(1 + e)$.

4. A particle is projected, at an angle θ to the horizontal, from a point *P* on a smooth inclined plane. After one bounce on the plane the particle returns to *P*. Prove that $\cot(\theta - \alpha) = (1 + e)\tan\alpha$, where α is the inclination of the plane and *e* is the coefficient of restitution between the particle and the plane.

EXERCISES 14*

1. A boy knows that the furthest distance he can throw a ball is 24 m. He is 16 m from a wicket: at what angles of projection can he throw the ball in order to hit the wicket?

2. A bullet is fired from the top of a cliff 120 m high, with a speed of 70 m/s, at an angle of elevation of $\sin^{-1}\frac{2}{5}$. Find how far the bullet travels in a horizontal direction before it enters the sea.

3. A particle is projected with speed *u* from a point *A*. It strikes at right angles at *B* a plane through *A* whose angle of inclination is α. Show that

(*a*) The vertical height of *B* above *A* is

$$(2u^2 \sin^2 \alpha)/(g(1 + 3\sin^2 \alpha)).$$

(*b*) The time of flight from *A* to *B* is

$$(2u)/(g\sqrt{(1 + 3\sin^2 \alpha)}).$$

4. A particle is projected from *A* with speed *u* to pass through a point *B*. The horizontal and vertical displacements of *B* with respect

---
* Exercises marked thus, †, have been metricized, *see* Preface.

EXERCISES

to A are b and a respectively. Show that if $u^2 > g[a + \sqrt{(a^2 + b^2)}]$ there are two possible angles of projection.

5. A particle is projected up and down a plane inclined at $\tan^{-1}(\frac{1}{3})$ to the horizontal, the speed of projection and the angle of projection being the same in each case. If the range up the plane is one-third of the range down the plane, find the angle of projection.

6. A gun is located on a coastal mountain at a height of 400 m above sea level. It fires a shell at 240 m/s at a stationary ship. If the shell takes 20 seconds to reach the ship, find the angle of projection and the horizontal displacement of the ship from the gun.

7. The range of a particle on a horizontal plane is R. Show that the greatest height h attained is given by the equation $16gh^2 - 8u^2h + gR^2 = 0$, u is the speed of projection. State the condition for this quadratic in h to have real roots and hence deduce the value of the maximum range.

8. A ball is projected from a point on the ground and just clears the top of a wall of height 4 m whose base is 6 m from the point of projection. If the angle of projection is 45 degrees, find the greatest height reached by the ball.

9. Given R is the maximum range of a particle up an inclined plane and T the corresponding time of flight, find a relation between R and T in terms of u, the speed of projection, and α, the inclination of the plane. If the projectile had been fired down the plane, would the same relation hold?

10. A particle is projected under gravity from a point at the foot of a fixed inclined plane so that its trajectory is in the vertical plane containing the line of greatest slope through the point. If T is the time of flight, prove that the vertical height of the particle above the plane at time $t\,(< T)$ is

$$\tfrac{1}{2}gt(T - t). \qquad \text{(London, part)}$$

11. Show that the range of a body projected up a plane inclined at an angle α to the horizontal is

$$(2V^2 \cos\theta \sin(\theta - \alpha))/(g \cos^2 \alpha)$$

where the velocity of projection is V at an angle θ to the horizontal.

If the range down the plane is double the range up the plane, for the same speed of projection, the angle θ being 45 degrees in each case, find the inclination of the plane to the horizontal. (London)

12. A cricketer threw a ball from ground level in the long field in such a way that it fell at the feet of the wicket-keeper 60 m away. Show that if u and v were the horizontal and vertical components, in m/s of the initial velocity of the ball, $uv = 294$.

By advancing 4 m towards the fieldsman the wicket-keeper could have taken the ball at a height of 1·5 m above the ground. Show that the ball was in the air for $2\tfrac{1}{7}$ seconds. (London)†

13. A particle is projected from a point A with velocity V at an angle α to the horizontal. When it has reached a point P its velocity is inclined at an angle β to the horizontal. Both α and β are taken to be positive when the particle is rising and negative when it is falling. Prove that:

(a) the vertical component of the velocity at P is $V \cos \alpha \tan \beta$;

(b) the height of P above the level of A is

$$\frac{V^2(\cos^2 \beta - \cos^2 \alpha)}{2g \cos^2 \beta};$$

(c) time to move from A to P is

$$\frac{V \sin (\alpha - \beta)}{g \cos \beta};$$

(d) Horizontal projection of AP is

$$\frac{V^2 \sin (\alpha - \beta) \cos \alpha}{g \cos \beta}.$$

(Oxford)

14. A particle is projected under gravity with velocity V in a direction inclined to the horizontal at an angle θ. Derive expressions for the horizontal and vertical displacements, x and y, at time t after projection, and deduce an equation for the path of the particle.

A vertical section of a valley is in the form of a parabola $x^2 = 4ay$, where a is a positive constant and the axis of y is vertically upwards. A gun placed at the origin fires a shell with velocity $\sqrt{(2gh)}$ at an angle θ to the horizontal. If the shell strikes the section at the point (x, y), prove that

$$x = (4ah \tan \theta)/(a + h + a \tan^2 \theta).$$

Deduce the greatest value of x as θ varies. (JMB)

15. A ball is projected under gravity from a point A on level ground directly towards a tall vertical mast 30 m away and strikes the ground at the foot of the mast. Show that during its flight the ball when viewed from A will *appear* to be descending the mast with uniform speed.

Viewed from A the ball appears at one moment to be passing a point P on the mast and half a second later to be passing a point Q 7 m below P. Find the initial velocity of the ball and its time of flight.

[Take g as 9·8 m/s^2.] (London)†

EXERCISES

16. A particle is projected under gravity in a horizontal direction from the highest point of a fixed sphere of radius a. If V is its speed of projection and the particle does not touch the sphere again, show that $V^2 > ag$.

17. A is the point of projection and B is a point on the parabolic path of a particle moving under gravity. A light source is placed at B. Show that while the particle travels from A to B, its shadow on a vertical wall moves with constant speed.

18. A missile is projected with an initial speed of 91 m/s at an angle of $\sin^{-1}(\frac{12}{13})$ with the horizontal. Find its range and time of flight. Three seconds later another missile is projected from the same point and hits the ground at the same time and in the same place as the first missile, find its speed and angle of projection.

19. A boat is moving directly away from a gun on the shore with speed V_1. The gun fires a shell with speed V_2 at an angle of elevation α and hits the boat. Find the distance of the boat from the gun at the moment it is fired.

20. From a point on a plane hillside of inclination α to the horizontal a particle is projected at right angles to the plane with speed V. The particle subsequently strikes horizontal ground. The distances from the foot of the hill of the point of impact with the ground and of the point of projection are each h. Show that.

$$2V^2 = gh \cot(\alpha/2).$$

If the speed just before impact is double the speed of projection, find the value of α. (London)

21. A bird flies in a straight line with uniform velocity u in an upward direction making an angle β with the horizontal. At the instant when the bird is at a height h vertically above a boy on the ground the boy throws a stone at an angle of elevation α. Show that, whatever the velocity of projection, the stone cannot hit the bird unless

$$\tan \alpha \geqslant (\sqrt{2gh}/u) \sec \beta + \tan \beta.$$

If the stone merely grazes the bird so that the motion of neither is appreciably disturbed, show that, in general, the bird will be hit again. (London)

22. A particle is projected from a point O with velocity V at an elevation α and strikes the horizontal plane through O at A. Find from first principles the distance OA.

Show that if the particle is projected from O with the same elevation to hit a target at a height h above A, then the velocity of projection must be

$$\frac{V^2 \sin \alpha}{(V^2 \sin^2 \alpha - \frac{1}{2}gh)^{1/2}}.$$

(WJEC)

23. A ball is thrown from a point P on a cliff height h above the seashore. It strikes the shore at a point Q where PQ is inclined at an angle α to the horizontal. If the angle of projection is also α, show that the speed of projection is $\sqrt{gh/(2\sin\alpha)}$, and that the ball strikes the shore at an angle $\tan^{-1}(3\tan\alpha)$ with the horizontal.

24. A shell is fired from a point O at an angle of elevation 60 degrees, with a speed of 40 m/s and strikes a horizontal plane through O, at a point A. The gun is fired a second time with the same angle of elevation and a different speed V. If it hits a target which starts to rise vertically from A with speed $9\sqrt{3}$ m/s at the same instant as the shell is fired, find V.

25. A shell is fired from a point O with speed u and angle of projection α to hit a target whose horizontal and vertical distances from O are h and v respectively. Show that
$$gh^2\tan^2\alpha - 2hu^2\tan\alpha + gh^2 + 2vu^2 = 0,$$
deduce that this equation has real roots in $\tan\alpha$ if $u^4 - 2gvu^2 - gh^2 \geq 0$ and hence that for the shell to hit the target $u^2 \geq gv + g\sqrt{h^2 + v^2}$.

26. A particle is projected horizontally with speed u from the top edge of a staircase in a direction which will take it directly down the staircase. The breadth of each stair from front to back is a and the height of each stair is h. Show that if $u^2 < ga^2/2h$, the particle strikes the first stair down.

The coefficient of restitution between the particle and each stair is e. Show that the further conditions that the second impact is on the second stair down are
$$a < u(1 + 2e)\sqrt{\frac{2h}{g}}$$
and
$$u\{1 + e + \sqrt{(1 + e^2)}\}\sqrt{\frac{2h}{g}} < 2a. \qquad \text{(London)}$$

27. A particle is projected under gravity from a point P with a velocity whose horizontal and vertical components are p and q. Find an expression for its range R on a horizontal plane in terms of p and q. If, before reaching the horizontal plane, the particle had hit a vertical wall whose distance from P was $(pq)/2g$ and rebounded to the point P, show that $e = \tfrac{1}{3}$ (where e is the coefficient of restitution between the particle and the wall).

28. A particle is projected under gravity from a point P with speed

EXERCISES

u at an angle of inclination of θ to the horizontal. It strikes a vertical wall and returns to P without bouncing. If e is the coefficient of restitution between the ball and the wall, find the vertical height of the point of impact with the wall above P if the distance of P from the wall is d.

29. A corridor has a horizontal floor and a ceiling 4·9 m high and is closed at one end by a vertical wall. A boy standing in the corridor throws a small ball from a height 1·3 m above the floor. If the greatest speed with which he can throw the ball is 14 m/s, show that he cannot hit the wall without first hitting the ceiling or the floor if he stands further from the wall than 20·8 m.

If he stands at half this distance from the wall, find the depth below the ceiling of the highest point of the wall which he can hit without first hitting the ceiling or the floor.

[Take g as 9·8 m/s^2]   (London)†

30. From a point on a horizontal floor a particle is projected which just clears the edge of a round table and strikes the centre. The table is of radius 1 m and height 1·25 m and the point of projection is at a *horizontal* distance of 2 m from the centre of the table. Find the velocity of projection.

Assuming the table to be smooth and inelastic, find the total horizontal distance travelled by the particle on striking the floor.

[Take g as 9·8 m/s^2]   (London)†

31. Two particles are projected simultaneously in the same vertical plane, i and j being unit horizontal and vertical vectors in that plane. The first particle is projected from the origin with velocity vector $nV\cos\alpha\,\mathbf{i} + nV\sin\alpha\,\mathbf{j}$, and the second particle is projected from a position $h\mathbf{i} + k\mathbf{j}$ (where $h > 0$, $k > 0$) with velocity vector $-V\cos\beta\,\mathbf{i} + V\sin\beta\,\mathbf{j}$. Write down the position vectors of each of the particles after time t has elapsed.

Show that the particles cannot collide unless $\sin\beta < n\sin\alpha$, and if they do collide, prove that $\sin(\beta + \gamma) = n\sin(\alpha - \gamma)$, where $\tan\gamma = k/h$.

Find the condition imposed on V if the point of collision is above the level of the origin.   (London)

32. If in Question 28, the particle strikes the wall at a height h above P, find the value of u in terms of d, h, g and e.

33. An elastic particle is projected from a point O on a smooth horizontal floor with horizontal and vertical components of velocity u and v respectively. When the particle is at the highest point of its path it strikes a vertical wall which is perpendicular to the plane of

its path. It rebounds from the wall and after rebounding *once* from the floor next strikes the floor at the point O. If the coefficient of restitution between the particle and both wall and floor is e, show that $e = \frac{1}{2}$.

Show that when it strikes the floor for the third time after reaching O again, the particle is at a distance $7uv/(16g)$ from O, and find the total time which has elapsed since it was first projected.

(WJEC)

34. Two equal particles are projected at the same instant from points A and B at the same level, the first from A towards B with velocity u at 45 degrees above AB, and the second from B towards A with velocity v at 60 degrees above BA. If the particles collide directly when each reaches its greatest height, find the ratio $v^2:u^2$ and prove that $u^2 = ga(3 - \sqrt{3})$, where a is the distance AB.

After the collision the first particle falls vertically. Show that the coefficient of restitution between the particles is $(\sqrt{3} - 1)/(\sqrt{3} + 1)$.

(JMB)

35. A particle is projected with speed u from a point on a plane of inclination α to the horizontal, motion taking place in a vertical plane through a line of greatest slope. The maximum ranges up and down the inclined plane are R_1 and R_2 respectively. Prove that R_1 and R_2 are the roots of the equation in R

$$g^2 R^2 \cos^2 \alpha - 2gu^2 R + u^4 = 0.$$

If R_3 is the maximum range on a horizontal plane, prove that

$$\frac{2}{R_3} = \frac{1}{R_1} + \frac{1}{R_2}.$$

(London)

36. A particle is projected upwards with speed V from a point A on an inclined plane. Its plane of projection meets the inclined plane in a line of greatest slope and its angle of projection (to the plane) is $\tan^{-1}(\frac{1}{2})$. If it meets the plane at right angles, find the angle of inclination of the plane. Show that the second point of impact is $4\sqrt{2}V^2(1 - e^2)/10g$ from A.

37. A particle is projected upwards with speed V from a point A on a smooth plane inclined at an angle α. Its plane of projection meets the inclined plane in a line of greatest slope and its angle of projection, measured to the plane is θ. If the particle strikes the plane at P when moving horizontally, show that $\tan \theta = \tan \alpha/(1 + 2\tan^2 \alpha)$. If the coefficient of restitution is $\frac{4}{5}$ and it rebounds and strikes the plane again at Q, find PQ.

EXERCISES

38. A particle is projected from the origin with initial velocity $28i + 100j$ towards an inclined plane whose line of greatest slope has the equation $r = 480i + \lambda(2i + j)$ where λ is a parameter. Show that the particle strikes the plane after 20 seconds and determine the distance along the line of greatest slope from the point where $\lambda = 0$ to the point of impact.

Show also that α, the acute angle between the direction of motion at impact and the inclined plane, is given by $5\sqrt{5} \cos \alpha = 2$.

(Velocity components are measured in m s^{-1} and displacements in m. Take g as 9.8 m s^{-2}.) (JMB)

39. Two particles P and Q are projected simultaneously out to sea from the same point on the top of a high cliff and in the same vertical plane. P is projected with velocity $V\sqrt{3}$ at 60 degrees to the horizontal and Q at 30 degrees to the horizontal. The particles remain one vertically above the other till one of them strikes the sea. Find the velocity of projection of Q and also the distance between the particles when (a) Q is at its greatest height, (b) P is at its greatest height, showing that one of these distances is three times the other. (London)

40. A particle is projected from a point O on a plane inclined at an angle α to the horizontal. The path of the particle lies in a vertical plane meeting the inclined plane in a line of greatest slope. If the particle returns to O after one bounce on the plane and θ is the angle the direction of projection makes with the plane, find the coefficient of restitution e in terms of θ and α.

41. Two particles, A and B, are projected simultaneously from the origin with velocities $-30i + 90j$ and $30i + 110j$ respectively. Both particles impinge on a plane which contains the origin, meeting the plane on a line of greatest slope which has the equation $r = \lambda(3i + j)$, where λ is a scalar. Show that the times of flight of A and B are equal and that the distance between the points of impact is $400\sqrt{10}$ m.

(Take g as 10 m/s^2) (JMB, part)

42. A particle is projected from a point O with initial speed $2\sqrt{(ga)}$ at an angle θ to the horizontal. It strikes a target which is at a horizontal distance a from O and at a vertical distance $3a/4$ above the level of O. Find the possible values of $\tan \theta$. Find also the least speed of projection of a particle from O if it is to strike the target. (London)

43. The muzzle speed of a gun is V and it is desired to hit a small target at a horizontal distance a away and at a height b above the gun. Show that this is impossible if
$$V^2(V^2 - 2gb) < g^2 a^2$$

but that if $V^2(V^2 - 2gb) > g^2a^2$ there are two possible elevations for the gun.

Show that, if $V^2 = 2ga$ and $b = \frac{3}{4}a$, there is only one possible elevation and find the time taken to hit the target. (Oxford)

44. A particle is projected with speed v m/s from a point A at an angle of elevation α and moves freely under gravity. The highest point in the path of the particle is B and AB is inclined at an acute angle θ to the horizontal. Show that $\tan \alpha = 2 \tan \theta$.

If $\alpha = 45°$ and the vertical height of B above A is 50 m, find the value of v. Find also the time that elapses from the instant of projection until the instant when the velocity of the particle is parallel to AB. (Take the acceleration due to gravity to be 10 m/s².)
(AEB)

15

CIRCULAR MOTION

15.1 INTRODUCTION

Newton's laws of motion show us that an unrestricted particle will continue to move *in a straight line* unless something deflects it from its path. For instance, in Chapter 14 we saw how a steady downwards force (its weight) on a projectile pulled it into a *parabolic* path.

It turns out that if we wish to make a moving particle move in a *circle* about a point O, we have to pull it always towards O. The magnitude of the force must not be too great or it will spiral into O, nor too little, but just enough to keep it moving in the required circle. Indeed, if a particle is *restricted* to motion in a circle by being on the end of a string or on the inside of a cylinder (say), then the tension in the string or the reaction on the particle will be inward and will automatically adjust to the necessary value. (Incidentally, should the string break when the particle is at a point P then the particle will move in a straight line in the direction of the tangent at P, *not* outwards.)

In the next sections we establish the exact value of the inward acceleration causing a particle to move in a given circle and hence we determine the force required.

15.2. THE ACCELERATION OF A PARTICLE MOVING IN A CIRCLE

WHEN a particle moves in a circle its direction is constantly changing. Hence its velocity changes and it must have an acceleration.

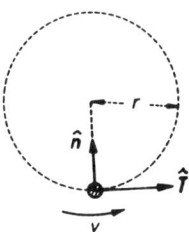

Figure 15.1

CIRCULAR MOTION

Consider the particle at an instant when its speed is v, and let $\hat{T}, \hat{n}$ be unit vectors along the tangent and radius respectively. (Refer to Figure 15.1.)

Then the velocity v of the particle is given by

$$v = v\hat{T},$$

and differentiating with respect to time t

$$\frac{dv}{dt} = \frac{dv}{dt}\hat{T} + v\frac{d\hat{T}}{dt}. \quad \ldots(i)$$

But $\hat{T}$ is a unit vector. So $d\hat{T}/dt$ is a vector at right angles to $\hat{T}$ directed towards the centre of the circle (refer to Section 3.7). Its magnitude is the rate of rotation of $\hat{T}$ which is the same as ω, the angular speed of the particle. Hence $d\hat{T}/dt = \omega\hat{n}$ and equation (i) becomes

$$\frac{dv}{dt} = \frac{dv}{dt}\hat{T} + v\omega\hat{n}.$$

That is, the acceleration of the particle has components dv/dt along the tangent and $v\omega$ along the radius towards the centre of the circle.

For a particle moving in a circle $\omega = v/r$ (refer to Section 4.5), so this last component can be written as v^2/r or $\omega^2 r$, which is more usual:

$$\frac{dv}{dt} = \frac{dv}{dt}\hat{T} + \frac{v^2}{r}\hat{n}.$$

Example 1. A particle moves in a circle so that after t seconds its speed is $(2t^2 + 4)$ m/s. The circle is of radius 9 m. Find the resultant acceleration of the particle after 1 second.

The tangential component of acceleration = dv/dt

$$= d(2t^2 + 4)/dt$$

$$= 4t \text{ m/s}^2.$$

The component towards the centre = v^2/r

$$= (2t^2 + 4)^2/9 \text{ m/s}^2.$$

When $t = 1$, these components are each 4 m/s². Hence

$$dv/dt = 4\hat{T} + 4\hat{n} \text{ m/s}^2,$$

which is a resultant acceleration of magnitude $4\sqrt{2}$ m/s², in the plane of the circle and inclined at 45 degrees to the radius to the particle.

FORCE AND MOTION IN A CIRCLE

Example 2. A particle moves at constant speed in a circle of radius 10 m, completing 30 revolutions in one minute. Find its resultant acceleration.

Since the speed is constant the tangential acceleration $dv/dt = 0$.

$$\text{The angular speed} = 30 \text{ rev/min}$$
$$= 30 \times 2\pi/60 \text{ rad/s}$$
$$= \pi \text{ rad/s}.$$

Therefore the component of acceleration towards the centre ($\omega^2 r$) is $\pi^2 10$ m/s².

Hence the resultant acceleration of the particle is towards the centre of the circle and of magnitude $10\pi^2$ m/s².

Exercises 15a

1. A particle moves in a circle of radius 3 m, so that after t seconds its speed is $(5t + 1)$ m/s. Find its acceleration after 1 second.

2. A car moves at a constant speed of 45 km/h round a circular bend of radius 200 m. Find its acceleration.

3. A particle moves in a circle of radius 1 m. What is its speed when the component of its acceleration towards the centre is 4 cm/s²?

4. Find the acceleration of a particle moving in a circle of radius 2 m with a constant angular speed of 5 rad/s.

5. A flywheel rotates uniformly at 300 rev/min. Find the acceleration of a point on the wheel 5 cm from the axle.

6. A stone on the end of a string 1 m long is made to describe circles in a vertical plane. If, when the string is horizontal, its speed is 2·1 m/s and is increasing at 9·8 m/s², find the acceleration of the stone.

7. A particle starts from rest and moves in a circle of radius r. When the radius to the particle has turned through an angle θ, the speed of the particle is $kr\theta$. Show that its acceleration makes an angle $\tan^{-1}(1/\theta)$ with the radius (k is a constant).

8. A particle starts from rest and moves in a circle of radius r. When the radius to the particle has turned through an angle θ, the speed of the particle is given by $V^2 = 2kr \sin \theta$ (k constant). Show that its acceleration is $k(\cos \theta \hat{T} + 2 \sin \theta \hat{n})$ and deduce the magnitude of this acceleration when $\theta = 30$ degrees.

9. A particle is projected with speed u so that it moves in a circle of radius r. The forces acting on it are such that its acceleration is always inclined at 45 degrees to the inward radius. Show that after time t its speed v is given by
$$v = (ur)/(r - ut)$$

297

CIRCULAR MOTION

15.3. FORCE AND MOTION IN A CIRCLE

Consider a particle of mass m moving in a circle of radius r at an instantaneous speed v. Then we have seen that its acceleration is $(dv/dt)\hat{T} + (v^2/r)\hat{n}$, where $\hat{T}, \hat{n}$ are unit vectors along tangent and radius. Hence, since $P = mf$, the resultant force acting on the particle must be $m[(dv/dt)\hat{T} + (v^2/r)\hat{n}]$. That is, it has components $m\,dv/dt$ along the tangent and mv^2/r towards the centre of the circle.

Thus, if a particle is to move in a circle, the forces acting on it must vary so that the sum of their components towards the centre is always mv^2/r. The rate of change of speed round the circle will depend on the tangential components of these forces.

We shall first consider cases when the speed is uniform.

15.4. UNIFORM MOTION IN A CIRCLE

If a particle moves in a circle at *uniform* speed, the tangential component of force must be zero and hence the resultant force must be towards the centre and of constant magnitude mv^2/r.

Example 1. A particle of mass 1 kg is attached to one end of a spring of natural length 50 cm and modulus 80 N. The other end is fastened to a point on a smooth horizontal table. If the particle is made to describe circles on the table at 72 rev/min, find the extension of the spring.

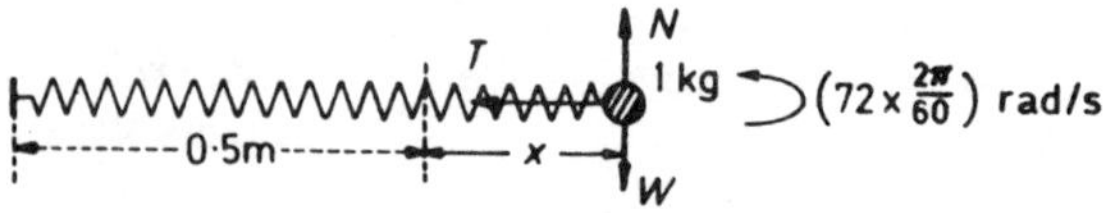

Figure 15.2

Let the extension of the spring be x (refer to *Figure 15.2*). The modulus of the spring is 80 N and hence, by Hooke's law, the tension

$$T = \lambda(x/l) = 80(x/\tfrac{1}{2}) = 160x \text{ N}.$$

Since the particle is moving in a circle its acceleration towards the centre

$$= \omega^2 r$$
$$= \left(72 \times \frac{2\pi}{60}\right)^2 \left(\frac{1}{2} + x\right)$$
$$= \frac{144\pi^2}{25}\left(\frac{1}{2} + x\right) \text{ m/s}^2.$$

UNIFORM MOTION IN A CIRCLE

Since $P = mf$, towards the centre of the circle

$$160x = 1 \times \frac{144\pi^2}{25}\left(\frac{1}{2} + x\right).$$

$$\therefore \quad x(160 - 144\pi^2/25) = (144\pi^2/25)1/2,$$

giving $\quad x = 0.2755$ m.

The extension of the spring is 27·6 cm.

Example 2. A particle of mass m is attached to one end of an inextensible string, length l, the other end of which is fastened to a fixed point. The particle can be made to describe horizontal circles at a uniform speed v with radius r. Show that $v^2\sqrt{l^2 - r^2} = gr^2$.

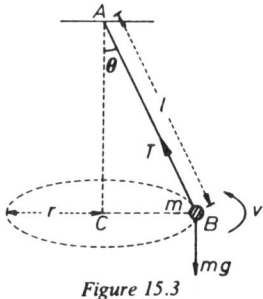

Figure 15.3

Let the string be inclined at an angle θ to the vertical (refer to Figure 15.3).

Since the particle is moving uniformly in a circle its acceleration is towards the centre and of magnitude v^2/r.

Since $P = mf$,

horizontally towards the centre
$$T \sin \theta = mv^2/r \qquad \ldots (i)$$

vertically $\qquad T \cos \theta - mg = 0 \qquad \ldots (ii)$

Rearranging (ii) and dividing (i) by (ii), we have

$$\tan \theta = v^2/rg.$$

But from triangle ABC, $\tan \theta = BC/AC = r/\sqrt{l^2 - r^2}$.

$$\therefore \quad r/\sqrt{l^2 - r^2} = (v^2/rg),$$

or $\qquad gr^2 = v^2 \sqrt{l^2 - r^2}.$

299

CIRCULAR MOTION

Exercises 15b

1. A car of mass 1 200 kg travels round a circular bend in the road at a constant speed of 27 km/h. The radius of the circle is 100 m and the road surface is horizontal. The driver turns the wheels of the car to bring into play a frictional force which will make the car go round the bend. What is the magnitude and direction of this force?

2. A particle of mass m moves with uniform speed v in a straight line. During its motion it passes a point O at a distance r from it. Give details of the force needed to make the particle (continuing at speed v) perform circles with O as centre.

3. One end of a light inextensible string, of length 50 cm, is fastened to a point O on a smooth horizontal table. To the other end a particle of mass 2 kg is attached which describes horizontal circles on the table about O. The particle moves with a uniform speed of 8 m/s. Find the tension in the string.

4. A light inextensible string, of length 2 m, is fixed at one end and carries a mass of 3 kg at the other. The mass is made to describe horizontal circles at a uniform angular speed of 4 rad/s. Find the tension in the string.

5. A smooth hemispherical bowl is fixed with its rim horizontal. A particle of mass 90 g describes horizontal circles on the inside of the bowl at 140 rev/min. If the internal radius of the bowl is 20 cm, find the depth of these circles below the rim and determine the reaction between the bowl and the particle.

6. A particle, of mass m, is attached to one end of an elastic thread of natural length l and modulus λ. The other end is fastened to a point on a smooth horizontal surface. The particle describes circles on the surface of radius $l'(l' > l)$. If ω is the angular speed of the particle, show that

$$\omega^2 = \frac{\lambda}{m}\left(\frac{1}{l} - \frac{1}{l'}\right).$$

7. A particle of mass 5 kg describes horizontal circles on the end of an elastic string of natural length 16 cm and modulus 40 g. N. If the string is extended by 4 cm, find the inclination of the string to the vertical and the speed with which the particle is rotating.

8. One end of a light inextensible string is attached to a point at a height of 4 m above a smooth horizontal surface. To the other end is attached a particle of mass 4 kg which describes circles of radius 3 m on the surface with the string taut. If it moves at a speed of 3 m/s, find the tension in the string and the reaction between the surface and the particle.

9. A light inextensible string of length 18 cm has its ends fastened to two points A and B. A is vertically above B and distant 12 cm from it. A small smooth ring P is threaded on the string and is made to perform circles at uniform speed with the string tight and BP horizontal. Find the number of revolutions completed in one minute.

10. A particle is placed on a rough horizontal rotating platform. In what direction must the frictional reaction act if the particle is to be carried round with the platform?

If the particle is distant r from the centre of rotation and the coefficient of friction is μ, find the maximum speed that can be given to the particle without it slipping.

11. The Moon is attracted to the Earth by a force of magnitude GMm/r^2, where M and m are the masses of the Earth and Moon respectively, and r is the distance between them. The Universal Gravitational Constant $G = 6\cdot66 \times 10^{-11}$ m^3/kg s^2.

Assuming that the Moon moves with a uniform speed in a circle around the Earth, completing one revolution in 27 days, and that the distance between them is 384 000 km, estimate the mass of the Earth.

15.5 MOTION IN A VERTICAL CIRCLE

Examples of motion in a vertical circle are: a ring sliding on a vertical circular wire; a particle sliding down the outside of a sphere; the motion of the bob of a simple pendulum.

In each case the weight of the particle will, in general, have a component along the tangent giving the particle an acceleration in that direction. So we can apply $P = mf$ along the tangent, as well as towards the centre, to obtain our information. However, rather than resolve along the tangent, it is usually more convenient to obtain a second equation from application of the Principle of Work.

Example 1. A small bead, of mass 2 g, is threaded on a smooth circular wire fixed with its plane vertical. The bead is released from rest in a position where the radius to the bead is horizontal. The radius of the wire is 6 cm. Find the reaction between the bead and the wire when the radius to the bead makes an angle of 60° degrees with the downward vertical.

Let the particle be moving with speed v at the instant when the radius to the bead makes an angle of 60 degrees to the vertical (refer to *Figure 15.4*). Then since it is moving in a circle the component of its acceleration towards the centre will be $v^2/0\cdot06$ m/s^2.

CIRCULAR MOTION

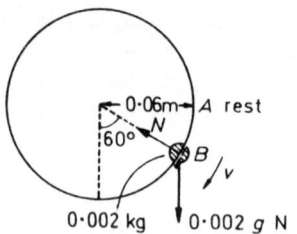

Figure 15.4

Since $P = mf$,

towards the centre $\quad N - 0{\cdot}002g \cos 60° = 0{\cdot}002(v^2/0{\cdot}06). \quad \ldots (i)$

Applying the principle of work between positions A and B,

$$\text{Final K.E.} - \text{Initial K.E.} = \text{Work done}$$

$\therefore \quad \tfrac{1}{2}(0{\cdot}002)v^2 - 0 = 0{\cdot}002g(0{\cdot}06 \cos 60°). \quad \ldots (ii)$

From (ii) $v^2 = 0{\cdot}06g$, and substituting this value of v^2 in equation (i) gives

$$N - 0{\cdot}001g = 0{\cdot}002g$$

$\therefore \quad N = 0{\cdot}003g \text{ N.}$

The reaction between the bead and the wire in this position is towards the centre and of magnitude $0{\cdot}003g$ N.

Example 2. A particle of mass m hangs at rest on the end of an inextensible string of length r. If the particle is projected horizontally with speed u, find the tension in the string when it makes an angle θ with the downward vertical. Deduce that the particle will make complete circles if $u^2 \geqslant 5gr$ and indicate the positions of maximum and minimum tension.

If $u^2 = 3\,gr$, determine the value of θ at the point where the string goes slack.

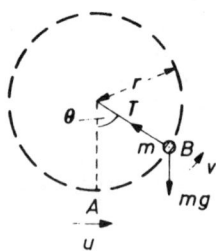

Figure 15.5

MOTION IN A VERTICAL CIRCLE

Consider the particle when the string is inclined at an angle θ to the vertical and it is moving with speed v (refer to *Figure 15.5*). Then, since the particle is moving in a circle the component of acceleration towards the centre is v^2/r.

Since $P = mf$,

towards the centre, $\quad T - mg\cos\theta = mv^2/r \quad \ldots\text{(i)}$

Applying the principle of work between positions A and B,

$$\text{Final K.E.} - \text{Initial K.E.} = \text{Work done}$$

$\therefore \quad\quad \tfrac{1}{2}mv^2 - \tfrac{1}{2}mu^2 = -mgr(1 - \cos\theta). \quad \ldots\text{(ii)}$

From equation (ii) $v^2 = u^2 - 2gr(1 - \cos\theta)$ and substituting this value in equation (i),

$$T - mg\cos\theta = mu^2/r - 2mg(1 - \cos\theta)$$

$\therefore \quad\quad T = mu^2/r - mg(2 - 3\cos\theta) \quad \ldots\text{(a)}$

which is the required formula for the tension.

For the particle to perform complete circles, the string must not go slack. That is,

$$T \geqslant 0 \quad \text{for all } \theta.$$

i.e. $\quad u^2/r - g(2 - 3\cos\theta) \geqslant 0 \quad \text{for all } \theta,$

or $\quad\quad u^2 \geqslant gr(2 - 3\cos\theta) \quad \text{for all } \theta.$

Now the right-hand side of this inequality will be greatest when $\theta = \pi$ and $\cos\theta = -1$. So the condition for performing complete circles is

$$u^2 \geqslant 5gr.$$

(Note that for a ring on a wire, say, the condition for complete circles is that the speed at the top is greater than zero. This is not a sufficient condition for a particle on a string. The speed at the top in this case must be still greater to ensure that the string remains taut.)

Inspection of equation (a) shows that the tension T is a maximum when $\cos\theta = 1$, i.e. when $\theta = 0$. And if the particle does make complete circles then T is a minimum when $\cos\theta = -1, \theta = \pi$. So the tension in the string is greatest at the bottom and least at the top.

If $u^2 = 3gr$, then equation (a) becomes $T = mg(1 + 3\cos\theta)$ and the condition for complete circles is not satisfied. At the moment when the string is about to go slack $T = 0$

$\therefore \quad\quad mg(1 + 3\cos\theta) = 0$
$\therefore \quad\quad \cos\theta = -\tfrac{1}{3}$
$\therefore \quad\quad \theta = 109°\,28'.$

The string goes slack, therefore, at the moment when it makes an angle of 70° 32' with the upward vertical.

Exercises 15c

1. A small stone, of mass 0·1 kg, hangs from a fixed point on the end of an inelastic string 0·2 m long. The stone is pulled to one side and held so that the string is taut and horizontal. If the stone is released from rest, find its speed and the tension in the string when the particle has fallen a vertical distance of 0·1 m.

2. A light rod is pivoted at one end O and carries a particle of mass 150 g at the other. The system is held at rest with the particle vertically above O and released gently. Find the stress in the rod when the rod makes an angle (*a*) 60 degrees (*b*) 120 degrees with the upward vertical.

3. A large cylinder, smooth on the inside and of internal radius r, is fixed with its axis horizontal. It contains a small particle lying at rest on the bottom. The particle is projected horizontally at right angles to the axis of the cylinder with speed u. Assuming that the particle does not lose contact with the cylinder, show that when the speed has been reduced to $u/2$, the angle θ between the radius to the particle and the downward vertical satisfies the relation

$$8gr(1 - \cos \theta) = 3u^2.$$

Find also an expression for the reaction between the particle and cylinder in this position in terms of m, u and r (m being the mass of the particle).

4. A particle is placed on the highest point of a fixed smooth sphere and given a gentle push. Find the angle between the radius to the particle and the upward vertical at the moment when the particle begins to leave the surface of the sphere.

5. A small particle hanging on the end of a light inelastic string 2 m long is projected horizontally. Calculate the least speed of projection needed to ensure that the particle performs complete circles.

If the speed of projection is 700 cm/s, find the position of the particle when the string goes slack.

6. A simple pendulum consists of a particle of mass 5 kg hanging on a string of length 10 m. If the particle is pulled aside so that the string makes an angle of 60 degrees with the downward vertical, and is released from rest, find the speed acquired and the tension in the string when it makes an angle $\cos^{-1}(\frac{2}{3})$ with the downward vertical.

EXERCISES

If the breaking strain of the string is 150 N, is there any danger of the string breaking in the subsequent motion?

7. A smooth circular wire, of radius a, is fixed with its plane vertical. A small ring, threaded on it, is projected with speed u from its lowest point. Show that the ring will describe complete circles if $u^2 > 4ga$.

If $u^2 = 2ga$, find the greatest height reached by the particle and, for a ring of mass m, the reaction between the ring and wire at that point.

8. A smooth cylinder of radius r is fixed with its axis horizontal. Two particles of mass m and $3m$ are connected by a light inelastic string and the system is hung over the cylinder with the string at right angles to the axis of the cylinder. The string is in contact with the cylinder and subtends an angle 2θ at its centre. Initially the particles are symmetrically placed on either side of the vertical and are released from rest. Find the speed acquired by the mass m when it reaches the top of the cylinder and find the reaction between the mass m and the cylinder in this position.

EXERCISES 15*

1. In an experiment to measure the ratio of charge (e) to mass (m) of an electron, a beam of electrons is passed through a magnetic field. The field is arranged so that it exerts a force on each electron which is always at right angles to its path and which can be adjusted so that the beam is deflected into a circular path whose radius r is measured.

If v is the speed of an electron and H the field strength, then the force exerted on the electron is Hev. Find a formula expressing e/m in terms of H, v and r.

2. A small ring can slide on a smooth circular wire fixed with its plane vertical. If the ring is released from the highest point on the wire, find the direction of its acceleration when it is on a level with the centre of the circle.

3. A small stone describes horizontal circles with angular speed ω on the end of a light inelastic string. If the depth of the circles below the point of suspension is h, find a formula expressing h in terms of ω.

Is it possible to whirl the stone round with the string horizontal?

4. A particle of mass 5 kg describes vertical circles on the end of a string 0·5 m long. The breaking strain of the string is 89 N. If the string breaks when the particle is in its lowest position, what is its speed at that instant?

* Exercises marked thus, †, have been metricized, *see* Preface.

✓5. A spring of natural length l has one end fastened to a smooth horizontal table. The other end is attached to a particle which describes circles on the table. When the particle moves with speed v the extension of the spring is x. If when the particle is suspended vertically on the spring its extension is e, show that
$$ev^2 = gx(l + x).$$

6. Inside a van a simple pendulum is hung from the roof. When the van rounds a bend at 54 km/h the pendulum makes an angle of 20 degrees with the vertical. Find the radius of the bend.

7. A particle describes vertical circles on the end of a light inelastic string 20 cm long. The ratio between the maximum and minimum tensions is 2:1. Find the speed of the particle at its highest point.

8. A right circular cone of semi-angle 60 degrees is fixed with its axis vertical and its vertex downwards. A small particle performs circles on the inside of the cone at a constant speed of 70 cm/s. Find the distance of the particle from the vertex.

9. Two particles, of mass $2m$ and m respectively, are connected by a light inextensible string of length a. The string passes through a fixed smooth ring and, while the $2m$ mass hangs in equilibrium a distance x below the ring, the other mass describes horizontal circles with angular velocity ω. Find the inclination of the moving string to the vertical, and show that
$$\omega^2(a - x) \doteq 2g.$$

10. A particle is projected from the top of a smooth sphere of radius a so that it slides down the outside of the sphere. If the particle leaves the sphere when it has fallen a distance $\tfrac{1}{4}a$, find the speed with which it was projected.

What will be its speed when it has fallen to the level of the centre of the sphere?

11. A light inextensible string of length 30 cm has one end fixed to a point A on a smooth vertical rod. To the other end a particle P of mass 20 g is attached. A second similar string (again 30 cm long) is attached to P and carries a small 30 g ring which slides on the rod. If P is made to describe horizontal circles at a speed of 420 cm/s, find the inclination of AP to the vertical.

12. A is a point on a smooth horizontal surface, B a point 1 m vertically above it. A light inelastic string of length 2 m has its ends fastened to A and B and carries a smooth small bead P of mass 1 kg threaded on it. If P is made to describe circles, centre A, on the surface at an angular speed of 8 rad/s, find the tension in the string.

EXERCISES

13. A particle P, lying on a smooth horizontal table, is connected to an identical particle Q by a light inextensible string. The string passes through a smooth hole in the table so that Q hangs freely. P is made to describe circles of radius 0·2 m on the table with speed u. Find u if (a) Q hangs at rest, (b) Q is made to perform circles with its portion of the string inclined at 60 degrees to the vertical.

In case (b), if the string is of length 0·4 m, find the angular speed with which Q is rotating.

✓14. A particle is attached to a fixed point by a light elastic string of natural length 20 cm and modulus twice the weight of the particle. If the particle is made to perform horizontal circles with angular speed ω show that ω must be greater than $\frac{7}{3}\sqrt{6}$ rad/s.

If $\omega = 7$ rad/s, find the radius of the circle in which the particle moves.

15. A particle of mass m describes complete circles in a vertical plane on the end of a light inextensible string. If the speed at the lowest point is twice the speed at the highest point, find the greatest and least tensions in the string.

Find also the height of the particle above the lowest point when the tension is $\frac{4}{3}mg$, the length of the string being $2a$.

16. A hemispherical bowl of radius 10 cm rotates uniformly about its axis which is vertical. A particle on the rough interior rotates with it in a position such that the radius to the particle makes an angle of 60 degrees with the vertical. If the coefficient of friction between particle and bowl is 1/2 and the particle is about to slip upwards, find the rate of rotation of the bowl in revolutions per minute.

17. A bead of mass m is threaded on a smooth circular wire of radius a fixed in a vertical plane. A light inextensible string attached to the bead passes through a smooth ring fixed at the centre of the wire and supports a particle of mass M hanging freely. The bead is projected with speed $\sqrt{(kga)}$ from the lowest point of the wire. Find the least value of k for the bead to reach the top of the wire.

Taking $k = 6$, show that the reaction between the bead and the wire vanishes at some point of the motion if M lies between m and $7m$.

(London)

18. A particle moves with constant speed v in a circle of radius r. Show that the acceleration of the particle is v^2/r directed towards the centre of the circle.

A rough horizontal plate rotates with constant angular velocity ω about a fixed vertical axis. A particle of mass m lies on the plate at a distance $5a/4$ from this axis. If the coefficient of friction between the plate and the particle is $\frac{1}{3}$ and the particle remains at rest relative to

the plate, show that
$$\omega \leq \sqrt{\left(\frac{4g}{15a}\right)}.$$

The particle is now connected to the axis by a horizontal light elastic string, of natural length a and modulus $3mg$. If the particle remains at rest relative to the plate and at a distance $5a/4$ from the axis, show that the greatest possible angular velocity of the plate is
$$\sqrt{\left(\frac{13g}{15a}\right)}$$
and find the least possible angular velocity. (JMB)

19. A light inextensible string ABC of length $5m$ has a particle of mass m attached at B, where $AB = 2m$. A ring of mass $3m$ is attached at C. The end A is fixed and the ring at C is free to move on a smooth vertical wire passing through A. The system rotates with constant angular velocity ω about the wire, with C below A. If the angle $BAC = 60$ degrees, show that $\omega^2 = g(4 + \sqrt{3/2})$ and find the reaction between the ring and the wire. (London)†

20. A, B and C are three points in order in a vertical straight line with A as the lowest point, $AB = 2a$ and $BC = \frac{1}{2}a$. A thin smooth wire $ADBEC$ passes through the three points and lies in one plane. ADB is a semi-circle on AB as diameter and BEC is a semi-circle on BC as diameter, D and E being on opposite sides of ABC. A small bead is threaded on the wire and is projected from A with speed $\sqrt{(5ag)}$. Show that the reaction between the wire and the bead equals the weight of the particle at three points on the wire and find the heights of these points above A. (WJEC)

21. A parcel rests on the horizontal back seat of a car, its coefficient of friction with the seat being 3/4. The car when travelling on a level road at 54 km/h is brought to rest by a uniform application of the brakes. Show that the parcel will not slide forwards on the seat if the stopping distance is more than about 15.3 m.

The car later rounds a circular bend on the road at a steady speed of 54 km/h. Show that the parcel will shift towards the side of the car if the radius of the bend is less than about 30·6 m. (London)†

22. A particle of mass 1 kg is attached to one end of a light inelastic string of length 0·5 m, the other end of which is attached to a fixed point O. The particle is held at a point on the same horizontal level as O and 0.4 m from it and is then released. Find the impulsive tension in the string when the string becomes taut.

If the string can sustain this impulsive tension find the speed of the particle and the tension of the string at the moment when the particle is vertically below O. (London)†

EXERCISES

23. A heavy particle is attached to a fixed point O by a light inextensible string of length a. When the particle is at rest vertically below O it is given a horizontal velocity u. In the ensuing motion the string becomes slack and then subsequently becomes taut again at the instant when it is horizontal. Prove that $u^2 = ga(2 + \frac{3}{2}\sqrt{3})$.
(Oxford)

24. A particle slides from rest at a point at an angular distance α from the highest point of a fixed smooth sphere of radius r. Show that while the particle is still in contact with the sphere the reaction between them is $mg(3\cos\theta - 2\cos\alpha)$, where θ is the angular distance of the particle from the highest point of the sphere. If $\cos\alpha = \frac{3}{4}$, show that the particle will leave the sphere when its angular distance from the highest point is 60 degrees. Show also that, in the ensuing motion, when the particle is at a distance $r\sqrt{3}$ from the vertical diameter of the sphere its depth below the centre of the sphere is $4r$. (J.M.B.)

25. A small bead P of mass m is threaded on a thin smooth wire bent into the shape of a circle of radius a fixed in a vertical plane. The lowest point of the circle is A. The bead is acted on by gravity and also by a force along the chord $\overrightarrow{PA}$ of magnitude mg/a times the length of the chord. Show that the motion of the particle is the same as if gravity alone were acting, but with g increased to $2g$.

If the particle is describing complete circles and its reaction on the wire at the top of the circle is outwards and one half its reaction on the wire at A, show that the speed V of the particle at A is given by $V^2 = 23ag$. (WJEC)

26. A particle of mass m, suspended from a point O by an inextensible string of length l, is projected horizontally from its lowest position with a speed $\sqrt{(2gh)}$. Find the greatest tension in the string.

If the string becomes slack during the subsequent motion, show (a) that $h > l$, (b) that slackness first occurs when the particle is at a point A which is higher than O by $\frac{2}{3}(h - l)$, (c) that the greatest height above A attained by the particle in the subsequent motion is $(h - l)(l + 2h)(5l - 2h)/27l^2$. (JMB)

27. A particle P is attached to two fixed points A and B in the same horizontal line by means of two light inextensible strings. It is projected with a velocity just sufficient to make it describe a circle in a vertical plane without the strings becoming slack, the angle PAB being α. When the particle is at its lowest point, the string BP breaks and the particle then describes a circle in a horizontal plane. Prove that $\cot\alpha = \sqrt{5}$. (Oxford)

28. One end of a light inelastic string of length a is attached to a fixed point O and the other end is attached to a particle P of mass

m. P is held with the string taut and horizontal and is then released. When the string becomes vertical it begins to wrap itself around a small smooth peg A at a depth b below O. Find the tension in the string when AP subsequently makes an angle θ with the downward vertical and show that the particle makes complete revolutions about A if $b > \frac{3}{5}a$. (WJEC)

29. A particle, of mass m, is suspended at one end B of a light inextensible string AB, of length $2a$, the other end of which is fixed. The particle is drawn aside, the string being taut, until AB makes an angle $\cos^{-1}(\frac{1}{4})$ with the downward vertical through A and the particle is then released from rest. When B has risen to a height such that AB makes an angle 45 degrees with the downward vertical, the mid-point of the string comes into contact with a small fixed horizontal peg C.

Determine (a) the speed of the particle at the moment the string first touches the peg, (b) the tension in the string at the moment when part of it is horizontal. (London)

30. A particle lies at rest on the smooth interior of a sphere of internal radius a. What is the least speed with which it must be projected horizontally to describe complete circles on the inside of the sphere?

If it is projected with speed $\sqrt{4ga}$, find the greatest height (above the base of the sphere) reached in the subsequent motion.

31. (a) An artificial satellite is moving in a circular orbit of radius R, with angular velocity ω, under the action of the Earth's gravitational field. Assuming that the force due to gravity varies as the inverse square of the distance from the centre of the Earth, find an expression for R in terms of ω, a and g, where a is the radius of the Earth and g is the acceleration due to gravity at the surface of the Earth. (C, part)

32. A small bead of mass m is suspended from a fixed point O by a light inelastic string. The point O is at a height $4a$ above a smooth horizontal table and the the bead moves on the table in a horizontal circle of radius $3a$ with constant speed $\sqrt{(2ga)}$. Find the reaction between the bead and the table.

Find also the reaction in the case when the beam moves in the same circle with the same speed as before, but is, instead, threaded on a smooth string of length $8a$ whose ends are fastened at O and at the point N on the table vertically below O. (C)

33. A smooth narrow tube is bent into the form of a circle, centre O and radius a, fixed in a vertical plane. Two particles A and B, of masses m and λm ($\lambda > 1$) respectively, are connected by a light inextensible string of length πa. The particles are initially at rest

EXERCISES

in the tube at opposite ends of the horizontal diameter. The string is taut and occupies the upper half of the tube.

The particles are released and after time t the diameter AOB makes an angle θ with the horizontal. Given that the string remains taut during the motion,

(i) show that $a\dot\theta^2 = \dfrac{2(\lambda - 1)g \sin \theta}{(\lambda + 1)}$

(ii) find in terms of θ
    (a) the force exerted by the tube on A,
    (b) the tension in the string. (AEB)

PART II—RIGID BODIES

A rigid body is one whose size and shape are, for all practical purposes, invariable. It will be regarded as a collection of particles very close together such that the distance between any two of them is invariable. For certain purposes however, e.g. finding centres of gravity, a rigid body will be treated as if it were a continuous distribution of matter.

16

MOTION OF A SYSTEM OF PARTICLES

16.1. CENTRE OF MASS

Definition—The *centre of mass* of a system of n particles, of masses $m_1, m_2, m_3, \ldots, m_n$, at points whose position vectors are $r_1, r_2, r_3, \ldots, r_n$ respectively, is the point whose position vector $\bar{r}$ is given by

$$\bar{r} = \frac{m_1 r_1 + m_2 r_2 + m_3 r_3 + \cdots + m_n r_n}{m_1 + m_2 + m_3 + \cdots + m_n}$$

i.e. $\bar{r} = \Sigma mr/\Sigma m$.

Its position is independent of the origin chosen which can be shown as follows:

Suppose its position is not independent of the origin and let G, G' be the centres of mass of a given system referred to origins O, O' respectively. Then $\overrightarrow{OG} = \Sigma mr/\Sigma m$ and $\overrightarrow{OG'} = \Sigma mr'/\Sigma m$, where r, r'

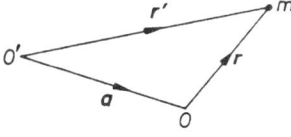

Figure 16.1

are the position vectors relative to O, O' respectively, of a typical member of the system. Then if $\overrightarrow{O'O} = a$ (say), it follows that $\overrightarrow{O'G} = a + \overrightarrow{OG}$ and $r' = a + r$ (refer to *Figure 16.1*). By definition

$$\overrightarrow{O'G'} = \Sigma mr'/\Sigma m$$
$$= \Sigma m(a + r)/\Sigma m$$
$$= (a\Sigma m/\Sigma m) + (\Sigma mr/\Sigma m)$$
$$= a + \overrightarrow{OG}$$
$$\overrightarrow{O'G'} = \overrightarrow{O'G}$$

315

MOTION OF A SYSTEM OF PARTICLES

i.e. G' and G have the same position vector relative to O' and are therefore the same point and the position of the centre of mass is independent of origin.

It will be shown later that the centre of mass coincides with the point through which the total weight of the system acts—its centre of gravity.

Example. *Masses of* 1, 2, 5 *and* 2 kg *lie in the plane* XOY *at points whose coordinates are* $(0, 1), (0, -1), (1, 2)$ *and* $(2, 2)$ *respectively. Find the coordinates of their centre of mass.*

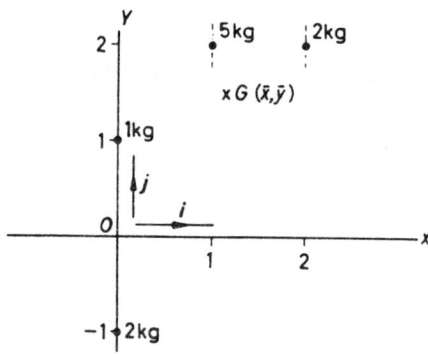

Figure 16.2

Taking unit vectors i, j parallel to OX, OY respectively (refer to Figure 16.2), the position vector of a point (x, y) will be $xi + yj$. Then, from our definition, the position vector of the centre of mass will be given by

$$\bar{r} = \Sigma mr / \Sigma m$$

$$\therefore \quad \bar{x}i + \bar{y}j = \frac{1(j) + 2(-j) + 5(i + 2j) + 2(2i + 2j)}{1 + 2 + 5 + 2}$$

$$= (9i + 13j)/10.$$

$$\therefore \quad \bar{x} = \tfrac{9}{10} \text{ and } \bar{y} = \tfrac{13}{10}.$$

The centre of mass of the system is at the point $(\tfrac{9}{10}, \tfrac{13}{10})$.

Exercises 16a

1. Four particles of mass $2m, m, m$ and $2m$ are placed at points whose position vectors are $a, 2a, b$ and $2b$ respectively. Find the position vector of their centre of mass.

2. Three particles of masses 1, 2 and 3 kg are situated at points whose position vectors are $-i + 2j + k$, $2i + 3k$ and $i - 2j + k$ respectively. Find the position vector of their centre of mass.

3. Four particles of masses 2, 2, 1 and 1 g are placed in a plane at the points whose cartesian coordinates are $(0, 3a)$, $(2a, 3a)$, $(2a, 0)$, and $(0, 0)$ respectively. Find the coordinates of their centre of mass.

If the masses of all four particles were doubled, what effect would this have on the position of their centre of mass?

4. Show that the centre of mass of three equal particles at the vertices of a triangle is at the intersection of the medians of the triangle.

5. A system of particles is in motion. A typical particle has mass m and position vector r, where r is a vector function of time t. Find a formula for the velocity of their centre of mass.

A bolas consists of three equal balls, each of mass m, connected by three light inextensible strings to the same point. The bolas is thrown so that the three balls always lie in the same horizontal plane. At a given instant the balls have velocities $40i + 60j$, $125i + 3j$, $-6i + 75j$ units respectively, where i and j are perpendicular horizontal unit vectors. Find the velocity of the centre of mass of the bolas at this instant.

6. A typical particle of a moving system has mass m and velocity v' *relative to the centre of mass*. Show that $\Sigma m v' = \mathbf{0}$.

16.2. MOTION OF THE CENTRE OF MASS

Consider again a set of particles such that a typical particle of mass m is instantaneously at a point whose position vector is r. Let the resultant force on this particle be P, so that

$$P = m\, d^2r/dt^2.$$

Adding vectorially for all such particles,

$$\Sigma P = \Sigma(m\, d^2r/dt^2) \qquad \ldots \text{(i)}$$

But if $\bar{r}$ is the position vector of the centre of mass of the system $\bar{r}\Sigma m = \Sigma mr$, and differentiating twice with respect to time

$$\frac{d^2\bar{r}}{dt^2}\Sigma m = \Sigma\left(m\frac{d^2r}{dt^2}\right) \qquad \ldots \text{(ii)}$$

Combining (i) and (ii)

$$\boxed{\Sigma P = \frac{d^2\bar{r}}{dt^2}\Sigma m}$$

MOTION OF A SYSTEM OF PARTICLES

Thus the vector sum of all the forces acting on the particles is equal to the product of the total mass and the acceleration of their centre of mass.

Now any internal forces between the masses (attractions, repulsions, tensions in strings, reactions etc.) will, by Newton's third law, occur in equal and opposite pairs. Hence only external forces will contribute to this vector sum ΣP, and equation (iii) expresses the important result that *the centre of mass of a system of particles moves as if it were a particle of mass Σm acted upon by all the external forces exerted on the system.*

Note that this justifies our earlier treatment of cars, engines, boxes etc. as if they were particles.

Example. A rigid framework of light rods forms a square ABCD. Masses of 1, 3, 2 and 4 kg are fastened to ABCD respectively, and the framework lies on a smooth horizontal surface. A force of $5\sqrt{2}$ N acts at A along AC, 6 N at B along BC, $6\sqrt{2}$ N at C along CA and 6 N at D along DA. Find the magnitude and direction of the acceleration of the centre of mass of the framework.

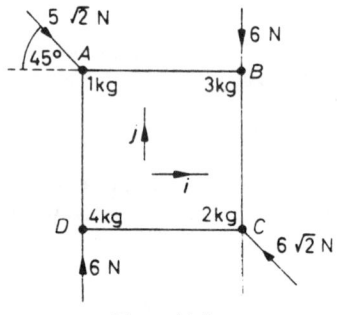

Figure 16.3

Reactions between the particles and the horizontal surface will arise so that the weights are counterbalanced and the vector sum of vertical forces is zero (and the vector sum of the internal stresses in the rods will also be zero). Hence, taking i, j as unit vectors parallel to $\overrightarrow{DC}$ and $\overrightarrow{DA}$ (refer to *Figure 16.3*), the vector sum of all the forces acting on the system is

$$\Sigma P = (5i - 5j) - 6j + (-6i + 6j) + 6j$$
$$= -i + j \text{ N}.$$

MOTION OF THE CENTRE OF MASS

The total mass $\Sigma m = 10$ kg.

Since
$$\Sigma P = (d^2\bar{r}/dt^2)\Sigma m$$
$$\therefore \quad -i + j = (d^2\bar{r}/dt^2)10$$
$$\therefore \quad d^2\bar{r}/dt^2 = \tfrac{1}{10}(j - i) \text{ m/s}^2.$$

The centre of mass of the framework has an acceleration of magnitude $\tfrac{1}{10}\sqrt{2}$ m/s² in a horizontal direction making an angle of 45 degrees with $\overrightarrow{CD}$ and $\overrightarrow{DA}$.

The acceleration of the centre of mass is clearly not affected by the lines of action of the forces nor by the distribution of the masses. Their motion *around* the centre of mass will, however, depend on these factors as will be seen later.

Conservation of momentum—Equation (i) at the beginning of this section can be stated in the form

$$\Sigma P = \Sigma(d(mv)/dt)$$

or
$$\Sigma P = d(\Sigma mv)dt.$$

If $\Sigma P = 0$, this gives $d(\Sigma mv)/dt = 0$ and hence $\Sigma(mv) = $ const.

This is the principle of conservation of linear momentum that if the vector sum of the external forces acting on a system is zero, its total momentum remains constant. Considerable use of this principle has been made in Chapter 13 (refer to Section 13.4 et seq).

Exercises 16b

1. Three forces F_1, F_2, F_3 act on a system of particles of total mass 2 kg. If $F_1 = 2i - 3j + k$, $F_2 = i + 4j - k$ and $F_3 = -i + j$ N find the acceleration they impart to the centre of mass.

2. Three particles are rigidly connected by light rods to form an equilateral triangle ABC. Particle A is a 5 g mass, B an 11 g mass and C a 9 g mass. The system lies at rest on a smooth horizontal table. If a force of 4 N is applied at A along AB and a force of 3 N at B perpendicular to AB (towards C), find the acceleration given to the centre of mass.

What would this acceleration have been if both forces had been applied at the mid-point of AB?

3. Two particles each of mass m lie at rest at points A, B on a smooth horizontal surface. A force of magnitude k mg is applied to A in a direction making an angle α with AB. Find the acceleration of the particle at A, and of the centre of mass of the system consisting of the two particles.

MOTION OF A SYSTEM OF PARTICLES

Would these accelerations have been the same if A and B had been connected by a light rod?

4. A man finds himself standing on very smooth ice unable to move. Use the principle of conservation of momentum to explain how he can set himself in motion by throwing away a cigarette packet.

5. A shell explodes into four pieces of masses 10, 50, 70 and 30 kg moving with velocities $6i + 16j + 27k, 20i + j - 30k, 14i - 3j + 17k$ and $-68i - 26k$ m/s respectively. Find the original velocity of the shell. What is the velocity of the centre of mass after the explosion?

16.3. MOTION ABOUT A FIXED POINT

We shall begin by discussing the moment of a force (or torque) and the moment of momentum (or angular momentum) of a system of particles.

As defined in Section 3.10, the moment of a force F about a point O is $r \times F$, where r is the position vector relative to O of a point on its line of action. This is also sometimes called the *torque* about O.

Again as in Section 3.10 $\hat{a} \cdot (r \times F)$ is the moment of F about an axis parallel to $\hat{a}$ and through O.

$$\text{SI unit of torque} \ldots \text{N m.}$$

Similarly, the moment of momentum of a particle about O is $r \times mv$, where m is its mass, v its velocity and r the position vector relative to O of a point on the line of action of v. This is also called the *angular momentum* of the particle about O.

$\hat{a} \cdot (r \times mv)$ is the angular momentum of the particle about an axis through O parallel to $\hat{a}$.

$$\text{SI unit of angular momentum} \ldots \text{kg m}^2/\text{s.}$$

We can gain some insight into the meaning of angular momentum in the following way.

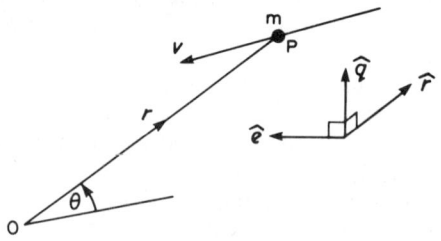

Figure 16.4

MOTION ABOUT A FIXED POINT

Suppose we wish to find the instantaneous angular momentum of a particle P about some point O (refer to *Figure 16.4*), and let $\omega (= d\theta/dt)$ be the instantaneous rate of rotation of OP. Then by the result of Section 3.7, $d\hat{r}/dt = \omega\hat{e}$ where $\hat{e}$ is a unit vector at right angles to *r* as shown.

Hence angular momentum $r \times mv = r \times m[dr/dt]$
$$= r \times m[d(r\hat{r})/dt]$$
$$= r \times m[(dr/dt)\hat{r} + r(d\hat{r}/dt)]$$
$$= r \times mr\omega\hat{e} \text{ since } (r \times \hat{r} = 0)$$
$$= mr^2\omega\hat{q}$$

where $\hat{q} = \hat{r} \times \hat{e}$.

Thus the magnitude of the angular momentum depends on the *mass*, the *distance from O* and the *angular speed* around O.

The angular momentum of a system of particles is defined as the vector sum of their individual angular momenta, i.e. $\Sigma(r \times mv)$.

We shall now obtain a relation between torque and rate of change of angular momentum for a system of particles analogous to the relation $(P = d(mv)/dt)$ between force and rate of change of momentum for a particle.

Consider a system of particles such that a typical particle has mass *m*, position vector *r* with respect to a fixed point O, and velocity *v*. Then if the resultant force on the particle is *P*, $P = d(mv)/dt$.

Hence $r \times P = r \times d(mv)/dt$

$\therefore \quad r \times P = d(r \times mv)/dt.$* $\left(\text{since }(dr/dt) \times v = v \times v = 0\right).$

Adding vectorially for all such particles,

$$\Sigma(r \times P) = \Sigma[d(r \times mv)/dt]$$
$\therefore \quad \Sigma(r \times P) = d(\Sigma[r \times mv])/dt. \qquad \ldots\text{(i)}$

Now the internal forces (if any) occur in equal and opposite pairs in the same line. Hence the sum of their moments about any point is zero. Thus equation (i) states that:

The sum of the moments of the external forces about a fixed point O, is equal to the rate of change of angular momentum about O.

---
* Since, as we have seen, $r \times mv$ is linked with the angular speed of the particle, the above equation shows that its angular acceleration depends on the moment of *P*. This gives some justification for regarding the moment of a force as its "turning effect".

MOTION OF A SYSTEM OF PARTICLES

A similar relation can be obtained for moments about an axis. Form the scalar product of both sides of equation (i) with $\hat{a}$. Then

$$\hat{a} \cdot \Sigma(r \times P) = \hat{a} \cdot d(\Sigma[r \times mv])/dt$$

$$\therefore \quad \Sigma[\hat{a} \cdot (r \times P)] = d(\Sigma[\hat{a} \cdot r \times mv])/dt.$$

i.e. *the sum of the moments of the external forces about a fixed axis is equal to the rate of change of angular momentum about that axis.*

If the sum of the moments of the external forces is zero, it follows that the angular momentum of the system remains constant. This is the *principle of conservation of angular momentum.*

Exercises 16c

1. The line of action of a force $2i - j + k$ N passes through the point whose position vector is $i + j$ metres, and the line of action of a force $i + 2j + k$ N passes through the point $j + k$ metres. Find the vector sum of the moments of these two forces about the point $i + j + k$ metres.

2. *ABCDEF* is a regular hexagon of side *a*. Forces of magnitude 1, 2, 3, 4 and 2 units act along *AB, BC, DC, DE* and *DB* respectively. Find the sum of their moments about an axis through the centre of the hexagon perpendicular to its plane.

3. Three particles *A, B, C* are each of mass 3 kg. At a certain instant *A* is at the point with position vector $2i + j - k$ m moving with speed $3i - 2j + 4k$ m/s. *B* is at $4i + j - 2k$ m moving at $3i + k$ m/s and *C* at $2i + k$ m moving at $3i - j + k$ m/s. Find the angular momentum of the system about the origin.

4. A light rod *AB* has a particle of mass *m* attached at *A* and another at *B*. The rod is made to rotate with uniform angular speed ω about an axis through *C*, a point which divides *AB* internally in the ratio of 1:2. If $AB = 3a$, find the angular momentum of the system about the axis of rotation.

What can be deduced about the moments of the external forces acting on this system?

5. Two particles *A, B*, each of mass 2 units are at rest at a point *O*. Variable forces F_1, F_2 are applied to *A, B* respectively, such that after time $t, F_1 = 4ti + 2j$ and $F_2 = 4i + 4tj - 2tk$ units. Find the velocity and displacement from *O* of each particle at time *t*.

Find also the angular momentum of the system at this time and verify that the sum of the moments of the forces is equal to the rate of change of momentum.

6. A thin uniform lamina rotates in its own plane about a point *O* in the lamina which is fixed. Show that its angular momentum at any instant has magnitude $k\omega$, where k is a constant and ω is the

MOTION RELATIVE TO THE CENTRE OF MASS

instantaneous rate of rotation. Hence, show that its angular acceleration is directly proportional to the magnitude of the torque of the external forces about O.

7. A wheel, rotating freely about its axis, is brought to rest in time T by the application of two brake blocks to its surface. These blocks produce constant forces of magnitude F acting tangentially to the rim of the wheel. If the disc is of radius a, find the initial angular momentum of the wheel about its axle.

8. A top consists of a flat disc spinning uniformly about a fixed vertical axis. A rough ring is lowered gently on to the disc and released so that the two rotate together about the same axis. Discuss any changes that take place in (a) the angular momentum of the combined system (b) the angular velocity of the top.

16.4. MOTION RELATIVE TO THE CENTRE OF MASS

Consider a system of particles whose centre of mass G has position vector $\bar{r}$ relative to a fixed point O; and let a typical particle have position vectors r, r' relative to O, G respectively. (Refer to *Figure 16.5*.) Then from triangle OQG, $r = \bar{r} + r'$.

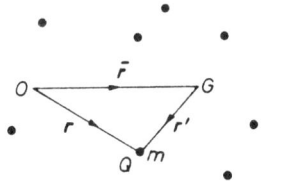

Figure 16.5

Now, from the result obtained in the previous section,

$$\Sigma(r \times P) = \frac{d}{dt}(\Sigma[r \times mv])$$

$$\therefore \quad \Sigma[(\bar{r} + r') \times P] = \frac{d}{dt}(\Sigma[(\bar{r} + r') \times mv])$$

$$\therefore \quad \Sigma(\bar{r} \times P) + \Sigma(r' \times P) = \frac{d}{dt}(\Sigma[\bar{r} \times mv]) + \frac{d}{dt}(\Sigma[r' \times mv])$$

$$\therefore \quad \bar{r} \times \Sigma P + \Sigma(r' \times P) = \bar{r} \times \Sigma\frac{d(mv)}{dt} + \frac{d}{dt}(\Sigma[r' \times mv]).$$

MOTION OF A SYSTEM OF PARTICLES

But $$\Sigma P = \Sigma \frac{d(mv)}{dt}$$

(refer to Section 16.2)

so that $$\bar{r} \times \Sigma P = \bar{r} \times \Sigma \frac{d(mv)}{dt}$$

∴ $$\Sigma(r' \times P) = \frac{d}{dt}(\Sigma[r' \times mv]).$$

Again internal forces "cancel out" and this equation (similar to equation (i) of Section 16.3) states that *the sum of the moments of the external forces about the centre of mass is equal to the rate of change of angular momentum about the centre of mass.*

From the scalar product of both sides of our equation with $\hat{a}$, we can show that

$$\Sigma[\hat{a} \cdot (r' \times P)] = d(\Sigma[\hat{a} \cdot r' \times mv])/dt.$$

Thus a similar relationship holds between moments about an axis through G and angular momentum about that axis.

Also the kinetic energy of a system of particles may be expressed as the sum of the K.E. of a particle of mass Σm moving with G and the K.E. of the system relative to G.

For since $r = \bar{r} + r'$, then $v = \bar{v} + v'$.

The total K.E. $= \Sigma(\tfrac{1}{2}mv^2) = \Sigma(\tfrac{1}{2}m[\bar{v} + v']^2)$

$\phantom{\text{The total K.E. }} = \Sigma(\tfrac{1}{2}m\bar{v}^2) + \Sigma(m\bar{v} \cdot v') + \Sigma(\tfrac{1}{2}mv'^2)$

$\phantom{\text{The total K.E. }} = \tfrac{1}{2}\bar{r}^2\Sigma m + \bar{v} \cdot \Sigma(mv') + \Sigma(\tfrac{1}{2}mv'^2).$

But $\Sigma(mv') = 0$, since $\Sigma(mv')/\Sigma m = v' = 0$.

∴ Total K.E. $= \tfrac{1}{2}\bar{v}^2\Sigma m + \Sigma(\tfrac{1}{2}mv'^2).$

Summary

In this chapter three important relationships have been obtained:

$$\Sigma P = (d^2\bar{r})/(dt^2)\Sigma m \quad \ldots\text{(i)}$$

and $$\Sigma(r \times P) = d\Sigma(r \times mv)/dt \quad \ldots\text{(iia)}$$

or $$\Sigma(r' \times P) = d\Sigma(r' \times mv)/dt \quad \ldots\text{(iib)}$$

(these last two relationships being also true for moments about an axis).

EXERCISES

Hence the motion of a system of particles depends on the vector sum of the external forces, and on the sum of their moments about a fixed point (or about the centre of mass). Indeed, the motion of a *rigid body*, whose particles are fixed relative to one another, is determined by these quantities.

In the succeeding chapters we shall make extensive use of these relations; sometimes taking moments about a fixed point (Chapters 17 and 21), sometimes about a fixed axis (Chapter 19) and sometimes about an axis through the centre of mass (Chapter 20).

EXERCISES 16

1. Seven equal masses are placed one at each of seven corners of a cube of side $2a$. Find the position of their centre of mass relative to the eighth corner O.

2. The point with position vector R_1 is the centre of mass of a set of particles each of mass m, while the point with position vector R_2 is the centre of mass of a second set of particles each of mass λm. Show that the centre of mass of the combined set has position vector

$$\frac{\lambda R_1 + R_2}{1 + \lambda}.$$

3. Two particles of masses 3 kg and 2 kg are moving so that at time t their velocities are $t\hat{a}$ and $2t^2\hat{b}$ m/s respectively, $\hat{a}$, $\hat{b}$ being constant unit vectors. Find the velocity of their centre of mass at time t and deduce the vector sum of the forces acting on them when $t = 2$ seconds.

4. A shower of n particles of masses $m, 2m, 3m, \ldots, nm$ is falling vertically under gravity. A horizontal cross wind exerts on each one a force of magnitude $\lambda(\text{mass})^2$. Find the horizontal and vertical components of the acceleration of their centre of mass. (Neglect air resistance.)

5. A moving particle P collides with and scatters a collection of three equal particles. After the collision, the first particle is brought to rest and the other three have velocities $-2i + j$, $4i + 2j$ and $i - 5j$ m/s respectively. Find the velocity of the centre of mass of the four particles immediately after the collision. If each of the three particles has a mass of 1 kg, determine the momentum of the first particle just before the collision.

6. A particle slides down the smooth face of a wedge which is itself free to move on a smooth horizontal surface. Without considering the forces between the particle and the wedge, show that the wedge must move.

Would this still be true if the face of the wedge in contact with the particle were rough.

7. Three particles of equal mass have position vectors $(5 + 3t)i + (2 - 5t)j$, $(-4ti + 2t)j$ and $(1 + t)i + (7 + 3t)j$ at time t. Find the position of their centre of mass. Deduce the total linear momentum of the system.

Is the angular momentum of the system constant?

8. Two particles of masses 3 kg and 2 kg move so that at time t their displacements from a fixed point O are $2i + 4tj$ and $(t^2 - 2)i + 4tj$ metres respectively. At the instant when $t = 5$ s find (a) the velocity of their centre of mass, (b) the angular momentum of the system (c) the vector sum of the forces acting on them (d) the vector sum of the moments of these forces about O.

9. A particle of mass m moving in a plane has coordinates (x, y) at time t. Show that its angular momentum A about a perpendicular axis through the origin is given by

$$A = m\left(x\frac{dy}{dt} - y\frac{dx}{dt}\right).$$

10. A cylinder of radius 0·1 m is free to rotate about its axis. A long string is wrapped round the cylinder and then pulled to make it rotate (the string does not slip). The cylinder starts from rest and the force in the string has the value $(3 + 2t)$ N after t seconds. Find the angular momentum about its axis, acquired by the cylinder in 2 seconds.

11. A top consists of a uniform disc spinning about a vertical axis. The top of the disc is rough. If a particle is gently dropped on the disc and carried round with it, how does this affect the angular speed of the top?

Does the change in angular speed depend on the distance of the particle from the axis?

12. A sphere rolls without slipping down an inclined plane. By considering its angular momentum about an axis through its centre of mass, show that it cannot move at uniform speed.

13. Three particles each of mass m have velocities $7i + 12j$, $-3i + 6j$, $2i - 3j$ cm/s respectively. Find the velocity of their mass centre G.

Hence find the velocities of the particles relative to G and verify that the total kinetic energy of the system is equal to that of a particle of mass $3m$ with the velocity of G together with the sum of the kinetic energies of the three particles relative to G.

14. Three particles, each of mass m, are attached to points A, B, C on a light rod of length $2a$. A and C are the ends of the rod and B

EXERCISES

the mid-point. The rod moves so that B has a uniform speed v and the rod rotates about B with angular speed ω. Find (a) the kinetic energy of the system (b) the angular momentum of the system about B.

What can be said about the forces acting on the system?

15. A top is spinning about a vertical axis on a smooth horizontal surface. A tangential horizontal frictional force is applied to its rim. What effect does this have on the top?

16. The velocity of the mass-centre of two particles, of masses m_1, m_2, moving in the same straight line is V and their relative velocity is V'; prove that the total kinetic energy of the two particles is

$$\tfrac{1}{2}MV^2 + \tfrac{1}{2}M'V'^2,$$

where $\qquad M = m_1 + m_2 \quad \text{and} \quad MM' = m_1 m_2.$

A shell of mass M, travelling with velocity V, is broken into two fragments of masses m_1, m_2 by an explosion which increases the kinetic energy by an amount E. Show that, if the two fragments initially move in the same line as the shell, their smallest relative velocity occurs when their masses are equal. (WJEC)

17. Define the vector product $a \times b$ of two vectors a, b and state what conclusions follow from $a \times b = 0$. The vector angular momentum H of a particle about the fixed point O is defined as the moment about O of its linear momentum. Write down an expression for H in terms of r, v and m, the position vector, velocity and mass, respectively, of the particle.

Show that the rate of change of angular momentum about O is equal to the moment about O of the resultant force on the particle. Show also that if the particle moves with constant angular speed ω in a circle about O, then $H = I\Omega$ where $I = mr^2$ and Ω is a vector of magnitude ω, perpendicular to both v and r. (JMB)

18. A system of n particles is confined to a plane Oxy, where Ox, Oy, Oz form a right-handed system of mutually perpendicular axes with origin O. The ith particle has mass m_i, position vector r_i (magnitude r_i) and is acted on by an external force F_i acting in the plane Oxy ($i = 1, 2, \ldots, n$). In addition there are internal interactions between the particles, the force of the jth particle on the ith being F_{ij} ($j = 1, 2, \ldots, n$), and F_{ii} is taken to be zero.

(i) Show that the sum of the moments of the two forces F_{ij} and F_{ji} about O is zero.
(ii) State what can be deduced about the sum of the moments of *all* the internal forces about O.

(iii) If L is the sum of the moments of *all* the external forces about O, show that

$$L = \frac{d}{dt}\left[\sum_i m_i r_i \times \dot{r}_i\right].$$

Consider the particular case in which each particle is moving in a circle, centre O, with the same variable angular speed ω in the sense Ox to Oy. Show that
(iv) at any instant $|\dot{r}_i| = \omega r_i$,
(v) $|r_i \times \dot{r}_i| = r_i^2 \omega$,
(vi) $L = I\omega k$ where I is the moment of inertia of the system about Oz and k is a unit vector along Oz. (JMB)

17

EQUIVALENT SYSTEMS OF FORCES

17.1. MEANING OF EQUIVALENCE

WE have seen, in the previous chapter, that the motion of a rigid body depends on the vector sum of the external forces acting on it, and on the vector sum of their moments. Accordingly we take two sets of localized forces $P_1, P_2, P_3 \ldots$ and $Q_1, Q_2, Q_3 \ldots$ to be *equivalent* if

$$\Sigma P = \Sigma Q$$

and
$$\Sigma(r \times P) = \Sigma(s \times Q)$$

where r, s are the position vectors of points on the lines of action of P, Q respectively, relative to some fixed point.

Example. The lines of action of two forces $F_1 = i + j - k$, $F_2 = j$ units pass through the points $(0, 0, 1)$ and $(1, 1, 1)$ respectively. Find the force, P which, together with a force $i + j$ units through the origin, forms a system equivalent to F_1 and F_2. Find also the point in which the line of action of P cuts the plane $z = 0$.

Equating the vector sums of the two systems,

$$(i + j - k) + j = (i + j) + P$$

$$\therefore \quad P = j - k$$

Let the line of action of P cut the plane $z = 0$ at the point $(a, b, 0)$. Then taking moments about the origin.

$$0 + (ai + bj) \times (j - k) = k \times (i + j - k) + (i + j + k) \times j$$

$$\therefore \quad ak + aj - bi = (j - i) + (k - i)$$

$$\therefore \quad -bi + aj + ak = -2i + j + k,$$

giving $b = 2$ and $a = 1$.

P is a force $j - k$ units passing through the point $(1, 2, 0)$.

17.2. COUPLES

A system of forces of particular interest consists of two forces equal in magnitude, opposite in direction and not in the same straight line. (Refer to *Figure 17.1*.) Such a system is called a *couple*.

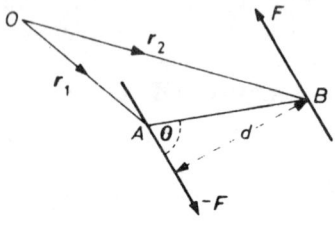

Figure 17.1

Since the vector sum of the two forces is zero, a couple can have no effect on the motion of the centre of mass of a body. However, the sum of the moments of the two forces cannot be zero so there will always be a turning effect.

A couple has the important property that the sum of the moments of the two forces is the same about all points. To prove this let A, B be points on the line of action of the two forces, and r_1, r_2 respectively their position vectors with respect to some origin O (refer to *Figure 17.1*). Then G the sum of the moments of the two forces is given by

$$G = (r_1 \times -F) + (r_2 \times F)$$
$$= (r_2 - r_1) \times F$$
$$= \overrightarrow{AB} \times F.$$

Thus G is independent of the position of O and $|G| = AB \sin \theta |F|$, i.e. the magnitude of G is the product of the magnitude of one force and the perpendicular distance d between them. G is called the moment of the couple.

There will be many different pairs of equal and opposite forces forming couples of the same moment G (say). They will all be equivalent to one another. For this reason a couple is usually referred to only by its moment, the constituent forces not being specified.

A set of couples G_1, G_2, G_3 ... can always be replaced by a single couple G such that

$$G = G_1 + G_2 + G_3 \dots$$

COPLANAR SYSTEMS

17.3. REPLACEMENT BY A SINGLE FORCE AND COUPLE

A simple equivalent system that can be found for *any* system of forces consists of a single force, whose line of action passes through a chosen point, and a couple. For we can, in general, find a force R such that $R = \Sigma P$, and, if the point chosen has position vector r', we can find a couple G such that $G + r' \times R = \Sigma(r \times P)$. (In special cases, of course, it may be that either or both of R and G are zero.)

Thus any system of forces may be replaced by a single force, acting through a given point, together with a couple.

Example. Forces λa, $2\lambda b$, $\lambda(c + b - a)$ act through the points with position vectors $0, 0, a - b$ respectively, a, b, c being constant vectors. Find an equivalent system consisting of a force acting at the point $(a + b)$ together with a couple.

Let the force be R and the couple of moment G. Then equating vector sums,
$$\lambda a + 2\lambda b + \lambda(c + b - a) = R$$
$$\therefore \quad R = \lambda(3b + c).$$

Taking moments about the origin,
$$0 + 0 + (a - b) \times \lambda(c + b - a) = (a + b) \times \lambda(3b + c) + G$$
$$\therefore \quad \lambda(a \times c + a \times b - b \times c + b \times a) = \lambda(a \times 3b + a \times c + b \times c) + G$$
$$\therefore \quad G = -\lambda(a \times 3b + 2b \times c)$$
$$\therefore \quad G = \lambda b \times (3a - 2c).$$

The required system is a force $\lambda(3b + c)$, acting through the point $a + b$, together with a couple of moment $\lambda b \times (3a - 2c)$.

17.4. COPLANAR SYSTEMS

The results of the preceding sections remain true, of course, for systems of forces all lying in the same plane, that is:

(a) If two systems are equivalent, the vector sums of the forces are equal, and the sums of moments about a fixed point are equal.

(b) Any system can be replaced by a single force, acting through a given point, together with a couple.

When the two systems of forces all lie in the same plane, the style of the solution can be modified. The vector sum condition can be met

EQUIVALENT SYSTEMS OF FORCES

by equating the sum of the resolved parts in two directions at right angles. Also, since the direction of all moments will be perpendicular to the plane of the forces, only their magnitudes need be considered—assigning + or − signs according as they have (say) anticlockwise or clockwise turning effects.

Example 1. *Forces of* 3, 4 *and* 8 N *act along the sides AB, BC, DA respectively of a rectangle ABCD.* $AB = 6$ m, $BC = 4$ m *and E is the mid-point of AB. Find the three forces acting along EC, CD and DE which are equivalent to the given forces.*

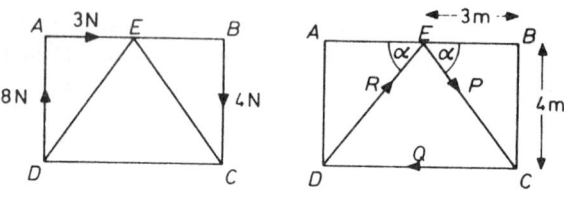

Figure 17.2

Resolving (refer to *Figure 17.2*),

parallel to DA $8 - 4 = R \sin \alpha - P \sin \alpha$, ...(i)
parallel to AB $3 = R \cos \alpha + P \cos \alpha - Q$.   ....(ii)

Taking moments about E (clockwise positive).

$$4 \times 3 + 8 \times 3 = Q \times 4 \qquad ....(iii)$$

From (iii) $Q = 9$ and substituting this value in (ii), we have $(P + R) \cos \alpha = 12$, i.e. since $\alpha = \tan^{-1} \frac{4}{3}$

$$P + R = 20.$$

Also from (i) $R - P = 5.$

Adding $2R = 25$

$\therefore$ $R = 12\frac{1}{2}$ and $P = 7\frac{1}{2}$.

The three forces required are $7\frac{1}{2}$, 9, and $12\frac{1}{2}$ N along EC, CD and DE respectively.

Example 2. ABCDEF is a regular hexagon of side 10 cm. *Forces of* 2, 4, 3 *and* 6 N *act along AB, BC, ED and FE respectively. Find the single force acting through the centre of the hexagon and the couple which together are equivalent to the given system.*

COPLANAR SYSTEMS

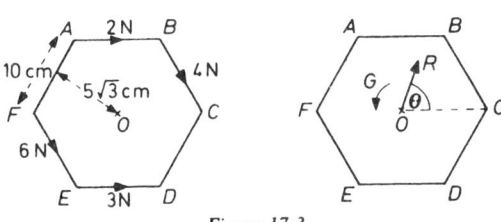

Figure 17.3

Let the required force and couple have magnitudes R and G, and let R be inclined at an angle θ to FC (refer to *Figure 17.3*).

Resolving,

parallel to EA $\qquad -4\sin 60° - 6\sin 60° = R\sin\theta \qquad$ .... (i)

parallel to AB $\quad 2 + 4\cos 60° + 3 + 6\cos 60° = R\cos\theta. \quad$ .... (ii)

Taking moments about O (anticlockwise positive),

$$-2 \times 5\sqrt{3} - 4 \times 5\sqrt{3} + 3 \times 5\sqrt{3} + 6 \times 5\sqrt{3} = G + O \dots \text{(iii)}$$

From (i) $\qquad R\sin\theta = -5\sqrt{3},$

and from (ii) $\qquad R\cos\theta = 10.$

Squaring and adding, $\quad R^2 = 75 + 100 \qquad \therefore \quad R = 5\sqrt{7}\,\text{N}.$

Dividing, $\qquad \tan\theta = -\dfrac{\sqrt{3}}{2}\,$ with $\sin\theta$ negative

$\therefore \qquad\qquad\qquad \theta = 360° - 40°\,54'$

$\therefore \qquad\qquad\qquad \theta = 319°\,6'.$

From (iii) $\qquad G = 5\sqrt{3}(-2 - 4 + 3 + 6)$

$\therefore \qquad\qquad G = 15\sqrt{3}\,\text{N cm} = 3\sqrt{3}/20\,\text{N m}.$

The required equivalent system consists of a force of $5\sqrt{7}\,\text{N}$, through O, making an angle of 40° 54' with OC on the same side as D together with an anticlockwise couple whose moment has magnitude $3\sqrt{3}/20\,\text{N m}$.

An additional property of a *coplanar* system of forces is that it can, in general, be replaced by a *single* equivalent force **R**. In this case, however, **R** cannot be made to act through a specified point. Its line of action is determined by the configuration of the system to which it is equivalent.

333

EQUIVALENT SYSTEMS OF FORCES

For we can, as before, find a force R such that $R = \Sigma P$. Then, since all the moments are perpendicular to the plane of the system, R can be positioned so that $r' \times R = \Sigma(r \times P)$, i.e. a point r' can be found such that the moment of R about the origin is equal to the sum of the moments of the forces about the origin. (Unless, of course, $\Sigma P = 0$, in which case the system is equivalent to a couple.)

Thus any system of coplanar forces can be replaced by a single force or a couple.

Example 3. *ABCD is a rectangle such that $AB = 5a$, $BC = 2a$. Forces of magnitude $2F, 6F, 4F$ and $5F$ act along the sides DA, BA, BC, DC respectively. Find the magnitude and direction of the single force equivalent to this system, and find the equation of its line of action referred to DC, DA as axes.*

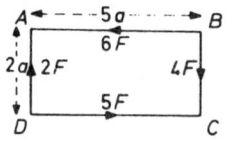

Figure 17.4

Let the single equivalent force be as shown in *Figure 17.4*. Resolving,

$$\text{parallel to } DA \quad 2F - 4F = R \sin \theta \quad \ldots \text{(i)},$$
$$\text{parallel to } AB \quad 5F - 6F = R \cos \theta \quad \ldots \text{(ii)}$$

Taking moments about D,

$$6F \times 2a - 4F \times 5a = R \sin \theta x \quad \ldots \text{(iii)}$$

From (i) $\quad R \sin \theta = -2F,$
and from (ii) $\quad R \cos \theta = -F.$
Squaring and adding, $\quad R^2 = 4F^2 + F^2 \quad \therefore \quad R = F\sqrt{5}$
Dividing, $\quad \tan \theta = 2$ with $\sin \theta$ negative
$\therefore \quad \theta \doteqdot 180° + 63° \, 26'$
$\therefore \quad \theta = 243° \, 26'.$

Substituting for $R \sin \theta$ in (iii),

$$-8Fa = -2Fx$$
$\therefore \quad x = 4a.$

334

COPLANAR SYSTEMS

Thus the line of action of R has slope 2 and passes through the point (4a, 0). Hence its equation is

$$y - 0 = 2(x - 4a)$$

or
$$y = 2x - 8a.$$

The required force is of magnitude $F\sqrt{5}$ making an angle of 63° 26′ with CD on the opposite side to A. It acts along the line $y = 2x - 8a$.

Note that it will be shown later that any system of *parallel* forces, coplanar or otherwise, can also be replaced by a single force or a couple (refer to Section 18.1).

Exercises 17a

1. The lines of action of two forces $F_1 = 2i$ and $F_2 = -i - k$ units pass through the points whose position vectors are k and $j + k$ respectively. Find the force P which, together with a force $-k$ units acting through the point whose position vector is j, forms a system equivalent to F_1 and F_2.

Show also that the line of action of P passes through the point $-j + k$.

2. A couple consists of two equal and opposite forces of 6 units acting parallel to Ox through the points (0, 0) and (0, 4). Find the magnitude and line of action of the force which, together with a force of 3 units parallel to the y-axis through (5, 1), forms an equivalent couple. Illustrate with a diagram.

3. i, j, k are unit vectors parallel to rectangular axes OX, OY, OZ respectively. Forces $i + j + k$, $i - 2j + k$ and $2i + j - k$ units act through the points (0, 0, 0), (1, 0, 1) and (0, -1, 1) respectively. Find the single force through the origin and the couple which are together equivalent to this system.

4. ABCDEF is a regular hexagon. Forces of magnitude $2F, 4F, 2F, 3F$ and $4F$ act along BA, BC, DC, DE and FE respectively. Find the three forces acting along AB, BD and DA which are equivalent to the given system.

5. ABCD is a rectangle with $AB = 4$ m, $BC = 3$ m. Forces of 6, 6, 15, 9 N act along AB, BC, CA and AD respectively. Replace the system by a single force acting through A with a couple.

6. ABCDEF is a regular hexagon. Forces of magnitude 2, 2, 4, 14, 14 and 4 N act along AB, BC, CD, ED, FE, FA respectively. Find the single force equivalent to this system.

7. A, B, C are the points (12, 0), (6, 8) and (0, 0) respectively referred to rectangular cartesian axes. Forces of magnitude $10F, 15F$ and $5F$

EQUIVALENT SYSTEMS OF FORCES

act along AB, BC and CA respectively. Find the magnitude of the single force equivalent to this system and the equation of its line of action.

8. $ABCD$ is a rhombus of side a such that angle ABC is 60 degrees. Forces of magnitude $2P$, $2P$, $4P$, $4P$ and $2P$ act along AB, BC, CD, DA and AC respectively. Show that the system is equivalent to a couple and find its moment.

9. If $ABCD$ is a parallelogram, show that

$$\lambda\overrightarrow{AB} + \mu\overrightarrow{BC} + \lambda\overrightarrow{CD} + \mu\overrightarrow{DA} = 0.$$

$ABCD$ is a parallelogram in which $AB = 4$ cm, $AD = 3$ cm and angle ABC is 30 degrees. Forces of 8, 6, 8 and 6 N act along AB, BC, CD and DA respectively. Show that the system reduces to a couple and find its moment.

10. A force of 10 N acts along the x-axis and parallel forces of 6 N and 4 N, in the same sense, act along the lines $y = 1$ and $y = 3$ respectively. Find the magnitude of the single force equivalent to this system and the equation of its line of action.

17.5. RESULTANTS

In this section an alternative method of reducing a set of forces to a simpler equivalent system will be developed. It will be shown that intersecting forces may be treated as if they were acting on a particle at the point of intersection and replaced by their resultant acting at that point.

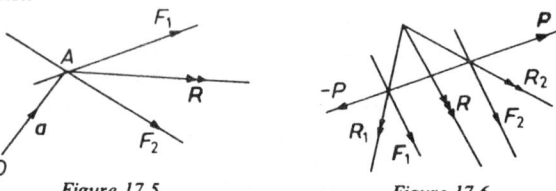

Figure 17.5   Figure 17.6

Consider two forces F_1, F_2 which intersect and their resultant R drawn through their point of intersection A (refer to *Figure 17.5*). Let A have position vector a with respect to some origin O. Then taking moments about O,

the sum of moments of F_1 and F_2 = $a \times F_1 + a \times F_2$
$$= a \times (F_1 + F_2)$$
$$= a \times R$$

$\therefore$ the sum of moments of F_1 and F_2 = moment of R.

RESULTANTS

Thus **R**, which by definition has the same vector sum as F_1 and F_2, also has the same moment about O. **R** is therefore equivalent to the system consisting of F_1 and F_2.

Parallel forces can also be reduced in this way (refer to *Figure 17.6*). Two equal and opposite forces **P** and $-$**P** (in the same straight line) can be introduced to intersect with the parallel forces F_1, F_2. Their presence will not alter the vector sum or the total moment of the system. We now have intersecting forces and the system may be reduced step by step taking them in pairs. The method breaks down, however, if F_1, F_2 form a couple.

Any system containing intersecting forces and/or suitable parallel forces may be reduced in this way.

If a system of forces can be reduced in this way to a single resultant **R**, then it follows that **R** is equivalent to the system.* For this reason the words 'resultant' and 'single equivalent force' are used interchangeably.

Example 1. ABCD is a square. Forces of 2, 3, 3, 2, and $\sqrt{2}$ kN act along the sides AB, CB, CD, AD and BD respectively. Find the magnitude, direction and line of action of their resultant.

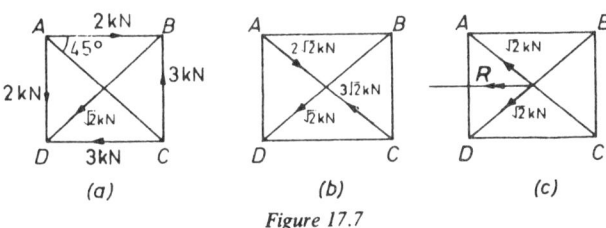

Figure 17.7

The forces along AB and AD have a resultant $2\sqrt{2}$ kN acting along AC. Similarly, the forces along CB, CD have a resultant $3\sqrt{2}$ kN acting along CA. Thus the given system of forces shown in *Figure 17.7(a)* can be reduced to that shown in *Figure 17.7(b)*.

This can then be further reduced to a force $\sqrt{2}$ kN acting along CA together with the given force of $\sqrt{2}$ kN along BD (refer to *Figure 17.7(c)*).

In turn the resultant **R** of these two forces is found acting at their point of intersection. Clearly **R** is of magnitude 2 kN acting parallel to CD.

* The statement that the sum of the moments of a set of forces about a fixed point O is equal to the moment of their resultant about O, is sometimes called the *principle of moments*.

That is, the resultant of the given system is a force of 2 kN acting parallel to CD through the point of intersection of AC and BD. (This result could, of course, have been obtained by resolving and taking moments as in Section 17.3.)

Example 2. Forces $\overrightarrow{BA}$, $3\overrightarrow{AC}$ and $2\overrightarrow{CB}$ act along the sides of a triangle ABC. Show that if M is the mid-point of BC their resultant is represented by $2\overrightarrow{AM}$. Show also that its line of action cuts BC at X where $BX:XC = 3:-1$.

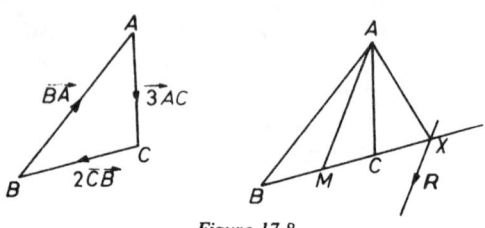

Figure 17.8

Referring to Figure 17.8, let $\boldsymbol{R}$ be the resultant of the given system. Then

$$\boldsymbol{R} = \overrightarrow{BA} + 3\overrightarrow{AC} + 2\overrightarrow{CB}$$
$$= (\overrightarrow{BA} + \overrightarrow{AC} + \overrightarrow{CB}) + (\overrightarrow{AC} + \overrightarrow{CB}) + \overrightarrow{AC}$$
$$= 0 + \overrightarrow{AB} + \overrightarrow{AC}.$$

But $\lambda\overrightarrow{AB} + \mu\overrightarrow{AC} = (\lambda + \mu)\overrightarrow{AD}$ where D divides BC in the ratio $\mu:\lambda$ (refer to Section 3.3).

∴ $\boldsymbol{R} = 2\overrightarrow{AM}$ where M is mid-point of BC.

Now, by this same theorem, the resultant of the forces $\overrightarrow{BA}$ and $3\overrightarrow{AC}$ is parallel to AX, where X divides BC externally in the ratio $3:1$. It must therefore lie along AX since the two forces intersect at A.

Hence the system can be reduced to two forces intersecting at X and thus $\boldsymbol{R}$ must act through X.

The resultant of the given system is a force $2\overrightarrow{AM}$ acting through the point X.

Example 3. $\boldsymbol{i}, \boldsymbol{j}, \boldsymbol{k}$ are unit vectors along rectangular axes OX, OY, OZ. A force $2\boldsymbol{i} + \boldsymbol{j} + \boldsymbol{k}$ units acts at the point $(4, -8, 8)$ and another force $\boldsymbol{i} - 2\boldsymbol{j} + 2\boldsymbol{k}$ units acts at the point $(a, -4, 0)$. Show that if these forces intersect, then $a = -3$ and find the vector equation of the line of action of their resultant.

RESULTANTS

The vector equation of a line through the point with position vector a and parallel to the vector b is $r = a + tb$, where t is a variable parameter (refer to Section 3.12).

Hence the equations of the lines of action of the two forces are

$$r = (4i - 8j + 8k) + t(2i + j + k)$$

and
$$r = (ai - 4j) + s(i - 2j + 2k).$$

If these two lines intersect, then values of s, t exist such that

$$(4i - 8j + 8k) + t(2i + j + k) = (ai - 4j) + s(i - 2j + 2k).$$

Equating components,

$$4 + 2t = a + s \quad \ldots \text{(i)}$$
$$-8 + t = -4 - 2s \quad \ldots \text{(ii)}$$
$$8 + t = 2s \quad \ldots \text{(iii)}$$

From (ii) and (iii) we obtain $t = -2$, $s = 3$ and substituting these values in (i) gives $a = -3$.

The point of intersection of the lines can be obtained by putting $t = -2$ in the equation of the line of action of the first force giving

$$r_0 = (4i - 8j + 8k) - 2(2i + j + k)$$
$$= -10j + 6k.$$

Now the resultant R of the two forces is such that

$$R = (2i + j + k) + (i - 2j + 2k)$$
$$= 3i - j + 3k.$$

Hence the equation of the line of action of R is

$$r = (-10j + 6k) + u(3i - j + 3k).$$

Exercises 17b

1. $ABCD$ is a square. Force of 1, 2, 2 and 1 N act along AB, CB, CD and AD respectively. Find the magnitude, direction and line of action of their resultant.

2. Forces represented by $\overrightarrow{AB}$, $\overrightarrow{CB}$ and $\overrightarrow{CA}$ act along the sides AB, CB and CA of a triangle ABC. Show that their resultant is represented by $2\overrightarrow{CB}$ and find the point in which its line of action cuts AC.

3. ABC is an equilateral triangle. Forces of 2, $2\sqrt{3}$ and 2 kN act along AC, CB and AB respectively. Find the magnitude and direction of their resultant and the point in which its line of action cuts BC.

EQUIVALENT SYSTEMS OF FORCES

4. ABC is a triangle. Three forces are *completely* represented by $4\overrightarrow{AB}, 2\overrightarrow{CB}$ and $\overrightarrow{CA}$. Show that their resultant is represented by $6\overrightarrow{MB}$ where M is the mid-point of AC. Find the points in which the line of action of this resultant cuts AB, AC and CB.

5. Three forces represented by $2\overrightarrow{AB}, 2\overrightarrow{BC}$ and $2\overrightarrow{CA}$ act at the mid-point of AC. Find their resultant.

6. Forces $3i + 2j + 3k$ and $i - 3k$ units act through points whose displacements are $9i - 3k$ and $3i - 2j + 3k$ units from a fixed point O. Show that the lines of action of these forces intersect and find the vector equation (with respect to O) of the line of action of their resultant.

EXERCISES 17*

1. $ABCD$ is a square of side a. Forces of magnitude $4P, 2P, 3P$ and P act along AB, CB, CD and AD respectively. Find the magnitude and direction of their resultant and the point in which its line of action cuts CD.

2. ABC is an equilateral triangle of side $2m$. Forces of 4, 2 and 3 N act along BA, BC and AC respectively. Find an equivalent system consisting of a force through B together with a couple.

3. A force $3i - 4j + 5k$ N acts at a point O and a second force $2i - j$ N acts through a point whose displacement from O is $j + 2k$ m. Find an equivalent system consisting of a force acting through the point whose displacement is i metres together with a couple.

4. ABC is an equilateral triangle and D, E, F are the mid-points of BC, CA, AB respectively. Forces of magnitude $P, 2P$ and $3P$ act along AB, BC and CA respectively. Find the three forces acting in DE, EF and FD which are equivalent to the given system.

5. Show that the line of action of the resultant of two like (in the same sense) parallel forces F_1, F_2 divides any line cutting them internally in the ratio $F_2 : F_1$.

What is the corresponding result for unlike parallel forces?

6. A trapezium $ABCD$ is such that AB is parallel to DC, angle ABC is a right angle, $AB = 4$ m, $BC = 3$ m and $CD = 7$ m. Forces of magnitude 16, 12, 21 and $9\sqrt{2}$ N act along BA, BC, DC and DA respectively. Find the magnitude of their resultant and show that it passes through the mid-point of AC.

7. A couple of moment $-4i + 7j - 5k$ units consists of forces acting at the points P_1, P_2 whose position vectors are $i + 5j + 4k$, $-3i + 2j + 3k$ respectively. Find the forces acting at P_1, P_2.

8. A force represented by $k\overrightarrow{BC}$ acts along the line BC. Express

* Exercises marked thus, †, have been metricized.

EXERCISES

the magnitude of the moment of this force about a point A in terms of the area of the triangle ABC.

Forces $4\overrightarrow{AB}$, $2\overrightarrow{CB}$ and $\overrightarrow{CA}$ act along the sides AB, CB, CA of a triangle ABC. Show that their resultant is represented by $3\overrightarrow{MB}$ where M is the mid-point of AC.

If the area of the triangle is Δ and $MB = l$, find the perpendicular distance of the line of action of the resultant from B.

9. $ABCD$ is a rectangle such that $AB = 14$ m, $BC = 8$ m and E is the point which divides AB internally in the ratio $3:4$. Forces of $1, 4, 7, 4$ and 10 kN act along BA, CB, DC, DA and ED respectively. Show that the system is equivalent to a couple and find its moment.

10. i, j, k are unit vectors parallel to the axes of x, y, z respectively. Three forces $2i + j + 3k$, $-i + 4j + 7k$ and $-i - 5j - 10k$ units act through the points $(3, 2, 4)$, $(2, -3, -6)$ and $(3, 11, 21)$ respectively. Show that they are in equilibrium.

11. $ABCD$ is a trapezium in which AB is parallel to DC. E, F are the mid-points of AB, DC respectively. Show that the resultant of forces completely represented by $\overrightarrow{AD}$ and $\overrightarrow{BC}$ is *completely* represented by $2\overrightarrow{EF}$.

12. $ABCD$ is a square of side a. Forces of magnitude $3P, 5P, 3P, 2P$ and $\sqrt{2}P$ act along BA, BC, DC, DA and CA respectively. Find the magnitude of their resultant and the equation of its line of action referred to DC, DA as axes.

13. $ABCD$ is a rectangle in which $AB = 2a$, $BC = a$ and E is the mid-point of AB. Forces $\sqrt{2}P, \sqrt{2}P, P, 3P$ act along AB, BC, CE and DE respectively. Find the magnitude and direction of the force which, when added to the system, makes it equivalent to a couple.

14. Two forces $F_1 = i + j + k$ and $F_2 = i + 2j - k$ act through points whose position vectors are $S_1 = i + j + 2k$ and $S_2 = pj + 5k$ respectively, relative to a fixed point and three mutually perpendicular unit vectors i, j and k. If the lines of action of F_1 and F_2 intersect, find p, and find the vector equation of the line of action of the resultant of F_1 and F_2. (London)

15. $ABCD$ is a quadrilateral. Forces represented by $\overrightarrow{AB}, \overrightarrow{AD}, \overrightarrow{BC}$ and $\overrightarrow{DC}$ act along the corresponding sides. Show that their resultant acts through the mid-point of DB and find its magnitude and direction.

16. ABC is a triangle such that $AB = 6$ m, $BC = 4$ m and $CA = 4$ m. Forces of 3, 2 and 2 N act along AB, BC and CA respectively. Show that the system is equivalent to a couple and find its moment.

Find the magnitude, direction and line of action of the resultant if the force in AB is reversed.

EQUIVALENT SYSTEMS OF FORCES

17. Forces X, Y act along rectangular axes of x and y respectively in the positive direction, together with an anticlockwise couple of moment G (in the same plane). Find the equation of the line of action of the single force equivalent to this system.

18. A system of forces in the plane of a triangle ABC has clockwise moments of $2G$, $-3G$ and $3G$ about the points A, B, C respectively. State why the system reduces to a single force F and not a couple. Find where the line of action of F meets BC and find its moment about the centroid of the triangle ABC.

19. A force has components X, Y parallel to rectangular axes Ox and Oy respectively and (x, y) is a point on the line of action. Show that the force is equivalent to an equal force at the origin O together with a couple. State the moment of the couple.

The components of three forces in the plane are given at time t by

$$(2P \cos \omega t, 0), \quad (P \cos \omega t, 2P \sin \omega t) \quad \text{and} \quad (3P \sin \omega t, P \cos \omega t),$$

and their lines of action pass respectively through O and the points (a, a) and $(-3a, 2a)$, where P, ω and a are constants. If the system is reduced to a force with components X', Y' at O and a couple G, find the values of X', Y' and G. Deduce the equation of the line of action of the resultant, and show that this line passes through a fixed point which is independent of t. (JMB)

20. $ABCD$ is a plane quadrilateral whose opposite sides meet in E and F. The mid-points of AC, BD and EF are X, Y and Z respectively. Show that
 (a) $AB + CD = CB + AD$,
 (b) $AB + CB + CD + AD = 4XY$,
and deduce that X, Y, Z are collinear.

If AB represents a force in magnitude, direction and position, show that the system of forces represented by

$$AB + BC + 2CD + 2DA + AC$$

is equivalent to a couple. (WJEC)

21. $ABCD$ is a square of side 6 m, F is a force coplanar with $ABCD$. The moments of F about A, B and C are 60, 24, -24 N m respectively. Show that the line of action of F passes through the mid-point of BC and find the magnitude of F. Find also where the line of action F meets CD, DA and AB.

22. Forces $2j$, $2j - 2i$ units act at points whose position vectors are $i + j + k$, and j respectively. Find the two forces, acting through the points whose position vectors are $i + j$ and $(i + k)/2$, which are together equivalent to the given system.

EXERCISES

23. Forces of magnitude 2, 1, 1 lb wt. act along the sides AB, AC, BC of an equilateral triangle ABC, in the directions indicated by the order of the letters. If E is the point where the perpendicular to BC at B meets CA produced, and F is the mid-point of AB, prove that the resultant is of magnitude $\sqrt{7}$ lb wt. and acts along EF. (Oxford)

(Note: a *lb wt.* is a unit of force.)

24. Forces of 1, 3, $2\sqrt{2}$ lb wt. act along the sides AB and CD and the diagonal BD respectively of a square $ABCD$, in the directions indicated by the order of the letters.

(a) Find the forces acting along the sides BC and AD and the diagonal CA which are equivalent to the given set of forces.

(b) Find the points where the resultant of the given set of forces meets AB and AD. (Oxford)

(Note: a *lb wt.* is a unit of force.)

25. $ABCD$ is a square of side 7 cm P, Q, R and S are points on AB, BC, CD and DA respectively such that

$$AP = BQ = CR = DS = 3 \text{ cm}$$

Forces 2, 3, 4 and 5 N act respectively along PQ, QR, RS and SP in the directions indicated by the order of the letters. Find the magnitude and direction of their resultant and the position of the point in which the resultant cuts AB (produced if necessary).

It is required to replace these forces by one force through the centre of the square $ABCD$ together with a couple. Find the magnitude of the force and the moment of the couple. (London)†

26. $ABCD$ is a square of side a. In the plane of the square, a force P acts through A and a perpendicular force $2P$ acts through B. Their resultant passes through D and through a point X on BC lying between B and C. Find the magnitude of the resultant and prove that $BX = a/4$.

Two additional forces are now introduced into the system: a force P acting through D parallel to the first force P, and a force $2P$ acting through C parallel to the second force $2P$. If the new forces have the same senses as the old ones, find the magnitude and the line of action of the resultant of the four forces.

If the new forces have opposite senses to the old ones, what will the resultant be? (London)

27. a, b, c are non-zero constant vectors and λ is a non-zero scalar. Three forces $\lambda(a + b + 2c)$, $\lambda(a - b + c)$ and $\lambda(a + b - c)$ act through points whose position vectors are a, b, c respectively.

343

Find the single force through the origin and the couple which are together equivalent to the given system.

State the condition that must be satisfied if the three forces are reducible to a single equivalent force.

28. (a) $OABC$ is a rectangle. O is the point $(0, 0)$, A the point $(2, 0)$ and B the point $(2, 1)$. Forces P, Q and R act respectively along $\overrightarrow{OA}$, $\overrightarrow{AB}$ and $\overrightarrow{BC}$ in the directions indicated by the order of the letters. Their resultant lies along the line $x + 2y = 7$. Find the magnitude of the resultant in terms of P. Find also the moment of a couple which when added to the system would transfer the resultant to the line $x + 2y = 9$.

(b) O is any point in the plane of a square $ABCD$ whose diagonals intersect at E. Four forces are represented completely by $3\overrightarrow{OA}$, $2\overrightarrow{OB}$, $3\overrightarrow{OC}$ and $2\overrightarrow{OD}$. Show that their resultant passes through E, and find its magnitude in terms of OE. (London)

29. Two forces represented by the vectors $\overrightarrow{AB}$ and $\overrightarrow{BD}$ act at the corner B of a plane quadrilateral $ABCD$ and two other forces represented by $q\overrightarrow{CA}$ and $r\overrightarrow{DC}$ act at the adjacent corner C. If the system is in equilibrium, find the angle between AD and BC and the values of q and r. (JMB, part)

30. The system S_1 consists of four forces in a plane. Fixed rectangular axes Ox, Oy are taken in the plane and the components of the four forces parallel to Ox are then respectively $-3, 2, 4, 1$ dynes, while the components parallel to Oy are respectively $1, -2, -1, 3$ dynes. The lines of action of the forces pass through the points $(1, 2)$, $(2, 4)$, $(3, 1)$, $(4, 5)$ respectively. Show that the line of action of the resultant of the system has equation $x - 4y + 5 = 0$.

If the system S_2, which is equivalent to S_1, consists of two parallel forces F_1 and F_2 with lines of action passing through the points $(-1, 0)$ and $(3, 0)$ respectively, find the components parallel to Ox and Oy of both F_1 and F_2. (WJEC)

(Note: a *dyne* is a unit of force.)

31. Four forces act, one along each side of a plane quadrilateral.

(a) If the forces all act the same way round the quadrilateral and each is represented in magnitude by the side in which it acts, show that the resultant is a couple whose moment is represented by twice the area of the quadrilateral.

(b) If the forces are in equilibrium and the quadrilateral is cyclic, show that the resultant of two adjacent forces acts along a diagonal and that each force is proportional to the opposite side.

(London)

EXERCISES

32. Forces F_1, F_2, F_3 act through points whose position vectors are r_1, r_2, r_3 where

$$F_1 = 8i + 8j + 4k$$
$$F_2 = 6i - 3j - 6k$$
$$F_3 = 5i - 10j + 10k$$
$$r_1 = 3i + 18j - 6k$$
$$r_2 = i + 7j - 16k$$
$$r_3 = 14i + 5j - 2k.$$

Find the single force through the origin and the couple which are together equivalent to this system.

Show that this system can be reduced to a single force and find the vector equation of its line of action.

33. Two forces $F_1 = i + 2j + k$ and $F_2 = 2i + j - k$ act at the points with position vectors $r_1 = i + k$ and $r_2 = j - 2k$ respectively. Find

(i) the angle between the forces F_1 and F_2,
(ii) a unit vector normal to F_1 and F_2,
(iii) the force R at the origin and the couple G to which the system reduces,
(iv) the magnitude of R,
(v) the moment G' of the forces F_1 and F_2 about a point with position vector $ai + bj + ck$.

Hence, or otherwise, show that $R.G'$ is independent of a, b and c. (AEB)

18
CENTRES OF GRAVITY

18.1. CENTRE OF PARALLEL FORCES

CONSIDER a set of parallel forces $F_1\hat{a}, F_2\hat{a}, F_3\hat{a}, \ldots$, $\hat{a}$ being a unit vector in their common direction, acting at points whose position vectors are $r_1, r_2, r_3 \ldots$ respectively.

Then if $\Sigma F \neq 0$* and if a single equivalent force R can be found, we have equating vector sums,

$$R = F_1\hat{a} + F_2\hat{a} + F_3\hat{a} + \cdots$$

i.e. $\qquad R = \hat{a}\Sigma F, \qquad \ldots\text{(i)}$

Also, if the point with position vector r lies on the line of action of R, then taking moments about the origin

$$\bar{r} \times R = r_1 \times F_1\hat{a} + r_2 \times F_2\hat{a} + \cdots$$

or $\qquad \bar{r} \times R = \Sigma(Fr) \times \hat{a}.$

Then substituting for R from equation (i),

$$\bar{r} \times \hat{a}\Sigma F = \Sigma(Fr) \times \hat{a}$$

$$\therefore \qquad \bar{r} \times \hat{a} = \frac{\Sigma(Fr)}{\Sigma F} \times \hat{a}. \qquad \ldots\text{(ii)}$$

Equation (ii) will always be satisfied if $\bar{r} = \Sigma(Fr)/\Sigma F$. Hence the line of action of R must pass through the point (G say) which has this position vector.

Again provided $\Sigma F \neq 0$,* such a point can always be found and its position is independent of $\hat{a}$. So we have shown that such a set of parallel forces can be reduced to a single force which always passes through G even if the direction of the parallel forces changes.

G is called the centre of the parallel forces.

* If $\Sigma F = 0$, then the set of parallel forces is equivalent to a couple. So this section shows that any system of parallel forces may be reduced to a single force or a couple.

CENTRE OF GRAVITY IN A RIGID BODY

18.2. CENTRE OF GRAVITY

The weights of a set of particles form a set of parallel forces. Hence, if the particles have position vectors $r_1, r_2, r_3, \ldots$ and weights of magnitude $W_1, W_2, W_3, \ldots$, their resultant weight will be of magnitude ΣW acting through the point

$$\bar{r} = \Sigma(Wr)/\Sigma W$$

called the *centre of gravity* of the system.

For a system of particles whose distances from the centre of the Earth are approximately the same, their weights are directly proportional to their masses. Hence $r = \Sigma(mr)/\Sigma m$ and their centre of gravity coincides with their centre of mass as defined in Section 16.1. The reader is referred to that section for calculations of the position of the centres of gravity of various sets of particles.

18.3. CENTRE OF GRAVITY OF A RIGID BODY

The particles that make up a rigid body are fixed relative to one another so that the centre of gravity is fixed relative to the body. Its position is independent of the alignment of the body relative to the Earth.

In calculating the position of the centre of gravity of a rigid body we shall, however, treat it as a continuous distribution of matter, dividing it into elements of weight δw. Taking the limit as $\delta w \to 0$, the position vector $\bar{r}$ of the centre of gravity will be given by $\bar{r} = \int r \, dw / \int dw$.

Rather than use this formula, however, it will be found more convenient to work from first principles. We shall take moments about suitable *axes* (refer to Section 3.10) using the fact that the moment of the resultant weight about a given axis is equal to the sum of the moments of the individual weights about that axis.

Example. A straight thin rod AB of length 2l has a uniform cross-section. Its density varies so that at a point distant x from A its weight per unit length is $\omega + \lambda x$. Find the position of its centre of gravity.

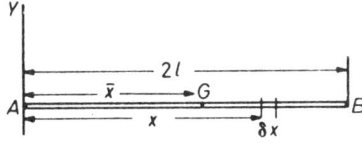

Figure 18.1

CENTRES OF GRAVITY

Let the centre of gravity G be at a distance $\bar{x}$ from A and consider an element of the rod of length δx distant x from A. (Refer to *Figure 18.1*.)

The weight of the element $\simeq (\omega + \lambda x)\delta x$, and the weight of the rod $\simeq \Sigma[(\omega + \lambda x)\delta x]$.

Now imagine the rod placed so that the weights act perpendicularly into the plane of the paper and take moments about a line AY perpendicular to the rod.

$$\therefore \quad \Sigma[(\omega + \lambda x)\delta x] \cdot \bar{x} \simeq \Sigma[(\omega + \lambda x)\delta x \cdot x].$$

Then in the limit as $\delta x \to 0$ (and the element becomes a particle),

$$\int_0^{2l} (\omega + \lambda x)\,dx \cdot \bar{x} = \int_0^{2l} (\omega + \lambda x)x\,dx$$

$$\therefore \quad \left[\omega x + \frac{\lambda x^2}{2}\right]_0^{2l} \cdot \bar{x} = \left[\omega \frac{x^2}{2} + \frac{\lambda x^3}{3}\right]_0^{2l}$$

$$\therefore \quad (2\omega l + 2\lambda l^2)\bar{x} = 2\omega l^2 + 8\lambda l^3/3$$

$$\therefore \quad \bar{x} = (3\omega l + 4\lambda l^2)/3(\omega + \lambda l).$$

The centre of gravity of the rod is at a point distant $(3\omega l + 4\lambda l^2)/3(\omega + \lambda l)$ from A. (Note that if $\lambda = 0$, so that the density is uniform, then $\bar{x} = l$, i.e. G is the mid-point of the rod as might be expected.)

18.4. CENTRES OF GRAVITY OF SOME STANDARD BODIES

It is not always necessary to consider a body as a set of elemental particles. It can often be split up into similar elemental masses whose centres of gravity are known and whose weights act through these centres of gravity.

Also, if the body has a line (or plane) of symmetry, these centres of gravity can be made to lie on it. Then taking moments about a suitable axis gives the centre of gravity of the whole body.

These principles will now be used to find the centres of gravity of some standard-shaped bodies. In most cases methods involving integration (as in Example 1, Section 18.3) will be used. However, it is sometimes possible to demonstrate the position of the centre of gravity without the use of calculus and we begin by establishing the position of the centre of gravity of a triangular lamina in this way.

CENTRES OF GRAVITY OF SOME STANDARD BODIES

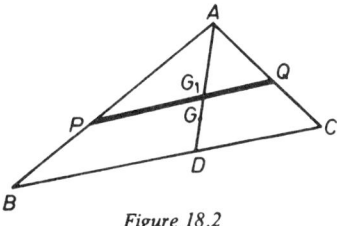

Figure 18.2

A uniform triangular lamina

Divide the triangle into n thin strips of equal widths (like PQ in *Figure 18.2*) parallel to one side. Then as n increases indefinitely, the strips become uniform thin rods with their centres of gravity at their mid-points. All these mid-points, such as G_1, lie on the median AD and hence the centre of gravity of the whole lamina must lie on AD. Similarly it must also lie on the other two medians of the triangle.

Thus the centre of gravity of a uniform triangular lamina is at the point of intersection of its medians, i.e., at a point G on AD such that $GD = \frac{1}{3}AD$.

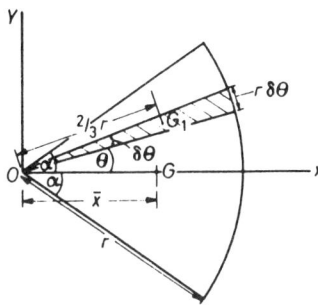

Figure 18.3

A uniform lamina in the form of a sector of a circle

Let the sector have radius r and subtend an angle 2α at the centre O of the circle (refer to *Figure 18.3*). From symmetry the centre of gravity G lies on OX the bisector of this angle.

Consider an elemental sector inclined at an angle θ to OX and subtending an angle $\delta\theta$ at O. It will be approximately a triangle with its centre of gravity G_1 distant $\frac{2}{3}r$ from O.

Then, if w is the weight per unit area of the lamina, the following table can be drawn up.

CENTRES OF GRAVITY

Body	Weight	Distance of C of G from OY
Whole sector	$w\frac{1}{2}r^2 2\alpha$	$\bar{x}$
Element	$w\frac{1}{2}r^2 \delta\theta$	$\frac{2}{3}r \cos\theta$

Again imagine the weights acting perpendicularly into the plane of the paper. Taking moments about OY,

$$(w\tfrac{1}{2}r^2 2\alpha)\bar{x} \simeq \Sigma(w\tfrac{1}{2}r^2 \delta\theta \cdot \tfrac{2}{3}r \cos\theta).$$

Then in the limit as $\delta\theta \to 0$,

$$w\tfrac{1}{2}r^2 2\alpha\bar{x} = \int_{-\alpha}^{\alpha} w\tfrac{1}{3}r^3 \cos\theta \, d\theta$$

$$\therefore \quad \alpha\bar{x} = \tfrac{1}{3}r[\sin\theta]_{-\alpha}^{\alpha}$$

$$\therefore \quad \bar{x} = \tfrac{2}{3}r(\sin\alpha/\alpha).$$

The centre of gravity of the sector lies on OX at a distance $\frac{2}{3}r(\sin\alpha/\alpha)$ from O.

If $\alpha = \pi/2$, so that the lamina has the form of a semicircular disc, then

$$\bar{x} = \frac{2}{3}r\frac{1}{\pi/2} = \frac{4r}{3\pi}.$$

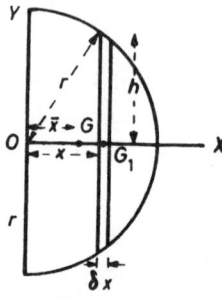

Figure 18.4

A uniform solid hemisphere

Referring to *Figure 18.4*, the hemisphere is of radius r and OX is the normal to the plane base through its centre. From symmetry the centre of gravity G lies on OX.

Consider an elemental disc, of thickness δx, with its plane parallel to the base of the hemisphere and distant x from it. If h is the radius

CENTRES OF GRAVITY OF SOME STANDARD BODIES

of this disc, then $h^2 = r^2 - x^2$ by Pythagoras theorem. Hence the volume of the disc $\simeq \pi h^2 \, \delta x = \pi(r^2 - x^2)\,\delta x$.

Let w be the weight per unit volume of the hemisphere.

Body	Weight	Distance of C of G from OY
Whole hemisphere	$w\tfrac{2}{3}\pi r^3$	$\bar{x}$
Element	$w\pi(r^2 - x^2)\,\delta x$	x

Taking moments about OY

$$(w\tfrac{2}{3}\pi r^3)\bar{x} \simeq \Sigma[w\pi(r^2 - x^2)\,\delta x \cdot x].$$

Then in the limit as $\delta x \to 0$,

$$w\tfrac{2}{3}\pi r^3 \bar{x} = \int_0^r w\pi(r^2 - x^2)x\,dx$$

$$\therefore \quad \tfrac{2}{3}r^3 \bar{x} = \left[r^2\frac{x^2}{2} - \frac{x^4}{4}\right]_0^r$$

$$\therefore \quad \tfrac{2}{3}r^3 \bar{x} = \frac{r^4}{2} - \frac{r^4}{4}$$

giving $\quad \bar{x} = \dfrac{3r}{8}.$

The centre of gravity of the hemisphere is a point on OX distant $\tfrac{3}{8}r$ from the base.

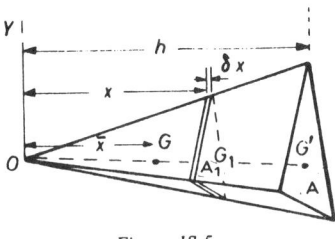

Figure 18.5

A uniform solid tetrahedron

Let the tetrahedron have height h and let the area of its base be A (refer to *Figure 18.5*).

Consider an elemental triangular lamina with its plane parallel to the base, of thickness δx and distant x from the vertex O of the tetrahedron.

CENTRES OF GRAVITY

Since such triangles are similar to the base triangle their centres of gravity all lie on OG'. Thus the centre of gravity G of the whole body lies on OG', the line joining the vertex to the centroid* of the base.

Again from similar triangles the area A_1 of the element is given by $A_1/A = x^2/h^2$ so that $A_1 = x^2 A/h^2$.

Let w be the weight per unit volume of the tetrahedron.

Body	Weight	Distance of C of G from OY
Whole body	$w\frac{1}{3}Ah$	$\bar{x}$
Element	$w(x^2 A/h^2)\delta x$	x

Taking moments about OY,

$$(w\tfrac{1}{3}Ah)\bar{x} \simeq \Sigma[w(x^2 A/h^2)\,\delta x . x].$$

Then in the limit as $\delta x \to 0$,

$$w\tfrac{1}{3}Ah\bar{x} = \int_0^h \frac{x^2}{h^2} Ax\,dx$$

$$\therefore \quad \tfrac{1}{3}h\bar{x} = \frac{1}{h^2}\int_0^h x^3\,dx$$

$$\therefore \quad \tfrac{1}{3}h\bar{x} = \frac{h^2}{4}$$

$$\therefore \quad \bar{x} = \frac{3h}{4}.$$

The centre of gravity of the tetrahedron lies one quarter of the way up a line joining the centroid of the base to the vertex.

A pyramid can always be split into tetrahedrons with a common vertex. If the number of sides of the base of the pyramid increases indefinitely, the perimeter of the base tends to a smooth curve and the pyramid becomes a cone. Hence our result can be extended to *any* pyramid or cone.

The centre of gravity of any uniform solid pyramid or cone lies one quarter of the way up a line joining the centroid of the base to the vertex.

---

* The centroid of a plane area coincides with the centre of gravity of a uniform lamina of the same shape. More general remarks about centroids are to be found in Section 18.5.

CENTRES OF GRAVITY OF SOME STANDARD BODIES

A frustum of a uniform thin spherical shell

Let the sphere be of radius r and let the rims of the frustum subtend angles 2α and 2β at the centre O respectively. From symmetry the centre of gravity G of the frustum lies on OX (refer to *Figure 18.6*).

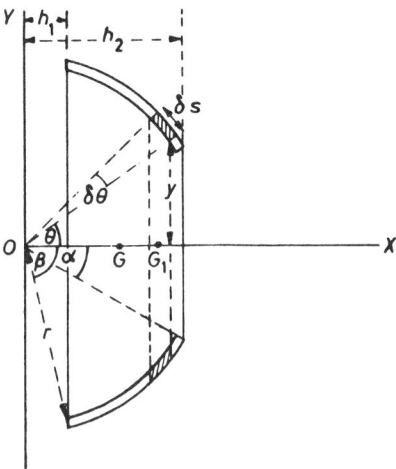

Figure 18.6

Consider an elemental frustum of radius y which subtends an angle 2θ at O and whose slant height is δs. Then the surface area of this element is approximately $2\pi y\, \delta s^* = 2\pi r \sin\theta\, r\, \delta\theta$.

Let w be the weight per unit surface area of the shell.

Body	Weight	Distance of C of G from OY
Whole frustum	$\Sigma(w 2\pi r^2 \sin\theta\, \delta\theta)$	$\bar{x}$
Element	$w 2\pi r^2 \sin\theta\, \delta\theta$	$r\cos\theta$

Taking moments about OY,

$$\Sigma(w 2\pi r^2 \sin\theta\, \delta\theta)\,.\, \bar{x} \simeq \Sigma(w 2\pi r^2 \sin\theta\, \delta\theta\,.\, r\cos\theta).$$

---

* In standard mathematical texts it is shown that the area of the curved surface of a frustum of a right pyramid or cone is given by $\frac{1}{2}$(sum of perimeters) × slant height. In this case surface area of the element = $\frac{1}{2}(2\pi y + 2\pi(y + \delta y))\delta s \simeq 2\pi y\, \delta s$ neglecting the second order small quantity $\delta y\, \delta s$.

353

CENTRES OF GRAVITY

Then in the limit as $\delta\theta \to 0$,

$$\bar{x}\int_\alpha^\beta w2\pi r^2 \sin\theta \, d\theta = \int_\alpha^\beta w2\pi r^3 \sin\theta \cos\theta \, d\theta$$

$$\therefore \quad \bar{x}\int_\alpha^\beta \sin\theta \, d\theta = r\int_\alpha^\beta \sin\theta \cos\theta \, d\theta$$

$$\therefore \quad \bar{x}[-\cos\theta]_\alpha^\beta = r[-\tfrac{1}{2}\cos^2\theta]_\alpha^\beta$$

$$\therefore \quad \bar{x}(\cos\beta - \cos\alpha) = \frac{r}{2}(\cos^2\beta - \cos^2\alpha)$$

$$\therefore \quad \bar{x} = \frac{r}{2}(\cos\beta + \cos\alpha)$$

$$\therefore \quad \bar{x} = \tfrac{1}{2}(h_1 + h_2).$$

The centre of gravity of the frustum lies on OX mid-way between the two planes which cut it off from the sphere.

If follows that the centre of gravity of a spherical cap lies half-way from its base to its vertex, and that the centre of gravity of a hemispherical shell is distant $\tfrac{1}{2}r$ from its base.

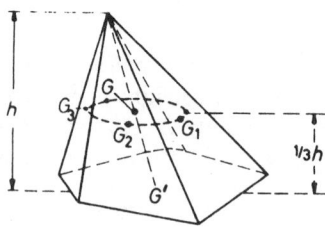

Figure 18.7

A uniform thin hollow pyramid (without base)

The surface of the pyramid consists of triangular laminae each of whose centre of gravity lies one-third of the way up its median. Thus the pyramid is equivalent to a set of particles at G_1, G_2, G_3 etc. all of which are at a height of $\tfrac{1}{3}h$ above the base of the pyramid (refer to *Figure 18.7*).

Hence the centre of gravity of the pyramid must be $\tfrac{1}{3}h$ above the base and, since cross-sections parallel to the base are similar, it must lie on the line joining the vertex to the centroid of its base.

If the number of sides of the base increases they will tend to a smooth curve and the pyramid will become a cone for which our results still hold.

Thus the centre of gravity of a hollow pyramid or cone lies one-third of the way up the line joining the centroid of the base to the vertex.

18.5. CENTROIDS

If, in the above examples, we had used elements of area (or volume) in our formulae instead of elements of weight, the points found would have been called *centroids* of area (or volume).

Hence for *uniform* bodies, centres of gravity, centres of mass and the centroids of the corresponding geometrical figures will coincide.

18.6. LIST OF STANDARD RESULTS

Note that the results below marked # have not been proved in the text but will be given as exercises later. It has been assumed, in each case, that the bodies are uniform and that their centres of gravity lie on lines of symmetry.

Body	*Position of C of G*
Straight rod	Mid-point
# Rod in form of circular arc of radius r subtending angle 2α at centre	$\dfrac{r \sin \alpha}{\alpha}$ from centre
# Lamina in form of parallelogram	Intersection of diagonals
Lamina in form of triangle	Intersection of medians
Lamina in form of sector of a circle of radius r subtending 2α at centre	$\dfrac{2r \sin \alpha}{3\alpha}$ from centre
Solid hemisphere of radius r	$\frac{3}{8}r$ from centre
Solid pyramid or cone	$\frac{1}{4}$ way up line from centroid of base to vertex
Frustum of spherical shell cut off by parallel planes	Mid-way between the planes
Hollow pyramid or cone	$\frac{1}{3}$ way up line from centroid of base to vertex

Exercises 18a

1. Show that the centre of gravity of a uniform lamina in the form of a parallelogram lies at the point of intersection of its diagonals.

CENTRES OF GRAVITY

2. A uniform thin rod is bent into the form of an arc of a circle of radius r subtending an angle 2α at the centre of the circle. Find the position of its centre of gravity.
What will be its position if the rod forms a semicircle?

3. Find the position of the centre of gravity of a uniform semi-circular disc by dividing it into strips parallel to its straight edge.

4. Find the position of the centroid of volume of a right circular cone by dividing it into discs parallel to its base.

5. A straight thin rod of length $2l$ is of variable density such that at a point distant x from one end its density is $\rho + \lambda x^2$. Find the position of its centre of gravity.

6. Find the position of the centre of gravity of a uniform thin shell in the form of the curved surface of a right circular cone by dividing it into rings parallel to its base.

7. Prove that the centre of gravity of a uniform lamina enclosed by the line $x = h$ and the curve $y^2 = 4ax$ is the point $(\frac{3}{5}h, 0)$.

8. Find the position of the centre of gravity of a uniform solid in the form of a cap of a sphere of radius r, the depth of the cap being h.

18.7. COMPOSITE BODIES

When a body consists of several parts, each of whose weight and centre of gravity are known, the centre of gravity of the body can be found by using the fact that the moment of the resultant weight is equal to the sum of the moments of the weights of the parts. That is, taking moments about some suitable axis,

$$(W_1 + W_2 + W_3 + \cdots)\bar{x} = W_1\bar{x}_1 + W_2\bar{x}_2 + W_3\bar{x}_3 \ldots$$

Example 1. A thin uniform wire is bent into the form of a triangle ABC such that $AB = AC = 13$ cm and $BC = 10$ cm. Find the position of its centre of gravity.

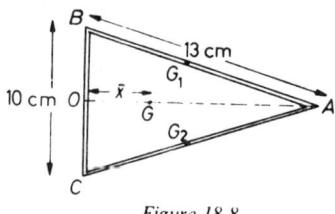

Figure 18.8

From symmetry the centre of gravity of the body lies on OA where O is the mid-point of BC (refer to *Figure 18.8*).

COMPOSITE BODIES

By Pythagoras' theorem $OA = \sqrt{AB^2 - OB^2} = 12$ cm. Hence the centres of gravity of the rods AB, AC at G_1, G_2 are 6 cm from BC. Let w be the weight per unit length of the wire.

Body	Weight	Distance of C of G from OB
Whole body	36w	$\bar{x}$
AB	13w	6 cm
AC	13w	6 cm
BC	10w	0

Taking moments about OB,

$$36w\bar{x} = 13w \times 6 + 13w \times 6 + 10w \times 0$$

$$\therefore \quad 36\bar{x} = 13 \times 12$$

$$\therefore \quad \bar{x} = 13/3.$$

The centre of gravity of the wire lies on the median OA distant $4\frac{1}{3}$ cm from O.

In many cases work can be simplified by "taking moments" with numbers proportional to the weights rather than the weights themselves, for if

$$(W_1 + W_2 + W_3 \ldots)\bar{x} = W_1\bar{x}_1 + W_2\bar{x}_2 \ldots$$

then

$$\left(\frac{W_1}{k} + \frac{W_2}{k} + \frac{W_3}{k} \ldots\right)\bar{x} = \frac{W_1}{k}\bar{x}_1 + \frac{W_2}{k}\bar{x}_2 \ldots$$

Example 2. A right circular cone and a hemisphere are joined so that their bases coincide. The cone is of height $2a$ and radius a, and the density of the material from which it is made is ρ. The hemisphere is of radius a and its density is 5ρ. Find the position of the centre of gravity of the combined body.

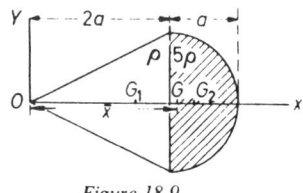

Figure 18.9

From symmetry all centres of gravity will lie on OX (refer to *Figure 18.9*).

CENTRES OF GRAVITY

The weights per unit volume of the two parts of the body $a\rho g$ and $5\rho g$ respectively.

Body	Weight	Relative weights	Distance of C of G from OY
Whole body	—	6	$\bar{x}$
Cone	$\frac{1}{3}\pi a^2 \cdot 2a \cdot \rho g$	1	$\frac{3}{4} \times 2a = \frac{3}{2}a$
Hemisphere	$\frac{2}{3}\pi a^3 \cdot 5\rho g$	5	$2a + \frac{3}{8}a = \frac{19}{8}a$

Taking moments about OY,

$$6\bar{x} = 1 \times \tfrac{3}{2}a + 5 \times \tfrac{19}{8}a$$

$$\therefore \quad \bar{x} = \tfrac{107}{48}a.$$

The centre of gravity of the combined body lies on OX distant $2\tfrac{11}{48}a$ from O.

When a standard shaped piece is removed from a standard body, the position of the centre of gravity of the remainder may be found in a similar manner. The piece removed is (in imagination) restored and the problem treated by the method of Examples 1 and 2 above.

Example 3. A frustum is cut from a solid right circular cone height h by a plane parallel to its base and distant $\tfrac{2}{3}h$ from it. Show that the position of the centre of gravity of the frustum is independent of the semi-vertical angle of the cone, and find its distance from the base.

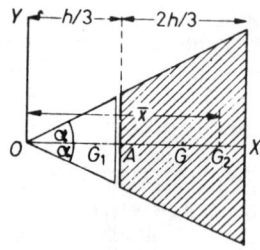

Figure 18.10

From symmetry all the centres of gravity lie on OX the axis of the cone (refer to *Figure 18.10*). G_1 is two-thirds of the way along OA, and G is two-thirds of the way along OX.

Let w be the weight per unit volume.

COMPOSITE BODIES

Body	Weight	Relative weights	Distance of C of G from OY
Whole cone	$\pi w \frac{1}{3} h^2 \tan^2 \alpha \cdot h$	27	$\frac{3}{4} h$
Small cone	$\pi w \frac{1}{3} \frac{h^2}{9} \tan^2 \alpha \cdot \frac{h}{3}$	1	$\frac{3}{4} \times \frac{h}{3} = \frac{h}{4}$
Remainder	—	26	$\bar{x}$

Taking moments about OY,

$$27 \times \tfrac{3}{4}h = 1 \times \frac{h}{4} + 26\bar{x}$$

$$81h = h + 104\bar{x}$$

$$\bar{x} = \tfrac{10}{13}h.$$

The position of the centre of gravity of the frustum does not depend on α and it is distant $\tfrac{3}{13}h$ from its base.

When a body has no line of symmetry, moments can be taken about more than one axis to fix the position of the centre of gravity.

Example 4. *A uniform lamina is in the form of a trapezium ABCD in which AB is parallel to DC and angle BCD is a right angle. $AB = 4$ m, $BC = 3$ m and $CD = 7$ m. Find the position of the centre of gravity of the lamina.*

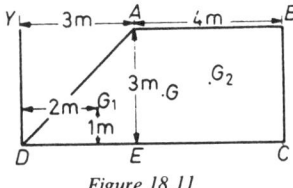

Figure 18.11

Divide the lamina into a rectangle and a right-angled triangle as shown in *Figure 18.11*. Then as G_1, the centre of gravity of the triangle, lies two-thirds of the way up the median from D, its distance from DY is 2 m. Similarly it is 1 m from DC.

Let w be the weight per square metre of the lamina.

Body	Weight	Distance of C of G from DY	from DC
Whole lamina	$33w/2$	$\bar{x}$	$\bar{y}$
Rectangle $ABCE$	$12w$	5 m	$\tfrac{3}{2}$ m
Triangle ADE	$9w/2$	2 m	1 m

359

Taking moments about DY,

$$\frac{33w}{2}\bar{x} = 12w \times 5 + \frac{9w}{2} \times 2$$

$$\therefore \quad \bar{x} = \frac{46}{11} \text{ m.}$$

Taking moments about DC,

$$\frac{33w}{2}\bar{y} = 12w \times \frac{3}{2} + \frac{9w}{2} \times 1$$

$$\therefore \quad \bar{y} = \frac{15}{11} \text{ m.}$$

The centre of gravity of the lamina is at a point distant $\frac{31}{11}$ m from BC and $\frac{15}{11}$ m from DC.

Exercises 18b

1. Three equal uniform rods are rigidly jointed to form three sides of a square of side $2a$. Find the position of the centre of gravity of the body so formed.

2. A thin uniform wire of length 4 m is bent into the shape of a right-angled isosceles triangle. Find the position of its centre of gravity.

3. A uniform thin lamina is in the shape of a square of side 8 cm with an isosceles triangle constructed on one side. The other two sides of the triangle are 5 cm in length. Find the position of the centre of gravity of the lamina.

4. A uniform solid body consists of a right circular cylinder, of height $2a$ and radius a, surmounted by a right circular cone of base radius a and height $4a$. Find the position of the centre of gravity.

Find also, the position of the centre of gravity when the cone is twice as dense as the cylinder.

5. A uniform solid body consists of a right circular cylinder, of height l and radius a, surmounted by a hemisphere of radius a. If the centre of gravity of the body lies on the common base of the cylinder and hemisphere, show that $2l = a\sqrt{2}$.

6. A circular lamina of radius 8 cm has a small circular hole drilled in it. The hole is of radius 2 cm with its centre 4 cm from the centre of the large circle. Find the position of the centre of gravity of the lamina.

7. The radii of two circular sections of a uniform solid right circular cone are a and $2a$. Show that the centre of gravity of the

EXERCISES

portion of the cone between these sections divides the join of their centres in the ratio 17:11. (London)

8. Show that the weight W of a triangular lamina may be replaced by the weights of three particles each of weight $\frac{1}{3}W$, one at each vertex.

Solve Example 4 of Section 18.7 by dividing the lamina into two triangles and replacing them by suitable particles at their vertices. Then take moments about two axes at right angles as before.

9. A piece of uniform thin card is in the form of a rectangle $ABCD$ in which $AB = 2a$, $BC = a$. E is the mid-point of CD and the triangle BCE is folded over, along the line BE, so that C coincides with the mid-point of AB. Find the distances of the centre of gravity of the folded card from AD and AB.

10. A uniform steel sheet has a rectangle $ABCD$ marked on it with $AB = 4$ m, $BC = 2$ m. Points E and F are also marked such that E is the mid-point of BC and F divides CD internally in the ratio of 1:3. The piece $ABEFD$ is then cut out. Find the distances of its centre of gravity from AB and AD.

EXERCISES 18*

1. Particles of weight 2, 3, 4 and 3 units are placed at points whose position vectors are $i + 2j + k$, $-i + 3j$, $3i - 3j + k$ and j respectively. Find the position vector of their centre of gravity.

2. $ABCD$ is a rectangle in which $AB = 10a$, $BC = 5a$. Particles of weight $3W$, W, $2W$ and $4W$ are placed at A, B, C and D respectively. Find the distances of their centre of gravity from AB and BC.

If, now, a particle of weight $5W$ is added at a point that divides AC internally in the ratio of 3:2, find the new position of the centre of gravity.

3. A thin uniform wire AB, 50 cm long, is bent at right angles at a point C 20 cm from A. Find the distances of the centre of gravity of the bent wire from AC and CB.

4. The weight per unit length of a rod AB falls exponentially along its length so that at a point distant x from A it is λe^{-x}. Show that the centre of gravity of the rod is at a distance

$$\frac{1 - e^{-l}(1 + l)}{1 - e^{-l}}$$

from A where l is the length of the rod.

* The exercises marked thus, †, have been metricized.

CENTRES OF GRAVITY

5. Show that the centre of gravity of three equal particles placed at the vertices of a triangle, coincides with the centre of gravity of three equal particles placed at the mid-points of the sides.

6. Three particles, each of weight w are placed at the vertices of a triangle ABC. A fourth particle also of weight w is placed at a point D so that the centre of mass of the four particles is at A. Show that the position of D is such that A is the centroid of triangle DBC.

7. A trapezium $ABCD$ is right-angled at D and the parallel sides are AB and DC. $AB = 6a$, $CD = 12a$ and $AD = 6b$. Find the position of the centre of gravity of a lamina of this shape.

8. A uniform thin wire is bent into a closed loop consisting of a semicircle of radius a with its diameter. Find the position of its centre of gravity.

9. A piece of thin sheet metal is in the form of a rectangle $ABCD$ such that $AB = 18$ cm, $BC = 12$ cm. A triangular piece ADE is cut away where E is the mid-point of DC. Find the position of the centre of gravity of the remainder.

10. A lamina consists of two triangles ABC, ABD fastened together along AB. The two triangles are of different material but have the same weight. ABC is an equilateral triangle of side $6a$, angle ABD is a right angle and $BD = 6a$. Find the position of the centre of gravity of the lamina.

11. $ABCD$ is a trapezium in which $AB = a$, $DC = b$, these being parallel sides distant h apart. Show that the centroid of the trapezium is at a distance

$$\frac{a + 2b}{3(a + b)}h$$

from AB and lies on the line joining the mid-points of AB and CD.

12. A spherical cavity of radius r is made inside a uniform solid sphere of radius R, in such a way that the two spherical surfaces touch at the point P. If G is the centre of mass of the remaining material, find the distance PG. (JMB, part)

13. From a uniform square lamina $ABCD$ of side a is cut a triangle CDE, with $CE = DE$. If the distance of E from CD is h, find the position of the centre of mass of the lamina $ABCED$.

(Oxford, part)

14. $\bar{x}$, $\bar{y}$ are the coordinates of the centroid of the area enclosed between the curve $y = f(x)$, the ordinates $x = a$, $x = b$ and the x-axis. Show that

$$\bar{x}\int_a^b y\,dx = \int_a^b xy\,dx \quad \text{and} \quad \bar{y}\int_a^b y\,dx = \tfrac{1}{2}\int_a^b y^2\,dx.$$

EXERCISES

Find the coordinates of the centroid of the area enclosed between the curve $y = x - x^2$ and the x-axis.

15. A uniform square lamina $ABCD$ is folded along the line joining E, F the mid-points of AB and BC so that the triangle EBF is at right angles to the plane $AEFCD$. If $AB = 2a$, find the distance of the centre of gravity of the folded lamina from the plane EBF.

16. Prove that the centre of mass of a uniform solid hemisphere of radius a is at a distance $\tfrac{3}{8}a$ from the centre of its base plane.

A uniform solid right circular cylinder of the same material as the hemisphere has radius a and height $\tfrac{4}{3}a$; a diameter of one of its ends is denoted by AB. The hemisphere and cylinder are joined so that the plane base of the hemisphere coincides with the end of the cylinder containing AB. Find the distance from AB of the centre of mass of the whole solid. (JMB, part)

17. A uniform solid body is in the form of a paraboloid obtained by revolving the area enclosed by the curve $y^2 = 4ax$ and the line $x = a$ about the axis of x. Show that the volume of the body is $2\pi a^3$ and find the position of its centre of gravity.

18. A body consists of two uniform solid cones of the same base radius, fixed base to base. One cone has height h_1, density ρ_1; the other has height h_2 and density ρ_2. Find the distance of the centre of gravity, of the composite body, from the common base of the two cones.

If the centre of gravity lies on this common base, find the ratio $\rho_1 : \rho_2$.

19. A uniform thin circular lamina of radius $2a$ is cut along a line whose distance from the centre of the circle is a. Find the distance of the centre of gravity of the smaller segment, thus created, from its straight edge.

20. A cylinder of height h has a right circular conical cavity of height $(2 - \sqrt{2})h$ cut from it. The base of the cone coincides with the base of the cylinder. Find the position of the centre of gravity of the remaining solid.

21. Find the position of the centre of mass of a uniform solid hemisphere of radius a.

Prove that the centre of mass of a uniform hemispherical shell, whose inner and outer radii are a and b, is at a distance

$$\frac{3}{8} \frac{(a+b)(a^2+b^2)}{a^2+ab+b^2}$$

from the centre and deduce the position of the centre of mass of a thin hemispherical shell. (London)

CENTRES OF GRAVITY

22. The centroid of the whole surface (including the base) of a solid right circular cone coincides with its centre of gravity. Find the semi-vertical angle of the cone.

23. $ABCD$ is a piece of thin plywood of uniform density in the form of a trapezium, BA being parallel to CD and each being perpendicular to BC. $AB = a$, $BC = b$ and $CD = c$. Find the distances of the centre of gravity of the plywood from BC and from AB.

The edges AB and CD are to be reinforced by thin strips of metal, each strip being uniform. Prove that this can be done without altering the centre of gravity of the plywood if the densities of the strips are in the ratio

$$c(c + 2a) : a(a + 2c).$$

(London)

24. A right circular cone of height $\sqrt{3}r$ and radius of base r has a sphere of the largest possible size cut from it. Find the position of the C of G of the remainder.

25. A uniform hemisphere, of radius $2a$, has a right cylindrical hole of radius a bored through it. The axis of the hole coincides with the axis of symmetry of the hemisphere. Find the position of the centre of gravity of the remaining body.

26. A uniform paper collar has the shape of the curved surface of a frustrum of a right circular cone. The radii of the top and bottom sections of the frustrum are 3 cm and $4\frac{1}{2}$ cm respectively and the planes of these sections are 2 cm apart. Find the position of the centre of gravity of the collar. (London, part)†

27. Prove that the centroid of a uniform solid right circular cone of height h is at a distance $\frac{1}{4}h$ from the centre of the base.

From the base of such a cone a right circular conical portion of height $h_1 (< \frac{1}{2}h)$, and with radius of base equal to that of the given cone, is hollowed out. The top conical portion, of height $\frac{1}{2}h$, of the given cone is cut off. Find the distance between the centre of the base of the original cone and the centroid of the remaining solid.

(London)

28. Find the position of the centre of mass of a thin hemispherical shell of mass m and radius a.

The density ρ at a distance r from the centre of a non-uniform solid sphere of mass $2M$ and radius a is given by the formula $\rho = \rho_0(1 - r^2/a^2)$, where ρ_0 is a constant. The sphere is divided into two equal parts by a plane through its centre. Find the centre of mass of either hemisphere.

(London, part)

19

ROTATION ABOUT A FIXED AXIS

19.1. EQUATIONS OF MOTION OF A RIGID BODY ROTATING ABOUT A FIXED AXIS

IN Section 16.3 it was shown that for a system of particles

$$\hat{a} \cdot \Sigma(r \times P) = d\{\Sigma[\hat{a} \cdot (r \times mv)]\}/dt,$$

i.e., that the moment of the external forces about an axis is equal to the rate of change of angular momentum about that axis. From this relation the equation of motion for rotation about a fixed axis can be derived. However, we obtain it below from first principles.

Consider a rigid body rotating about a fixed axis and in particular a constituent particle A, of mass m, distant r, from the axis.

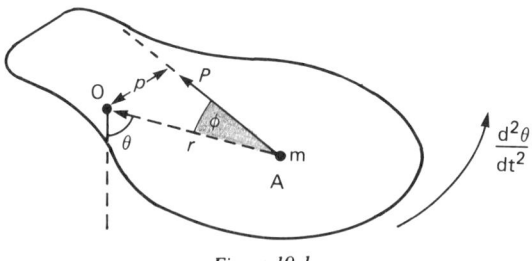

Figure 19.1

Figure 19.1 shows a section through A perpendicular to the axis of rotation and cutting it at O. Then, since A moves in a circle centre O, the resultant force P on it must lie in this plane making an angle ϕ say with OA.

Since $P = mf$, we have that perpendicular to OA

$$P \sin \phi = mr(d^2\theta/dt^2) \quad \text{(refer to Section 5.7)}$$
$$rP \sin \phi = mr^2(d^2\theta/dt^2)$$
$$pP = mr^2(d^2\theta/dt^2)$$

[where p is the perpendicular distance from O to the force P.]

ROTATION ABOUT A FIXED AXIS

Now the force P is at right angles to the axis of rotation and hence the quantity pP is the moment of P about this axis (refer to Section 3.10). Summing over the whole body

$$\Sigma(pP) = \Sigma[mr^2(d^2\theta/dt^2)].$$

But $d^2\theta/dt^2$ is the same for each particle in a rigid body.

∴ $$\Sigma(pP) = (\Sigma mr^2)\,d^2\theta/dt^2,$$

or $$L = I\alpha.$$

L is the sum of the moments of the external forces about the axis of rotation (the internal forces occur in equal and opposite pairs and do not affect the sum). $\alpha = d^2\theta/dt^2$ is the angular acceleration of the body about its axis.

$I = \Sigma mr^2$ is called the *moment of inertia* of the body about the given axis. It is a measure of both the mass of the body and its distribution about the axis.

19.2. KINETIC ENERGY OF A ROTATING BODY

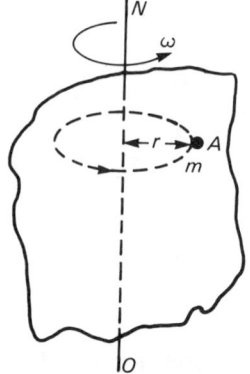

Figure 19.2

Consider a rigid body rotating with angular speed ω about a fixed axis ON (refer to *Figure 19.2*). Let m be the mass of a typical particle A moving in a circle of radius r. If v is the speed with which A is moving then

K.E. of particle $= \tfrac{1}{2}mv^2$ [refer to Section 12.5].

And since A moves in a circle, $v = \omega r$ (refer to Section 4.5).

Hence
$$\text{K.E.} = \tfrac{1}{2}m(\omega r)^2$$
$$= \tfrac{1}{2}m\omega^2 r^2.$$

CALCULATION OF MOMENTS OF INERTIA

∴ Total K.E. of the body $= \Sigma \frac{1}{2}m\omega^2 r^2$.

Since ω^2 is independent of the position of A in the body,

total K.E. $= \frac{1}{2}\omega^2 \Sigma mr^2$.

We again have the quantity $I = \Sigma mr^2$, which is the moment of inertia of the body and which first arose in Section 19.1. Before proceeding further we shall consider moments of inertia in detail.

19.3. CALCULATION OF MOMENTS OF INERTIA

Considering the body as a continuous distribution of matter, the mass of the constituent particle m is replaced by δm and

$$I = \lim_{\delta m \to 0} \Sigma r^2 \, \delta m$$

$$= \int r^2 \, dm$$

where the integration is taken throughout the whole body.

SI unit of moment of inertia ... kg m².

Example 1. Find the moment of inertia of a thin uniform rod, of mass M and length 2a, about an axis through its centre perpendicular to its length.

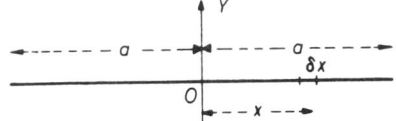

Figure 19.3

Take the centre of the rod as the origin, Ox along the rod, Oy perpendicular to the rod (refer to *Figure 19.3*). The mass per unit length is $M/2a$.

Consider a length δx of the rod distant x from O, its mass δm is $\delta x \cdot M/2a$

∴ moment of inertia of elemental piece $\delta I = \delta x M x^2 / 2a$

∴ moment of inertia of the rod about $Oy = \int_{-a}^{a} (Mx^2/2a) \, dx$

$$= (M/2a)[x^3/3]_{-a}^{a}$$

$$= Ma^2/3.$$

367

ROTATION ABOUT A FIXED AXIS

The moment of inertia can be written as $M(a/\sqrt{3})^2$ and, in general, moments of inertia can be written as Mk^2, where M is the mass of the body. The quantity k is known as the *radius of gyration*. The moment of inertia of the body is the same as if the whole of its mass had been concentrated at a distance k from the axis of rotation.

Example 2. *Find the radius of gyration of a thin rod AB, of length l, about an axis through one end A perpendicular to its length. Its mass per unit length being $k(l^2 - x^2)$, where x is the distance from A.*

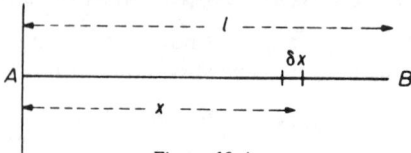

Figure 19.4

$$\delta m = \delta x\, k(l^2 - x^2) \quad [\text{refer to } \textit{Figure 19.4}]$$

∴ moment of inertia $\delta I = \delta x\, k(l^2 - x^2)x^2$

∴ moment of inertia of rod $= \displaystyle\int_0^l k(l^2 - x^2)x^2\, dx$

$\qquad\qquad\qquad\qquad\qquad = k[(l^2 x^3/3) - (x^5/5)]_0^l$

$\qquad\qquad\qquad\qquad\qquad = 2kl^5/15 \qquad\qquad\qquad \ldots\ldots (i)$

The mass of the rod $= \displaystyle\int_0^l dm = \int_0^l k(l^2 - x^2)\, dx$

$\qquad\qquad\qquad\qquad = k[l^2 x - x^3/3]_0^l$

i.e. $\qquad\qquad M = 2kl^3/3. \qquad\qquad\qquad \ldots\ldots (ii)$

From equations (i) and (ii),

$$\text{moment of inertia} = Ml^2/5$$

$$\text{radius of gyration} = l/\sqrt{5}.$$

Example 3. *Find the moment of inertia of a circular hoop, of mass M and radius a, about an axis through its centre perpendicular to its plane.*

Figure 19.5

368

CALCULATION OF MOMENTS OF INERTIA

Since the whole of the hoop lies at a constant distance a from the axis (refer to *Figure 19.5*),

$$\text{moment of inertia of the loop} = \int dm a^2 = a^2 \int dm = Ma^2.$$

Suppose two sets of particles, set A and set B, have moments of inertia I_A, I_B about the same axis. Then the moment of inertia I of the combined set about the *same* axis is given by

$$\therefore \quad I = \Sigma_A mr^2 + \Sigma_B mr^2,$$
$$I = I_A + I_B.$$

In general, $I = I_A + I_B + I_C + \cdots$. For a continuous body where δI is the moment of inertia of a typical element,

$$I = \lim \Sigma \, \delta I = \int dI.$$

Thus moments of inertia may be added provided they are about the same axis. This fact is used in the examples that follow.

Example 4. Find the moment of inertia of a thin uniform circular disc, of mass M and radius a, about an axis through its centre perpendicular to its plane.

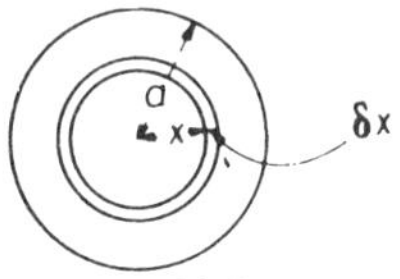

Figure 19.6

The mass per unit area is $M/\pi a^2$. The disc can be divided by means of concentric circles into circular rings. Consider one of these rings, thickness δx, radius x (refer to *Figure 19.6*).

$$\text{Its area} = 2\pi x \, \delta x.$$
$$\therefore \quad \text{its mass} = 2\pi x \, \delta x M / \pi a^2.$$

Its radius is x. Therefore, by the result of Example 3,

$$\text{its moment of inertia } \delta I = 2\pi x \, \delta x M x^2 / \pi a^2$$

$$\therefore \quad \text{moment of inertia of the lamina} = \int_0^a 2\pi x^3 M / \pi a^2 \, dx$$

$$= M 2\pi / \pi a^2 \left[\frac{x^4}{4} \right]_0^a$$

$$= Ma^2/2.$$

Example 5. *Find the moment of inertia of a thin uniform hollow sphere, of mass M and radius a, about a diameter.*

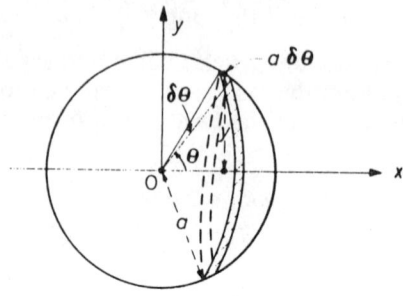

Figure 19.7

Let the centre of the sphere, O, be the origin of coordinates and the diameter be Ox. Divide the shell into circular rings by planes perpendicular to Ox (refer to *Figure 19.7*). The radius of a typical ring will be y and its slant thickness $a\,\delta\theta$.

∴ area of a typical ring $= 2\pi y a\,\delta\theta$.

The mass per unit area is $(M/4\pi a^2)$. ∴ the mass of a typical ring is $2\pi y a\,\delta\theta(M/4\pi a^2)$ and by the result of Example 3,

the moment of inertia $\delta I = (2\pi y a\,\delta\theta M/4\pi a^2)y^2$

∴ moment of inertia of the shell, $I = \int_{\theta=0}^{\theta=\pi} (2\pi y^3 a M/4\pi a^2)\,d\theta$.

Since $y = a\sin\theta$,

$$I = \int_0^\pi (2\pi a^4 \sin^3\theta M/4\pi a^2)\,d\theta$$

$$= \frac{Ma^2}{2}\int_0^\pi \sin^3\theta\,d\theta$$

$$= \frac{Ma^2}{2}\cdot\frac{2}{3}\int_0^\pi \sin\theta\,d\theta \quad \text{(by reduction formula)}$$

$$= \frac{Ma^2}{3}[-\cos\theta]_0^\pi.$$

Therefore $I = \dfrac{2Ma^2}{3}$.

Exercises 19a

1. Find the radius of gyration of a rod, of length 6 m, about an axis through one end of the rod A and perpendicular to the rod. The mass per unit length of the rod is $(12 - x)$ kg/m, where x m is the distance along the rod from A.

2. Find the moment of inertia of a thin uniform rod, mass M, length $2a$, about an axis through the centre inclined at an angle θ to the rod.

3. Four small balls, each of mass M, are rigidly jointed by four light rods to form a rectangle $ABCD$, the length of whose sides are $2a$ and $2b$. Find the moment of inertia of the system about

   (a) an axis through the centre of the rectangle perpendicular to the plane,

   (b) an axis in the plane of the rectangle perpendicular to the sides of length $2a$,

   (c) an axis in the plane of the rectangle perpendicular to the sides of length $2b$,

   (d) a diagonal of the rectangle.

4. Find the moment of inertia of a rectangular lamina, of mass M and sides $2a$ and $2b$, about an axis in the plane of the rectangle passing through its centre (a) perpendicular to the sides of length $2a$, (b) perpendicular to the sides of length $2b$.

5. Find the moment of inertia of the following bodies each assumed to be of mass M:

   (a) A thin hollow cylinder radius a about its axis.

   (b) A homogeneous solid cylinder radius a about its axis.

   (c) A homogeneous solid sphere radius a about a diameter.

   (d) A circular annulus inner and outer radii a and b, respectively about its axis.

   (e) A lamina in the form of an isosceles triangle height h, base $2a$ about the axis of symmetry.

6. A thin circular disc, of radius 0·2 m and weight 0·5 kg, has two small weights of 0·02 kg at the opposite ends of a diameter. Find the moment of inertia about an axis through the centre perpendicular to the plane of the disc.

7. Find the moment of inertia of a cube, of mass M and side $2a$, about an axis through its centre parallel to one of its edges.

19.4. PARALLEL AXIS THEOREM

We have so far calculated moments of inertia for axes through the centre of mass. These results can be extended to give moments of inertia about parallel axes to those through the centre of mass.

Theorem

If the moment of inertia of a body, of mass M, about an axis through its centre of mass is Mk^2, then its moment of inertia about a parallel axis distant a away is $M(k^2 + a^2)$.

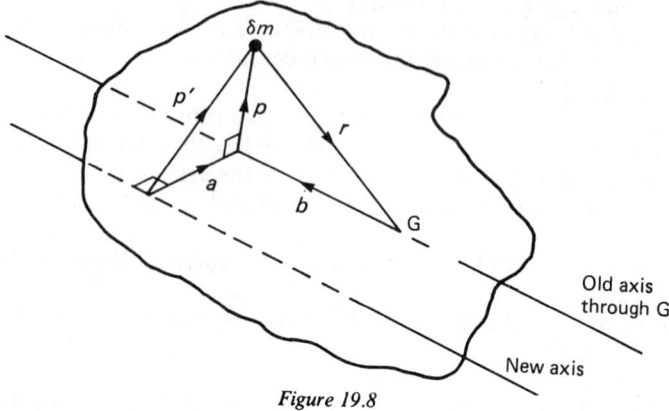

Figure 19.8

Let δm be a typical element of the body and p and p' the perpendiculars from δm to the axis through the centre of mass G and a new parallel axis respectively (refer to *Figure 19.8*).

Then
$$p' = p + a$$

∴ moment of inertia about the new axis

$$I' = \int_M (p')^2 \, dm = \int_M (p + a)^2 \, dm$$

$$= \int_M p^2 \, dm + 2\int_M p \cdot a \, dm + \int_M a^2 \, dm$$

$$= Mk^2 + 2a \cdot \int_M p \, dm + a^2 \int_M dm.$$

Now $\int_M r \, dm / \int_M dm$ gives the position vector of the centre of mass, and since r is measured from G this means that $\int_M r \, dm = \mathbf{0}$. Also $r = b + p$ (refer again to *Figure 19.8*) so $\int_M b \, dm + \int_M p \, dm = \mathbf{0}$. But since b and p are at right angles these integrals must be separately zero, in particular $2a \cdot \int_M p \, dm = 0$.

Hence
$$I' = Mk^2 + Ma^2.$$

The moment of inertia about the new parallel axis is $M(k^2 + a^2)$.

PARALLEL AXIS THEOREM

Example 1. Find the moment of inertia of a rod, of mass M and length 2a, about an axis perpendicular to the rod through one end.

By the result of Example 1, Section 19.4, the moment of inertia about a parallel axis through the centre of mass is $Ma^2/3$ and the distance between these axes is a. Therefore, by the parallel axis theorem, the required moment of inertia is

$$I = M\left(\frac{a^2}{3} + a^2\right) = M\frac{4a^2}{3}.$$

Example 2. Find the moment of inertia of a solid homogeneous cylinder, of mass M, radius of base a and height h, about a diameter of its base. Given that the moment of inertia of a circular disc about a diameter is mass × (radius)²/4, deduce the moment of inertia about a parallel axis through the centre of mass.

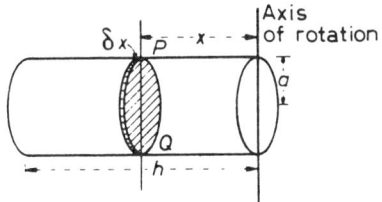

Figure 19.9

The mass per unit volume is $M/\pi a^2 h$. Consider the cylinder divided into elemental circular discs by planes parallel to the base. Let PQ (refer to *Figure 19.9*) be a typical disc width δx distant x from the axis.

$$\text{Volume of the disc} = \pi a^2 \, \delta x$$

$$\therefore \quad \text{mass of the disc} = \pi a^2 \, \delta x M/\pi a^2 h$$

$$= M \, \delta x/h.$$

The moment of inertia of a circular disc about a diameter is mass × (radius)²/4. Therefore, by the parallel axis theorem, the moment of inertia of the disc about the required axis is

$$\text{mass} \times \left[\frac{(\text{radius})^2}{4} + x^2\right]$$

ROTATION ABOUT A FIXED AXIS

$$\delta I = M\delta x[a^2/4 + x^2]/h$$

moment of inertia of cylinder, $I = \int_0^h \dfrac{M(a^2/4 + x^2)}{h} dx$

$$= M[(a^2 x)/4 + (x^3/3)]_0^h / h$$
$$= M[(a^2 h)/4 + (h^3/3)]/h$$
$$I = M[(a^2/4) + (h^2/3)].$$

If Mk^2 is the moment of inertia of the cylinder about a parallel axis through the centre of mass, and since the distance between the axes is $h/2$, then

$$M[k^2 + (h/2)^2] = M[(a^2/4) + (h^2/3)]$$
$$\therefore \quad Mk^2 = M[(a^2/4) + (h^2/12)].$$

19.5. PERPENDICULAR AXIS THEOREM FOR A LAMINA

Another useful theorem which can be applied to laminas but *not* solid bodies is the following: Given that the moments of inertia of a lamina about two perpendicular axes in its plane are I_A and I_B. If the two axes meet at O then the moment of inertia of the lamina about an axis through O perpendicular to the plane of the lamina is $I_A + I_B$.

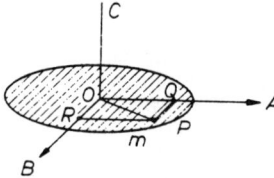

Figure 19.10

Let OA, OB be the two given axes in the plane of the lamina and OC perpendicular to the plane of the lamina (refer to *Figure 19.10*). Consider a constituent particle of mass m at a point P, where $OP = r$. The moment of inertia of the lamina about OC (I_C) is

$$I_C = \Sigma m r^2 = \Sigma m(PQ^2 + PR^2)$$
$$= \Sigma m PQ^2 + \Sigma m PR^2$$
$$\therefore \quad I_C = I_A + I_B.$$

MOMENTS OF INERTIA OF STANDARD BODIES

Example. By the result of Question 4, Exercises 19a, the moments of inertia of a rectangular lamina, sides 2a, 2b, about axes in the plane of the lamina passing through the centre and perpendicular to the sides are $Ma^2/3$ and $Mb^2/3$. By the perpendicular axis theorem the moment of inertia about an axis through the centre perpendicular to the plane of the lamina is

$$Ma^2/3 + Mb^2/3 = M(a^2 + b^2)/3.$$

19.6. MOMENTS OF INERTIA OF STANDARD BODIES

The following results are given for the purpose of quick reference. They apply to bodies of uniform density and mass M. Moments of inertia about other axes may be calculated by means of the Parallel Axis Theorem and (in the case of laminas) by the Perpendicular Axes Theorem.

Body (all uniform)	Dimensions	Axis (through the centre of mass)	Moment of inertia M = mass of body
Rod	length $2l$	perpendicular to rod	$Ml^2/3$
Rectangle	length $2l$	perpendicular to the length $2l$	$Ml^2/3$
Circular disc	radius a	a diameter	$Ma^2/4$
Sphere	radius a	a diameter	$2Ma^2/5$
Hollow sphere	radius a	a diameter	$2Ma^2/3$
Solid cylinder	radius a	the axis	$Ma^2/2$
Hollow cylinder	radius a	the axis	Ma^2

The values of the moment of inertia in several cases can be remembered by *Routh's Rule* which states that: the moment of inertia I of a *solid* body about an axis of symmetry is given by

$$I = \text{Mass} \left[\frac{\text{sum of squares of perpendicular semi-axes}}{3, 4 \text{ or } 5} \right].$$

The divisor to be 3, 4 or 5 according as the body is rectangular, elliptical or ellipsoidal respectively.

For a circular disc (a special case of an ellipse) radius a about a diameter

$$I = M[(a^2 + 0)/4] = (Ma^2)/4.$$

For a rectangular lamina sides $2a$, $2b$ about an axis through its centre perpendicular to its plane

$$I = M[(a^2 + b^2)/3].$$

ROTATION ABOUT A FIXED AXIS

Exercises 19b

1. Find the moment of inertia of a uniform circular disc (mass M, radius a) about a tangent.

2. By means of the perpendicular axis theorem, find the moment of inertia of a circular ring about a diameter.

3. Find the moment of inertia of a cube (mass M, edge $2a$) about a line through its centre perpendicular to one face. By means of the parallel axis theorem, find the radius of gyration of the cube about one edge.

4. Show that the moment of inertia of a rectangular lamina (mass M, sides $2a$ and $2b$) about an axis through one corner parallel to a diagonal is $14(M/3)[(a^2b^2)/(a^2 + b^2)]$.

Find the perpendicular distance from the centre of the rectangle to the axis, and deduce the moment of inertia of the rectangle about a diagonal.

5. Use Routh's Rule to find the moment of inertia of (*a*) a sphere about a diameter (*b*) an elliptical lamina about its major and minor axes.

6. Show that the square of the radius of gyration of a solid cone, height h, radius of base a, about an axis through the apex parallel to a diameter of its base is $3h^2/5 + 3a^2/20$.

7. Use the result of Question 6 to deduce the square of the radius of gyration of a cone about a diameter of its base.

8. Find the moment of inertia of a solid cone, of mass M and radius of base a, about its axis.

9. A governor consists of three solid metal spheres each of radius 0·02 m and mass 20 kg. Their centres rotate in the same plane in a circle of radius 0·15 m about a vertical axis. Find the moment of inertia of the three spheres about the axis.

10. A uniform circular plate, of radius a, has a mass M. A hole of radius $a/5$ is punched in the plate, the centre of the hole being a distance of $3a/5$ from the centre of the plate. Find the moment of inertia of the plate about (*a*) a diameter through the centre of the hole, (*b*) a line through the centre of the disc perpendicular to its plane.

19.7. UNIFORM ANGULAR ACCELERATION

Referring back to Section 19.1, we have that for a rigid body rotating about a fixed axis

$$L = I\alpha$$

where L is the sum of the magnitude of the moments of the external forces about the axis, α is the angular acceleration and I is the

UNIFORM ANGULAR ACCELERATION

moment of inertia about the axis. If L is constant, then α is constant and since

$$\frac{d\omega}{dt} = \alpha \quad \text{(const.)} \qquad \ldots\text{(i)}$$

$$\omega = \alpha t + C.$$

If $\omega = \omega_0$, when $t = 0$, $\omega_0 = C$.

$\therefore$
$$\omega = \omega_0 + \alpha t$$

$\therefore$
$$\frac{d\theta}{dt} = \omega_0 + \alpha t$$

$\therefore$
$$\theta = \omega_0 t + \tfrac{1}{2}\alpha t^2 + D.$$

If $\theta = 0$, when $t = 0$, $D = 0$

$\therefore$
$$\theta = \omega_0 t + \tfrac{1}{2}\alpha t^2.$$

Also from (i),

$$\frac{d\theta}{dt} \cdot \frac{d\omega}{d\theta} = \alpha,$$

i.e.
$$\omega \frac{d\omega}{d\theta} = \alpha$$

$\therefore$
$$\tfrac{1}{2}\omega^2 = \alpha\theta + E.$$

Now when $\theta = 0$, $\omega = \omega_0$

$\therefore$
$$E = \tfrac{1}{2}\omega_0^2$$

$\therefore$
$$\tfrac{1}{2}\omega^2 = \alpha\theta + \tfrac{1}{2}\omega_0^2$$

$\therefore$
$$\omega^2 = \omega_0^2 + 2\alpha\theta.$$

Summarizing these results we have that, for a body rotating with constant angular acceleration α,

$$\omega = \omega_0 + \alpha t$$

$$\theta = \omega_0 t + \tfrac{1}{2}\alpha t^2$$

$$\omega^2 = \omega_0^2 + 2\alpha\theta.$$

These equations should be compared with those summarized in Section 5.3 for a body moving with constant linear acceleration f.

Example 1. A flywheel is subject to a constant torque which imparts a constant angular acceleration of 1 rad/s^2. *After* 2 min *it attains full* speed. *It then moves uniformly for* 11 *min, after which it is brought to rest by a constant retarding force. If the flywheel rotates through* 3 600 rad *while it is being brought to rest, find its maximum speed, its angular retardation and the total number of revolutions it makes.*

Let ω be its maximum angular speed.

First part of the motion:

$$\omega = \omega_0 + \alpha t$$
$$= 0 + 1 \times 120$$
$$= 120 \text{ rad/s}$$
$$\theta = \omega_0 t + \tfrac{1}{2}\alpha t^2$$
$$= 0 + \tfrac{1}{2} \times 1 \times 120^2 \qquad = 7\,200 \text{ rad}$$

Second part of the motion:
The acceleration is zero,

$\therefore$
$$\theta = \omega_0 t + \tfrac{1}{2}\alpha t^2$$
$$= 120 \times 660 + 0 \qquad = 79\,200 \text{ rad}$$

Third part of the motion:
The angle turned through is $\qquad\qquad\qquad$ 3 600 rad

$\therefore\quad$ since $\qquad \omega^2 = \omega_0^2 + 2\alpha\theta,$
$$0 = 120^2 + 2\,.\,\alpha\,.\,3\,600$$
$\therefore \qquad\qquad\qquad \alpha = -2 \text{ rad/s}.$
$\overline{}$

Total angular distance 90 000 rad

$\therefore \qquad$ Total number of revolutions $= 90\,000/2\pi.$

Example 2. *A uniform circular wheel, of mass* 4·8 *kg and radius* 0·5 *m, rotates under the action of a constant couple about a fixed axis through its centre perpendicular to its plane. In* 12 *seconds the speed of rotation increases from* 15 rev/min *to* 33 rev/min. *Find the moment of the couple on the disc and the increase in kinetic energy of the disc.*

The moment of inertia of the flywheel is given by
$$I = Ma^2/2$$
$$= 4\cdot 8 \times (0\cdot 5)^2/2$$
$$= 0\cdot 6 \text{ kg m}^2.$$

UNIFORM ANGULAR ACCELERATION

Applying $L = I\alpha$ about the axis of rotation,

$$L = 0 \cdot 6\alpha$$

$$\therefore \quad \alpha = L/0 \cdot 6$$

and if the couple L is constant, α is constant so that

$$\omega = \omega_0 + \alpha t$$

$$\therefore \quad 33 \times 2\pi/60 = 15 \times 2\pi/60 + (L/0 \cdot 6)12,$$

giving $\quad L = 3\pi/100$ N m.

The increase in kinetic energy is

$$\tfrac{1}{2}I(\omega_2^2 - \omega_1^2) = \tfrac{1}{2} 0 \cdot 6[(33 \times 2\pi/60)^2 - (15 \times 2\pi/60)^2]$$
$$= \tfrac{1}{2} 0 \cdot 6 \times 4\pi^2/60^2[33^2 - 15^2]$$
$$= \tfrac{1}{2} 0 \cdot 6 \times 4\pi^2/60^2 \times 48 \times 18$$
$$= 288\pi^2/1\,000 \text{ J}.$$

Example 3. A uniform disc, of mass m and radius a, is free to rotate without friction about an axis through its centre perpendicular to its plane. One end of a light string is attached to a point on the circumference of the disc and part of the string is wound on the circumference. The other end carries a particle of mass 4m which hangs freely. If the system is released from rest, find the angular acceleration of the disc and the speed of the particle after the disc has turned through an angle θ radians. Assume that part of the string is wound on the circumference of the disc throughout the motion.

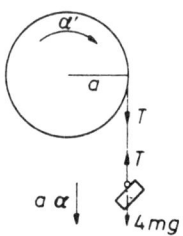

Figure 19.11

Let T be the tension in the string and α the angular acceleration of the disc. Therefore $a\alpha$ is the acceleration of the particle (refer to Figure 19.11).

The tension T in the string acts on the disc and has a moment Ta about the axis of the disc.

For the disc $L = I\alpha$ about its axis

$$Ta = m(a^2/2)\alpha$$

∴ $$T = ma\alpha/2. \quad \ldots (i)$$

For the particle $P = Mf$

∴ $$4mg - T = 4ma\alpha. \quad \ldots (ii)$$

Substituting from equation (i) in equation (ii),

$$4mg - ma\alpha/2 = 4ma\alpha$$

∴ $$8g = 9a\alpha$$

$$\alpha = 8g/9a.$$

Since this acceleration is constant, the angular speed ω of the wheel is given by

$$\omega^2 = \omega_0^2 + 2\alpha\theta$$

∴ $$\omega^2 = 0 + 2 \cdot (8g/9a)\theta$$

∴ $$\omega = \frac{4}{3}\sqrt{\frac{g\theta}{a}}.$$

The speed of the particle is

$$a\omega = a\frac{4}{3}\sqrt{\frac{g\theta}{a}}$$

$$= 4/3\sqrt{ag\theta}.$$

Exercises 19c

1. A flywheel has a moment of inertia of 1 kg m² about its axis. It starts from rest under the action of a constant couple L and reaches a speed of 30 rev/min in 10 s, find L.

2. A flywheel moves with a uniform angular acceleration. In the third and fourth seconds afterwards, it rotates through 70 and 94 revolutions respectively. Find its initial angular speed and its angular acceleration.

3. A wheel is rotating with an angular speed of 130 rad/s. It is subjected to a constant retardation of 6 rad/s. Find the time taken to complete $600/\pi$ revolutions.

ANGULAR SIMPLE HARMONIC MOTION

4. A disc increases its speed of rotation uniformly from 600 rev/min to 2 100 rev/min in 2·5 s. Find the angular acceleration and the number of revolutions made in this time.

5. A uniform circular flywheel of mass 64 kg and radius 1 m starts from rest and rotates under the action of a constant couple. After making 400 revolutions its angular speed is 40 rev/s. Find the moment of the couple.

6. A uniform disc (of mass km, radius a) is free to rotate without friction about an axis through its centre perpendicular to its plane. One end of a light string is attached to a point on the circumference of the disc and part of the string is wound round the circumference. The other end of the string is pulled with a force mg. The disc starts from rest. Find the length of string which unwinds from the disc in time t.

7. A flywheel has a light string coiled round its axle. The string is pulled with a force of 40 N until 40 cm of the string has unwound. The string then slackens and drops off and the flywheel is found to be rotating at $8/\pi$ times a second. Find the moment of inertia of the flywheel.

8. A flywheel is rotating about its axis at 210 rev/min. It is acted on by a constant frictional couple and after 20 seconds it is rotating at 60 rev/min. Find how many more revolutions it will make before it is brought to rest. If the moment of the couple is 3 N m, find the moment of inertia of the flywheel.

9. A wheel and axle has a moment of inertia of 4 kg m^2 rotating about its axis of symmetry. It is subject to a frictional couple of 0·15 N m. A light string is wound several times around the axle which has a radius of 6 cm. To the free end of the string is attached a weight of 3 kg. The system is released from rest. How long will it take for the weight to drop 1 m?

10. A wheel and axle has total mass M and radius of gyration k, it is free to rotate about its axis. The radii of the wheel and axle are a and b respectively. Masses m, m' ($m > m'$) are suspended by means of light strings wound round the circumference of the wheel and axle respectively. The strings are free to unwind without slipping and their tensions have opposite turning moments on the system. If the system is free to move, show that the mass m has an acceleration of $ga(ma - m'b)/(ma^2 + m'b^2 + Mk^2)$.

19.8. ANGULAR SIMPLE HARMONIC MOTION

(Compound Pendulum)

In the previous section we considered the equation of motion, $L = I\alpha$, when L was constant. L can vary in many ways, one is when

ROTATION ABOUT A FIXED AXIS

it varies as θ the angle turned through and $L = -k\theta$ (k a positive constant). The equation of motion is $-k\theta = I\alpha$.

or
$$\frac{d^2\theta}{dt^2} = -\frac{k}{I}\theta$$

which can be written

$$d^2\theta/dt^2 = -n^2\theta \qquad \ldots\text{(i)}$$

which is the equation of simple harmonic motion (refer to Section 5.6). In a similar manner to that section, if $\omega = d\theta/dt$

$$(d\theta/dt) \cdot (d\omega/d\theta) = -n^2\theta$$
$$\omega\, d\omega/d\theta = -n^2\theta$$
$$\tfrac{1}{2}\omega^2 = -\tfrac{1}{2}n^2\theta^2 + C.$$

If $\omega = 0$, when $\theta = \theta_0$, $C = \tfrac{1}{2}n^2\theta_0^2$

$$\therefore \qquad \omega^2 = n^2(\theta_0^2 - \theta^2)$$
$$\omega = n\sqrt{(\theta_0^2 - \theta^2)} \qquad \ldots\text{(ii)}$$

i.e.
$$d\theta/dt = n\sqrt{(\theta_0^2 - \theta^2)}$$

$$\therefore \qquad \int \frac{d\theta}{\sqrt{(\theta_0^2 - \theta^2)}} = \int n\, dt$$

$$\sin^{-1}(\theta/\theta_0) = nt + \varepsilon$$

$$\theta = \theta_0 \sin(nt + \varepsilon). \qquad \ldots\text{(iii)}$$

If in this equation we increase t to $t + (2\pi/n)$,

$$\theta = \theta_0 \sin[n(t + 2\pi/n) + \varepsilon]$$
$$= \theta_0 \sin(nt + \varepsilon)$$

which, as in Section 5.6, gives the periodic time

$$T = 2\pi/n. \qquad \ldots\text{(iv)}$$

Differentiating equation (iii)

$$\omega = \frac{d\theta}{dt} = \theta_0 n \cos(nt + \varepsilon) \qquad \ldots\text{(v)}$$

ANGULAR SIMPLE HARMONIC MOTION

Summarizing we have

If
$$\frac{d^2\theta}{dt^2} = -n^2\theta$$
$$\omega = n\sqrt{(\theta_0^2 - \theta^2)}$$
$$\theta = \theta_0 \sin(nt + \varepsilon)$$
$$T = 2\pi/n$$
$$\omega = \theta_0 n \cos(nt + \varepsilon)$$

Example 1. *A rigid body of mass M is rotating under gravity about a horizontal fixed axis. (This system is known as a Compound Pendulum). The distance of the centre of gravity G from the axis is h and the radius of gyration of the body about an axis through G parallel to the fixed axis is k. Show that the period of small oscillations is* $2\pi\sqrt{[(k^2 + h^2)/hg]}$.

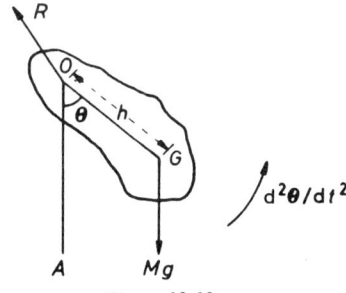

Figure 19.12

Consider the vertical plane through G, the centre of gravity cutting the axis at right angles in O, then $OG = h$. Let AO be the vertical through O and $\angle AOG = \theta$. Then, by the parallel axis theorem, the moment of inertia about the axis through O is $M(k^2 + h^2)$. Taking moments about O (refer to *Figure 19.12*) since

$$L = I\alpha$$
$$Mgh \sin\theta = -M(k^2 + h^2)\,d^2\theta/dt^2.$$

The negative sign occurs because the torque is in the direction of θ decreasing

$$\therefore \quad \frac{d^2\theta}{dt^2} = -\frac{gh}{k^2 + h^2}\sin\theta$$

ROTATION ABOUT A FIXED AXIS

If θ is small, $\sin\theta \simeq \theta$

$$\therefore \quad \frac{d^2\theta}{dt^2} = -\left(\frac{gh}{k^2+h^2}\right)\theta.$$

Comparing with $d^2\theta/dt^2 = -n^2\theta$, we have $n = \sqrt{[gh/(k^2+h^2)]}$, and the body oscillates with simple harmonic motion of period $2\pi\sqrt{(k^2+h^2)/gh}$. A simple pendulum of length l has the period $2\pi\sqrt{l/g}$. Therefore, by comparison, the body has the same period of oscillation as a simple pendulum of length $l = (k^2+h^2)/h$ known as the *simple equivalent pendulum*.

Example 2. A uniform solid sphere of radius 28 cm is pivoted about a horizontal axis distant 14 cm from its centre. A particle P, whose mass is one-third that of the sphere, is attached to the surface of the sphere at its lowest point. If P is now pulled aside so that the diameter through P makes an angle of 10 degrees with the downward vertical and is released, find (a) the time taken till it makes 5 degrees with the vertical, (b) the maximum angular speed acquired.

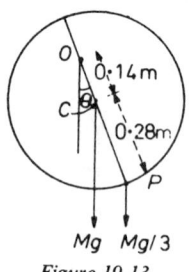

Figure 19.13

Let the vertical plane through P and the centre of the sphere C meet the horizontal axis O. Referring to *Figure 19.13*, the moment of inertia of the sphere about O is (by the parallel axis theorem)

$$M(2a^2/5 + OC^2) = M[(2(0\cdot28)^2/5 + (0\cdot14)^2].$$

Therefore, the moment of inertia of the sphere and the particle about O is given by

$M[2(0\cdot28)^2/5 + (0\cdot14)^2] + M(0\cdot42)^2/3$

$\quad = M[6(0\cdot28)^2 + 15(0\cdot14)^2 + 5(0\cdot42)^2]/15$

$\quad = M(0\cdot14)^2[24 + 15 + 45]/15$

$\quad = M(0\cdot0196 \times 28)/5$

ANGULAR SIMPLE HARMONIC MOTION

Taking moments about O and since $L = I\alpha$,

$$Mg(0\cdot 14)\sin\theta + (Mg/3)(0\cdot 42)\sin\theta = -M(0\cdot 0196 \times 28/5)\,d^2\theta/dt^2$$

$$\therefore \quad \frac{d^2\theta}{dt^2} = \frac{-(5\times 0\cdot 28\times 9\cdot 8)}{(0\cdot 0196\times 28)}\sin\theta$$

$$= -25\sin\theta.$$

Now if θ is small $\sin\theta \simeq \theta$

$$\therefore \quad d^2\theta/dt^2 = -25\theta.$$

Referring back to our summary of equations for simple harmonic motion, and comparing with $d^2\theta/dt^2 = -n^2\theta$, we have that $n = 5$.

$$\therefore \quad \theta = \theta_0 \sin(5t + \varepsilon).$$

When $t = 0$, $\theta = \theta_0 = 10° \times \pi/180 = \pi/18$

$$\therefore \quad \pi/18 = \pi/18 \sin(\varepsilon)$$

$$\therefore \quad \varepsilon = \pi/2$$

and $\quad\quad \theta = (\pi/18)\sin(5t + \pi/2)$

$\therefore$ when $\quad\quad \theta = 5° = \pi/36$ radians

$$\pi/36 = (\pi/18)\sin(5t + (\pi/2))$$

$$\therefore \quad 5t + (\pi/2) = \pi/6 \quad\text{or}\quad 5\pi/6.$$

Hence $\quad\quad 5t = \pi/3.$

$$t = (\pi/15)\,\text{s}.$$

To find the maximum angular speed we use the equation

$$\omega = n\sqrt{(\theta_0^2 - \theta^2)}$$

which is a maximum when $\theta = 0$

Hence $\quad\quad \omega_{\max} = n\theta_0$

$$= 5\,.\,(\pi/18)\,\text{rad/s}.$$

Example 3. A uniform rod AB, of mass M and length 2a, is smoothly hinged to a fixed point at one end A. The other end B is attached to a light elastic string (modulus Mg/2, natural length a) connected to a point C vertically below A. If CA = 4a, find the period of small oscillations about a vertical line through A.

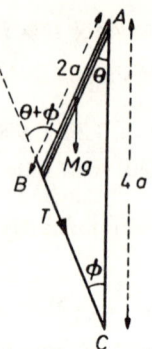

Figure 19.14

Referring to *Figure 19.14* since θ is small
$$AB + BC \simeq AC \quad \text{thus} \quad BC \simeq 2a = AB$$
and
$$\theta = \phi \qquad \ldots \text{(i)}$$

The moment of inertia of the rod about A is
$$M(a^2/3 + a^2) = \tfrac{4}{3}Ma^2.$$

The tension in the string is given by
$$T = \lambda \frac{\text{extension}}{\text{original length}}$$
$$= \tfrac{1}{2}Mg\left(\frac{BC - a}{a}\right).$$

By the cosine rule
$$BC = \sqrt{(4a^2 + 16a^2 - 16a^2 \cos\theta)}$$
$$= 2a\sqrt{(5 - 4\cos\theta)}.$$

$\therefore \qquad T = \tfrac{1}{2}Mg\left[\dfrac{2a\sqrt{(5 - 4\cos\theta)} - a}{a}\right]. \qquad \ldots \text{(ii)}$

Taking moments about A (refer to *Figure 19.14*),
$$Mga \sin\theta + T4a \sin\phi = -\tfrac{4}{3}Ma^2 \, d^2\theta/dt^2.$$

Therefore from equations (i) and (ii)
$$Mga \sin\theta + \tfrac{1}{2}Mg\left[\frac{2a\sqrt{(5 - 4\cos\theta)} - a}{a}\right]4a \sin\theta = -\tfrac{4}{3}Ma^2 \frac{d^2\theta}{dt^2}.$$

For small oscillations $\sin\theta \simeq \theta$ and $\cos\theta \simeq 1$,

$$ga\theta + \tfrac{1}{2}g[(2a - a)/a]4a\theta = -\tfrac{4}{3}a^2\, d^2\theta/dt^2$$

∴ $$3g\theta = -\tfrac{4}{3}a\, d^2\theta/dt^2$$

i.e. $$d^2\theta/dt^2 = -(9g/4a)\theta$$

which is the equation of simple harmonic motion. Hence, the period

$$T = 2\pi/\sqrt{(9g/4a)}$$
$$= (4\pi/3)\sqrt{(a/g)}.$$

In the case of a simple pendulum $T = 2\pi\sqrt{l/g}$ and the length l of the simple equivalent pendulum is given by $\tfrac{2}{3}\sqrt{a/g} = \sqrt{l/g}$ thus

$$l = 4a/9.$$

19.9. REACTION AT THE AXIS OF A ROTATING BODY

Refer again to *Figure 19.12*. Consider a rigid body, of mass M, rotating about a fixed axis under the action of a system of forces. Let the plane through G, the centre of gravity, meet the axis in O. Let $OG = h$, OA be the initial line in the plane and $\angle AOG = \theta$.

If Mk^2 is the moment of inertia of the body about a parallel axis through G (the centre of gravity), the moment of inertia about the axis of rotation is $M(k^2 + h^2)$. Taking moments about O,

$$L = M(k^2 + h^2)\frac{d^2\theta}{dt^2} \qquad \ldots \text{(i)}$$

where L is the sum of the moments about the axis. Now by the result of Section 16.2, the centre of gravity moves as if it were a particle of mass M acted on by all the external forces on the system (including the reaction at the axis)

∴ $$\Sigma P = M\frac{d^2\bar{r}}{dt^2}$$

where $\bar{r}$ is the position vector of the centre of gravity. Equating components

$$\Sigma P_x = M(d^2\bar{x}/dt^2) \qquad \ldots \text{(ii)}$$
$$\Sigma P_y = M(d^2\bar{y}/dt^2) \qquad \ldots \text{(iii)}$$
$$\Sigma P_z = M(d^2\bar{z}/dt^2). \qquad \ldots \text{(iv)}$$

ROTATION ABOUT A FIXED AXIS

These four equations [three in the case of motion in two dimensions] enable us to find the reaction at the axis.

Example. A thin uniform rod AB, of mass m and length 2a, is free to rotate in a vertical plane about a horizontal axis at A. A mass 2 m is attached to the end B. The rod is held horizontal and released. Find the horizontal and vertical components of the reaction at the axis when the rod makes an angle θ with the downward vertical.

Let R, the reaction at the axis, have components X and Y along and perpendicular to the rod whose centre is C.

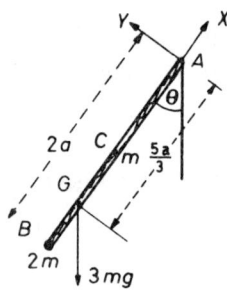

Figure 19.15

Let G be the centre of mass of the rod AB and the attached particle at AB. Then G divides CB in the ratio $2:1$ so that $AG = 5a/3$ (refer to *Figure 19.15*).

Taking moments about A, we have
$$3mg \cdot (5a/3) \sin \theta = -I(d^2\theta/dt^2).$$
The moment of inertia of the system about the axis through A is
$$m(a^2/3) + a^2) + 2m(2a)^2 = 28ma^2/3.$$
Therefore, from the above equations,
$$5mga \sin \theta = -(28ma^2/3)(d^2\theta/dt^2)$$
$$\therefore \quad d^2\theta/dt^2 = -(15g/28a) \sin \theta. \quad \ldots \text{(i)}$$
Since G describes a circle about O, the components of its acceleration along and perpendicular to the rod are respectively, $5a(d\theta/dt)^2/3$ and $5a(d^2\theta/dt^2)/3$. The total mass of the system is $3m$. Resolving along AB,
$$3m \cdot (5a/3)(d\theta/dt)^2 = X - 3mg \cos \theta \quad \ldots \text{(ii)}$$

388

REACTION AT THE AXIS OF A ROTATING BODY

Resolving perpendicular to AB,

$$3m \cdot (5a/3)(d^2\theta/dt^2) = Y - 3mg \sin\theta \qquad \ldots \text{(iii)}$$

From equations (i) and (iii),

$$5ma((-15g/28a)\sin\theta) = Y - 3mg \sin\theta$$

$$\therefore \quad Y = 9mg \sin\theta/28. \qquad \ldots \text{(a)}$$

Integrating equation (i),

$$\tfrac{1}{2}(d\theta/dt)^2 = C + 15g \cos\theta/28a.$$

But $d\theta/dt = 0$ when $\theta = \pi/2$, $\therefore C = 0$. Substituting in equation (ii),

$$5ma(15g \cos\theta/14a) = X - 3mg \cos\theta$$

$$\therefore \quad X = 117mg \cos\theta/14. \qquad \ldots \text{(b)}$$

Horizontal component $= X \sin\theta - Y \cos\theta$
$= 117mg \cos\theta \sin\theta/14$
$\quad - 9mg \sin\theta \cos\theta/28$
$= 225mg \sin 2\theta/56.$

Vertical component $= X \cos\theta + Y \sin\theta$
$= 117mg \cos^2\theta/14 + 9mg \sin^2\theta/28$
$= (mg/28)(234 \cos^2\theta + 9 \sin^2\theta).$

Exercises 19d

1. A rod AB, of negligible weight and length $6a$, has two weights $2m$ and $3m$ attached at B and C respectively, where $AC = 4a$. It is free to rotate in a vertical plane about a horizontal axis through A. Find the period of small oscillations.

2. If, in Question 1, the rod AB had been held horizontally and released, find the components of the reaction at the axis when the rod made an angle θ with the downward vertical.

3. A uniform solid cylinder, of mass M and radius a, is free to rotate with one of its generators as axis. If the axis is horizontal and the cylinder makes small oscillations under gravity, find the length of the simple equivalent pendulum.

4. A uniform rod of mass 0·3 kg and length 1 m, is free to rotate in a vertical plane about a horizontal axis distant 0·1 m from one end. A mass of 4 kg is attached to the other end. Find the length of the simple equivalent pendulum.

5. Three uniform rods AB, BC, CA (each of mass m, length $2a$) are rigidly jointed to form an equilateral triangle ABC. The system can rotate freely about a horizontal axis through A perpendicular

to the plane of the rods. Find the length of the simple equivalent pendulum for oscillations of the system about the axis through A.

6. A uniform rod, of mass 10 kg and length 2·5 m, has a uniform circular disc of radius 25 cm, mass 100 kg, rigidly attached at its centre to one end of the rod. The rod is free to rotate in a vertical plane about a horizontal axis through the other end, find the period of small oscillations.

7. A compound pendulum has radius of gyration k about an axis through its centre of mass. It rotates about a parallel axis making small vertical oscillations. Show that, if the position of this axis is chosen to make the period of oscillation a minimum, then the length of the simple equivalent pendulum is $2k$.

8. A heavy uniform rod (of mass m, length a) has a mass m attached at one end. It is free to rotate in a vertical plane about a horizontal axis through the other end. (a) Find the period of small oscillations about the axis. (b) If the rod is slightly disturbed from the position where the mass is at its highest point, find the reactions at the axis in the subsequent motion.

9. A circular disc (radius a, mass m) is free to rotate in a vertical plane about its centre O, which is fixed. A point P on the rim is attached by a light elastic string (modulus kmg, natural length $a/2$) to a point Q in the plane of the disc vertically below O. If $OQ = 2a$, find the period of small oscillations of the system.

10. A rigid body can rotate about a horizontal axis. The vertical plane through the centre of mass G meets the axis in O and $OG = h$. By the result of Example 1, the period of oscillation is $2\pi\sqrt{[(k^2 + h^2)/hg]}$, where k is the radius of gyration of the body about a parallel axis through G. If OG is extended to C where $OC = (k^2 + h^2)/h$ show that if the body oscillates about a parallel axis through C, its periodic time is unaltered.

11. A uniform rod AB, of mass m and length $2a$, is smoothly hinged to a fixed point at A. A particle of mass $2m$ is firmly attached at B. A light elastic string, of modulus mg and natural length a, joins B to a point C vertically below A. If $CA = 4a$ and the system performs small oscillations in the vertical plane containing A and C, find the length of the simple equivalent pendulum.

19.10. ENERGY METHODS

We have seen in Section 16.4 that the principle of work can be extended from a single particle to a set of particles so that:

Final K.E. − Initial K.E. = Work done by forces on particles.

ENERGY METHODS

Hence it applies to a rigid body and, since the internal forces are in equal and opposite pairs acting *in the same straight line*, we need only consider the work done by external forces.

Example 1. *A uniform thin ring of radius a is free to rotate about a horizontal axis perpendicular to the plane of the ring through a point O on the ring itself. P is the opposite end of the diameter through O. If OP is held horizontal and then P is projected downwards with speed u, find its speed when OP is first vertical.*

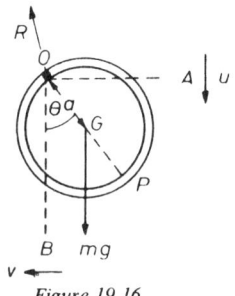

Figure 19.16

The moment of inertia of the ring about a perpendicular axis through G is ma^2 (m being the mass of the ring) (refer to *Figure 19.16*).

Hence its moment of inertia about the axis through O is $m(a^2 + a^2) = 2ma^2$ (using the parallel axis theorem).

Applying the principle of work between A and B,

$$\tfrac{1}{2}I(\omega_2^2 - \omega_1^2) = \text{Work done by weight} \quad (R \text{ does no work}).$$

$$\tfrac{1}{2}2ma^2\left\{\left(\frac{v}{2a}\right)^2 - \left(\frac{u}{2a}\right)^2\right\} = mga,$$

giving
$$v = \sqrt{u^2 + 4ga}.$$

In some problems it is more convenient to determine the work done from the moment of the forces (or couples). Taking moments about the axis of rotation we have

$$L = I\omega \, d\omega/d\theta.$$

Integrating with respect to θ from $\theta = \theta_1$ to $\theta = \theta_2$

$$\int_{\theta_1}^{\theta_2} L \, d\theta = I \int_{\omega_1}^{\omega_2} \omega \, d\omega$$

∴
$$\int_{\theta_1}^{\theta_2} L \, d\theta = \tfrac{1}{2}I(\omega_2^2 - \omega_1^2).$$

ROTATION ABOUT A FIXED AXIS

Hence $\int_{\theta_1}^{\theta_2} L \, d\theta$ is equal to the increase in kinetic energy and must be the work done by the external forces producing the moment L.

If these forces have constant moment L, then

$$\int_{\theta_1}^{\theta_2} L \, d\theta = L(\theta_2 - \theta_1),$$

i.e. work done is moment of the forces × angle turned through. This result will apply in particular to a couple of constant moment G. The work done by such a couple will be $G\theta$ where θ is the angle through which it turns.

Example 2. A constant force, of magnitude F, is applied to the circumference of a stationary circular flywheel of mass 200 kg and radius 1 m. After 25 revolutions, the flywheel is rotating with an angular speed of 1 rev/s. Find the magnitude of F.

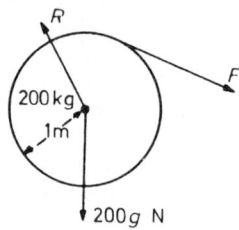

Figure 19.17

The moment of inertia of the flywheel (refer to *Figure 19.17*)

$$= Ma^2/2$$
$$= 200 \times 1^2/2$$
$$= 100 \text{ kg m}^2.$$

The work done by F as the wheel makes 25 revolutions ($= 50\pi$ rad) is

moment of F × angle turned through $= (F \times 1)50\pi$.

The weight and reaction acting at the centre O do no work. Hence by the principle of work,

Final K.E. − Initial K.E. = Work done

∴ $\quad \tfrac{1}{2}100(2\pi)^2 - 0 = F \times 50\pi$

$$F = (100 \times 4\pi^2)/(2 \times 50\pi)$$
$$F = 4\pi \text{ N}.$$

ENERGY METHODS

Example 3. A uniform rod AB, of mass m and length 2a, can turn freely in a vertical plane about a horizontal axis through its centre O. Particles of mass 3 m and 4 m are attached at A and B respectively. The system is released from rest in a horizontal position. Find its angular speed when it is vertical.

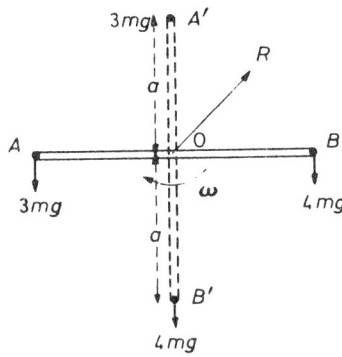

Figure 19.18

The moment of inertia of the rod about the axis O is $ma^2/3$, and of the two masses at A and B, $3ma^2$ and $4ma^2$ respectively. Therefore the total moment of inertia of the system is $22ma^2/3$.

In the vertical position the net work done by gravity is $4mga - 3mga$ (refer to *Figure 19.18*) i.e. mga.

Final K.E. − Initial K.E. = Work done

$$\therefore \quad \tfrac{1}{2} \cdot 22ma^2(\omega^2 - 0)/3 = mga$$
$$\therefore \quad \omega^2 = 3g/11a.$$
$$\omega = \sqrt{(3g/11a)} \text{ rad/s}.$$

Exercises 19e

1. A uniform rod, of mass $3m$ and length $2a$, is free to rotate in a vertical plane about a horizontal axis through one end A. A mass m is attached to the other end B. The system is released from rest in a horizontal position. Find its angular speed about A when vertical.

2. A wheel, whose moment of inertia about its axis is 100 kg m^2, makes 5 rev/s. It is brought to rest after 20 seconds by a constant couple. Find the magnitude of the couple.

3. A uniform circular disc, of mass m and radius a, can rotate in a vertical plane about a horizontal axis through a point O on its circumference. The disc is held with the diameter OA through O

horizontal and then released. If a constant frictional couple $mga/2\pi$ opposes the motion, find the speed of rotation about the axis when OA is vertical.

4. A uniform circular disc, of mass m and radius a, can rotate in a vertical plane about a horizontal axis through A (a point on the circumference of the disc). AB is a diameter of the disc. At B is rigidly attached, by its centre, a rod of length $6a$, mass $3m/2$. The rod is at right angles to AB and in the plane of the disc. The system is released from rest with AB vertical and B uppermost. If at any subsequent time AB makes an angle θ with the upward vertical, find an expression for its angular speed.

5. A circular disc, of mass m and radius a, is free to rotate in a vertical plane about a horizontal axis through a point A on the circumference of the disc. A particle B, of mass km, is attached to the disc at the opposite end of the diameter through A. The system is slightly disturbed from a position of rest with B vertically above A. Show that the speed v of the particle when passing through the lowest point is given by $(3 + 8k)v^2 = 32ga(1 + 2k)$.

6. A uniform rod AB, of mass $3m$ and length $2a$, can turn freely in a vertical plane about a horizontal axis through its centre O. Particles of mass $2m$ and $3m$ are attached to the rod at A and B respectively. The system is released from rest with the rod horizontal, and on passing through the lowest position the heavier particle falls off. Find the angular speed of the rod when it subsequently passes through the horizontal position.

7. A rod AB, of mass m and length $2a$, is free to rotate in a vertical plane Π about its end A. A light elastic string, modulus $mg/4$, original length $2a$, is attached to a point C on the same level as A in the plane Π. The other end of the string is attached to the end B of the rod. The rod is released from rest in a position where $\angle BAC = 60$ degrees. Find the angular velocity of the rod when passing through the lowest point if $AC = 2a$.

19.11. IMPULSIVE MOTION ABOUT A FIXED AXIS

When impulsive forces act on a rigid body, we have very large forces acting over a very small interval of time. During this time comparatively small forces such as weight and frictional couples can be ignored. We have, for the equation of motion about the fixed axis,

$$L = I\alpha$$

or $$L = I(d\omega/dt).$$

IMPULSIVE MOTION ABOUT A FIXED AXIS

Integrating with respect to t over our small time interval

$$\int_{t_1}^{t_2} L \, dt = \int_{\omega_1}^{\omega_2} I \, d\omega$$

$$\int_{t_1}^{t_2} L \, dt = I\omega_2 - I\omega_1 \qquad \ldots (i)$$

Now $L = \Sigma pF$ and over our small time interval, F, though varying in magnitude, acts along a constant line of action, hence p is constant

$$\therefore \quad \int L \, dt = \int (\Sigma pF) \, dt$$

$$= \Sigma \left(\int pF \, dt \right)$$

$$= \Sigma \left(p \int F \, dt \right)$$

$$= \Sigma pJ$$

where J is the impulse of the force F (refer to Section 13.2). Thus equation (i) becomes

$$\Sigma pJ = I\omega_2 - I\omega_1$$

and once again, since the internal impulses will occur in equal and opposite pairs, the summation can be taken over the external forces only. This equation is equivalent to the statement that, the sum of the moments of the external impulses about the axis is equal to the increase of angular momentum.

Example 1. A thin uniform rectangular lamina ABCD can rotate freely about the line AB which is horizontal. The mass of the lamina is 0.5 kg and $BC = 1$ m. An impulse of 4 N s is applied to the mid-point of CD in a plane at right angles to the lamina. Find the angular speed of the lamina immediately after the blow.

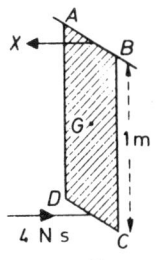
Force diagram

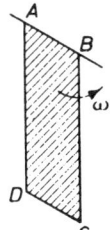
Velocity diagram

Figure 19.19

395

ROTATION ABOUT A FIXED AXIS

Let X be the magnitude of the reaction of the axis and ω the instantaneous angular speed of rotation about AB (refer to *Figure 19.19*). Since the moment of the external impulses about the axis is equal to the increase of angular momentum, taking moments about AB to eliminate X

$$4 \times 1 = mk^2\omega.$$

For a rectangle about one edge $k^2 = 4a^2/3$.
Therefore in this case $k^2 = 4(\tfrac{1}{2})^2/3 = \tfrac{1}{3}$

∴ $$4 \times 1 = 0{\cdot}5 \times \tfrac{1}{3}\omega$$

∴ $$\omega = 24 \text{ rad/s}.$$

Example 2. *A rod AB, of mass m, and length $2a$, can turn freely about a horizontal axis through A. It is hanging vertically at rest with B below A. A horizontal impulse I is given to the rod, at a distance x from A. Show that, if $x = 4a/3$, there is no impulsive reaction at the axis, and in this case, find the initial kinetic energy given to the rod.*

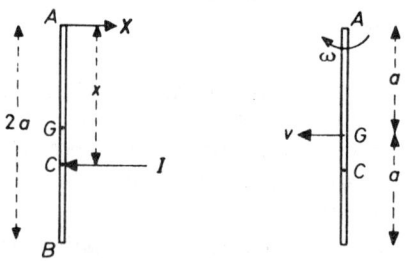

Force diagram   Velocity diagram
Figure 19.20

Let X be the impulsive reaction at the axis. Since I is horizontal, then X will be horizontal. Also let ω be the instantaneous angular speed of rotation about A. Referring to *Figure 19.20*

$$\text{Speed of } G \qquad v = a\omega \qquad \ldots\text{(i)}$$

and since

$$\text{impulse} = \text{change in momentum,}$$
$$I - X = m(a\omega - 0)$$
∴ $$I - X = ma\omega. \qquad \ldots\text{(ii)}$$

Also
the moment of the impulse = change of angular momentum.
Taking moments about A (to eliminate X),
$$Ix = mk^2\omega - 0.$$
For a rod about one end $k^2 = 4a^2/3$.
$$\therefore \qquad I = 4ma^2\omega/3x. \qquad \ldots\text{(iii)}$$
From equations (ii) and (iii)
$$X = I - ma\omega$$
$$= 4ma^2\omega/(3x) - ma\omega$$
$$= ma\omega((4a/3x) - 1)$$
$$\therefore \qquad X = 0, \quad \text{if} \quad 4a/3x = 1$$
i.e. if $\qquad x = 4a/3.$

Thus, when $AC = 4a/3$, there is no impulsive reaction at the axis and C is known as the *centre of percussion*.

When $X = 0$, equation (ii) becomes
$$I = ma\omega$$
$$\therefore \qquad \omega = I/ma. \qquad \ldots\text{(iv)}$$

The motion of the rod is one of rotation about the axis at A.
$$\therefore \qquad \text{K.E.} = \tfrac{1}{2}mk^2\omega^2$$
$$= \tfrac{1}{2}m\frac{4a^2}{3}\omega^2$$
$$= \frac{2ma^2}{3} \cdot \frac{I^2}{m^2a^2} \quad \text{(from equation (iv))}$$
$$= 2I^2/3m.$$

Exercises 19f

1. A uniform rod, of mass 4 kg and length 2 m, lies at rest on a smooth horizontal surface. An impulse of 8 N s is applied at one end in a direction making an angle of 30 degrees with the rod. Find the kinetic energy created by the blow.

2. A uniform rod, of mass M, has an impulse applied at right angles to one end. If the other end begins to move with speed v, find the magnitude of the impulse.

3. A is a point on the rim of a uniform circular disc, centre O, lying at rest on a smooth horizontal surface. An impulse J is applied at A

ROTATION ABOUT A FIXED AXIS

which causes A to move with speed v, in a direction making an angle θ with AO. Find the magnitude and direction of J.

4. A solid body, of mass m, is free to rotate about a horizontal axis. A perpendicular plane Π passes through G (the centre of gravity of the body) and meets the axis in a point O, where $OG = h$. The body is given a horizontal impulse of magnitude I whose line of action lies in the plane Π. I is at right angles to OG and meets OG in the point X where $OX = x$. Show that the initial angular speed about the axis through O is $Ix/m(k^2 + h^2)$, where k is the radius of gyration of the body about an axis through G parallel to the axis of rotation. Find R the magnitude of the impulsive reaction at the axis and show that if $x = (k^2 + h^2)/h$ [the length of the simple equivalent pendulum (refer to Section 19.7)] then $R = 0$.

5. A uniform rod, of mass m and length $2a$, which is free to rotate about a horizontal axis through a point distance x above its centre, hangs in equilibrium. It is given a horizontal impulse at a point distance x below its centre. If there is no impulsive reaction at the axis, find x.

6. A square lamina, of mass m and length of side $2a$, is free to rotate, with its plane vertical about a horizontal axis through one corner A. An impulse I is given to the lamina at the diagonally opposite corner when it is hanging in equilibrium. If I is in the plane of the lamina and in a horizontal direction, find the kinetic energy imparted to the lamina.

7. A uniform rod, of mass m and length $2a$, is free to rotate in a vertical plane about a horizontal axis through A. It is released from rest in a horizontal position. Find its angular speed when passing through the vertical position. At this moment a horizontal impulse I is applied to the centre of the rod which reverses and halves its angular speed, find the magnitude of I.

8. A uniform circular lamina, of mass m and radius a, is free to rotate in a vertical plane about a horizontal axis through a point P on its circumference. It is released from rest with the diameter through P horizontal and is brought to rest by a small fixed inelastic peg distant a vertically below P. Find the impulsive reaction at the moment of impact.

EXERCISES 19*

1. Find the moment of inertia of a solid hemisphere (radius a, mass M) about a diameter of its plane face. Deduce its moment of inertia about a tangent at the vertex.

* Exercises marked thus, †, have been metricized.

EXERCISES

2. A uniform circular disc, of mass 4·9 kg and radius 0·5 m, can turn freely about a fixed axis through its centre of mass perpendicular to its plane. A constant couple of 0·2 kg m acts on the body in a plane perpendicular to the axis of rotation. Find the time taken for the speed of rotation to increase from 4 to 20 rad/s and the angle turned through in this time.

3. A light inextensible string passes over a pulley of radius a and moment of inertia $ma^2/2$. Masses $2m$ and m are attached to the ends of the string. Find the acceleration of the masses if a constant frictional couple G acts on the wheel.

4. A uniform circular disc centre O of mass M and radius a, is free to turn in a vertical plane about a horizontal axis through its centre. A particle (of mass m) is attached to a point P on its circumference, and PO makes an angle α with the upward vertical. The system is released from rest. Find the angular speed of the disc as it passes through the lowest point.

5. Show that the moment of inertia of a hollow sphere (density ρ, external and internal radii R and r respectively), about a diameter is $8\pi\rho(R^5 - r^5)/15$.

6. Show that the moment of inertia of a uniform straight rod of mass M about an axis (a) through the centre perpendicular to the rod, and (b) through one end perpendicular to the rod is the same as the moment of inertia of three particles, one of mass $2m/3$ in the centre of the rod, and one of mass $m/6$ at each end of the rod. [This is an example of two systems whose moments of inertia about *all* axes are equal. They are known as *equimomental systems*.]

7. Four rods each of mass m and length $2a$, are rigidly jointed at their ends to form a square. The system makes small oscillations in a vertical plane about a horizontal axis through the mid-point of one side. Find the length of the simple equivalent pendulum.

8. A circular disc, of mass m and radius a, is free to rotate in a vertical plane about its centre O which is fixed. Two light elastic strings AB, CD each of modulus kmg and original length $a/2$, are attached one to each end of a diameter BOC of the disc. The other ends of the string are attached to two points A and D in the plane of the disc on the same horizontal level as O. $AO = 2a = OD$. The system is slightly disturbed from its position of stable equilibrium. Find the time of small oscillations.

9. A circular cylinder, of mass m and radius a, is free to turn about its axis which is horizontal. A light inextensible string is wound around the cylinder its free end is attached to a freely hanging mass m. The string unwinds not slipping on the cylinder. Find the tension in the string when the system is moving.

10. Two particles, of mass $2m$ and m, are connected by a light inextensible string passing over a rough circular pulley of mass m and radius a. The pulley is free to turn about its axis. Find the acceleration of the particles, assuming the string does not slip on the pulley.

11. Twelve equal rods, each of mass m and length $2a$, are rigidly jointed together to form a cube. Find the moment of inertia of the cube about a line through its centre parallel to one of the rods.

12. CD is a uniform rod of mass m and length $2a$. It is rigidly jointed at right angles at C to the mid-point of a similar rod AB. The system makes small oscillations in a vertical plane about a horizontal axis through a point P on AB, where $AP = a/2$. Find the length of the simple equivalent pendulum.

13. A rod AB is free to rotate about a horizontal axis through A. It is given simultaneously two parallel impulses of magnitudes I and J in a direction perpendicular to AB. I acts at the mid-point of the rod and J at B. If there is no impulsive reaction at the axis through A, find the ratio $I:J$.

14. A uniform straight rod mass m has the same moment of inertia about all axes as three particles, one of $2m/3$ at its centre and two of $m/6$, one at each end of the rod. [Equimomental Systems.] Deduce that a uniform parallelogram mass M is equimomental with: a particle of mass $4M/9$ at its centre of mass, four particles each of mass $M/36$ at the vertices, and four particles each of mass $M/9$ at the mid-points of the sides.

15. Two particles of mass $4m$ and $3m$ are connected by a light inextensible string passing over a rough circular pulley (of mass $2m$ and radius a). The pulley is free to turn about its axis. Find the accelerations of the particles assuming that the string does not slip on the pulley.

16. A circular lamina, of mass m, radius a and centre O, rotates in a vertical plane about a horizontal axis through a point P on its circumference. The axis is perpendicular to the plane of the lamina. A particle, of mass m, is attached to the lamina at a point B on the diameter through P, where $PB = 3a/2$. Show that if, at any time t, PO makes an angle θ with the downward vertical, then $3a\dot\theta^2 - 4g\cos\theta$ is constant. By differentiating this equation with respect to t, or otherwise, find the length of the simple equivalent pendulum.

17. Find the moment of inertia of a triangular lamina (mass M, perpendicular height h) about a line through the vertex parallel to the base. Hence by the parallel axis theorem, find its moment of inertia about the base. Verify that both these results are the same as

EXERCISES

the moments of inertia of three particles each of mass $M/3$ situated at the mid-points of the sides of the triangle.

18. Find the moment of inertia of a uniform solid right circular cone of mass M, height h and base radius r, about (a) the axis, (b) a line through the vertex parallel to the base.

If these results are equal, show that $r = 2h$ and that the minimum moment of inertia about a line parallel to the base is $51Mh^2/80$.

(London)

19. If in Question 16 the particle of mass m is removed from the circular lamina, show that the length of the simple equivalent pendulum is unaltered.

20. Four thin rods each of mass m, length $2a$ are rigidly jointed to form a square $ABCD$. The system can swing in its own plane about a horizontal axis through A, the diagonal AC never rising above the horizontal. Show that the magnitude of the resultant reaction on the axis varies between $8mg/5$ and $44mg/5$.

21. A lamina is in the form of a trapezium whose parallel sides are of lengths a and b and are at a distance h apart. If the mass of the lamina is M, use the equimomental system for a triangular lamina (refer to Question 17) to find its moment of inertia about the side of length a.

22. A uniform cube of edge $2a$ is placed with one edge AB on a perfectly rough plane. The cube is in unstable equilibrium with the diagonally opposite edge to AB in the same vertical plane as AB. If the cube is gently pushed, find its angular speed when one of its faces hits the plane.

23. A wheel consists of a circular disc of radius a with four circular holes each of radius $a/4$. The centres of the holes form a square and each centre is distant $a/2$ from the centre of the disc. If m is the mass of the wheel after the holes have been punched in it, find the moment of inertia of the wheel about its axis.

A light elastic string of modulus mg and original length $8a/9$ is connected at one end to a point P on the circumference of the wheel and at the other end to a point Q. If Q is in the plane of the wheel and $OQ = 3a$ where O is the centre of the wheel, find the time of small oscillations of the system about its equilibrium position.

24. A thin uniform rod AB, of mass m and length $2a$, has one end A resting on a rough horizontal plane. The rod is gently disturbed from the vertical position and rotates about the end A which begins to slip when the rod makes an angle of 45 degrees with the vertical. Find the coefficient of friction between the rod and the plane.

25. A uniform rod of length $20\,m$ is held on a horizontal table perpendicular to one edge. $14\,m$ of the rod projects over the edge.

The rod is allowed to turn about the edge of the table. If $\mu = 0.61$, show that the rod will begin to slip when it has turned through an angle $\tan^{-1} 0.25$.

26. A uniform circular disc is of mass M and radius a. Find its moment of inertia about a horizontal axis perpendicular to its plane which passes through a point O on the rim of the disc.

The disc is freely pivoted about this axis and is held with the diameter through O horizontal. It is then released. Find the horizontal and vertical components of the thrust on the axis when the disc has turned through an angle θ.

Hence or otherwise, show that the total thrust on the axis has $\frac{1}{3}Mg$ as a minimum and $\frac{7}{3}Mg$ as a maximum value. (London)

27. A rigid body can turn freely about a fixed axis through its centre of gravity and the moment of inertia of the body about this axis is I. A constant couple of magnitude N acts on the body in a plane perpendicular to the axis of rotation. Find (a) the time taken for the angular velocity of the body to increase from ω_1 to ω_2 (b) the angle through which the body rotates in this time.

A uniform circular disc of mass 20 kg and radius 2 m rotates, under the action of a constant couple, about a fixed axis through its centre perpendicular to its plane. If the speed of rotation changes from 10 rev/min to 35 rev/min in 15 seconds, find the moment of the couple on the disc.

Find also the angular acceleration and the number of revolutions made by the disc in this time. (London)†

28. A non-uniform rod AB of length $8a$ has mass M and can swing freely in a vertical plane about one end A. The radius of gyration of the rod about A is $6a$ and the distance of its centre of mass from A is $5a$. When the rod is hanging in equilibrium, it is struck by a horizontal impulse of magnitude I at a point P distant $5\frac{1}{7}a$ from A. If in the subsequent motion the rod just reaches the upward vertical, prove that $I = 7M\sqrt{5ag/3}$.

29. A point P moves in a circle of radius r and centre O. The angle between OP and a fixed direction is denoted by θ. Prove that the tangential and normal components of the acceleration of P are $r\,d^2\theta/dt^2$ and $r(d\theta/dt)^2$ respectively. (In your proof do not assume any known expressions for these components.)

A uniform rough rod of mass m and length $2a$ is freely pivoted to a fixed point at its centre O. A small ring of mass m is threaded on the rod, and the rod and ring are released from rest with the rod horizontal and the ring distant $r\,(<a)$ from O. The inclination of the rod to the horizontal is denoted by θ. Show that, before the ring slips,
$$(a^2 + 3r^2)(d\theta/dt)^2 = 6gr\sin\theta,$$

and find $d^2\theta/dt^2$ in terms of θ. Show that the ring will slip when θ reaches the value given by

$$\tan\theta = \frac{\mu a^2}{a^2 + 9r^2}$$

where μ is the coefficient of friction between the ring and the rod.
(JMB)

30. A uniform rod of mass m and length $2a$ lies on a smooth horizontal table and is rotating freely in the plane of the table about one end O which is fixed. The rod strikes a particle of mass m at rest at a distance x from O, whereupon the particle acquires a speed u. Find the impulsive reaction at O and show that this reaction vanishes if $x = 4a/3$.

If the particle adheres to the rod on impact and there is no impulsive reaction at O, find what fraction is lost of (a) the angular velocity of the rod (b) the kinetic energy of the system. (London)

31. A uniform circular lamina, of radius a and mass m, is free to rotate about a tangent which is fixed in a horizontal position. It is released from rest with its plane horizontal and is brought to rest by an inelastic peg fixed at a distance a, vertically below the point of contact of the tangent. Find the impulsive reaction at the moment of impact.

32. A rectangular closed box $ABCDPQRS$ is made from a uniform sheet of metal. The ends $ABCD$ and $PQRS$ are squares of side a and the length of the box is $2a$. Two adjacent sides $APQB$ and $APSD$ are removed and the remainder, which is of mass M, can rotate freely about AP which is horizontal. Find (a) its moment of inertia about AP (b) the period of small oscillations. (London)

33. Find the moment of inertia of a uniform circular disc about an axis through its centre perpendicular to its plane.

A light inextensible string is connected at its ends to two particles of masses m_1 and m_2 ($m_1 > m_2$) and passes over a uniform circular pulley of mass M which can rotate freely about a fixed horizontal axis through its centre. The particles hang freely and the system is released from rest. If the pulley is sufficiently rough to prevent the string slipping, find the acceleration of either particle.

Show that the motion is identical with that obtained by neglecting the inertia of the pulley and increasing the masses of the particles by equal amounts. (London)

34. Prove that the period of small oscillations of a rigid body moving under gravity about a smooth fixed horizontal axis is

$$2\pi\sqrt{\left(\frac{I}{Mgh}\right)},$$

where M is the mass of the body, I its moment of inertia about the axis and h the distance of its centre of mass from the axis.

Three equal uniform rods AB, BC, CD, each of mass m and length $2a$, are rigidly joined together at B and C so as to form three sides of a square, and the whole can rotate freely about a horizontal axis through A perpendicular to the plane of the rods. Find the moment if inertia of the system about the axis of rotation and the period of small oscillations about that axis.

(Proofs of formulae relating to moments of inertia are not required.)

(JMB)

35. A piece of heavy uniform wire of mass M and length $20a$ is bent in the form of an isosceles trapezium $ABCD$ in which the equal sides BC and AD are each of length $5a$ and CD is of length $8a$. Prove that the moment of inertia of the wire about an axis through N, the mid-point of CD, perpendicular to the plane of $ABCD$ is $149Ma^2/15$.

The wire is freely suspended so that it can oscillate in a vertical plane about the same axis through N. Calculate the period of small oscillations about the position of stable equilibrium. (London)

36. A uniform circular disc mass M can rotate freely about its axis which is horizontal. A light inextensible string is wrapped round the circumference of the disc, one end being attached to the disc the other supporting a freely hanging particle P of mass m. The system is released from rest. Find, by the principle of conservation of energy, the speed of P when it has fallen a distance d, and the tension in the string.

37. Four uniform rods each of length $2a$ and mass m are rigidly jointed together at their ends to form a square $ABCD$. The system can rotate freely about a horizontal axis through A perpendicular to the plane of $ABCD$. The system is gently disturbed from its position of unstable equilibrium with C vertically above A. Find its angular speed when C is vertically below A.

38. Find the moment of inertia of a uniform rod OA of mass m and length $2a$ about an axis through O perpendicular to the rod.

The rod is free to rotate in a vertical plane about a fixed pivot at O. A particle of mass $2m$ is attached to the rod at A with A vertically above O and the system is slightly disturbed from its position of rest. Show that when the rod makes an angle θ with the downward vertical

$$14a\dot\theta^2 = 15g(1 + \cos\theta)$$

and find the speed of the particle when it reaches its lowest point.

At this point the particle drops off. Show that A subsequently rises to a height $6a/7$ above the horizontal plane through O.

(London)

20
MOTION IN TWO DIMENSIONS

20.1. GENERAL EQUATIONS OF MOTION

A rigid body may be considered as a collection of particles rigidly joined together. In Sections 16.2 and 16.4 we saw that the motion of a system of particles (of total mass M say) may be considered in two parts:

(a) a translational motion, the centre of mass G moving as if it were a particle of mass M under the action of the external forces;

(b) a rotational motion about G, depending on the sum of the moments of these external forces.

Hence, to investigate the motion of a rigid body, we shall apply $P = mf$ at G and, for two-dimensional motion, $L = I\alpha$ about an axis through G.

Example 1. A sphere, of mass M and radius a, is rolling down a plane which is inclined at an angle α to the horizontal. Find the acceleration of the centre of mass and show that for rolling $\mu \geqslant 2 \tan \alpha / 7$ where μ is the coefficient of friction.

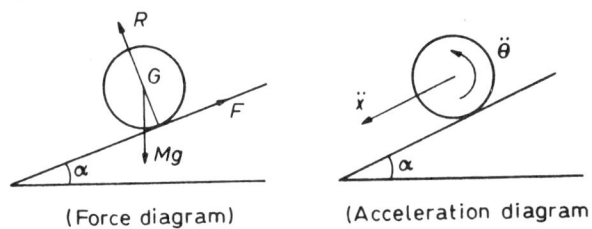

(Force diagram)   (Acceleration diagram)

Figure 20.1

Refer to *Figure 20.1*.
Applying $P = mf$ at the centre of mass G,

along the plane $\qquad Mg \sin \alpha - F = M\ddot{x}^*$ ....(i)

* Dots are a conventional way of indicating differentiation with respect to time, thus

$$\dot{x} = \frac{dx}{dt}, \qquad \ddot{x} = \frac{d^2 x}{dt^2}, \qquad \ddot{\theta} = \frac{d^2 \theta}{dt^2} \quad \text{etc.}$$

MOTION IN TWO DIMENSIONS

perpendicular to the plane $\quad Mg\cos\alpha - R = 0.$ $\quad\ldots$ (ii)

Taking moments about the C.G., $\quad L = I\alpha$

$$Fa = Mk^2\ddot{\theta}. \quad\ldots\text{(iii)}$$

There are four unknowns $\ddot{x}$, $\ddot{\theta}$, F, R and one other equation is required, this is found from the geometry of the body. Since it is rolling without slipping,

$$x = a\theta$$

$$\therefore \quad \ddot{x} = a\ddot{\theta}. \quad\ldots\text{(iv)}$$

From equations (iii) and (iv), since $k^2 = 2a^2/5$ for a sphere rotating about a diameter,

$$F = \tfrac{2}{5}M\ddot{x} \quad\ldots\text{(a)}$$

Substituting in equation (i)

$$Mg\sin\alpha - \tfrac{2}{5}M\ddot{x} = M\ddot{x}$$

$$\therefore \quad \ddot{x} = \tfrac{5}{7}g\sin\alpha. \quad\ldots\text{(b)}$$

Therefore the body moves with constant acceleration.

We have assumed that F is large enough to prevent slipping at the point of contact, that is, $F \leqslant \mu R$. Hence from equations (ii) and (a)

$$\tfrac{2}{5}M\ddot{x} \leqslant \mu \cdot Mg\cos\alpha.$$

Substituting for $\ddot{x}$ from equation (b),

$$\tfrac{2}{5}M\tfrac{5}{7}g\sin\alpha \leqslant \mu Mg\cos\alpha.$$

$$\therefore \quad \mu \geqslant 2\tan\alpha/7.$$

If μ is less than this value $F(=\mu R)$ is insufficient to stop the sphere slipping at its point of contact with the plane. In this case $\ddot{x} \neq a\ddot{\theta}$ and equation (i) becomes

$$Mg\sin\alpha - \mu R = M\ddot{x}.$$

From equation (ii) $\quad R = Mg\cos\alpha$

$$\therefore \quad \ddot{x} = g(\sin\alpha - \mu\cos\alpha).$$

Example 2. A car, of mass M, is moving with speed v around a horizontal circular track of radius r. a is the height of the centre of gravity assumed to be in a central position above the track, and d, the distance between the inner and outer wheels. Show that, if $v^2 > grd/2a$, the car will overturn (assume that the force of friction is large enough to prevent sliding).

406

GENERAL EQUATIONS OF MOTION

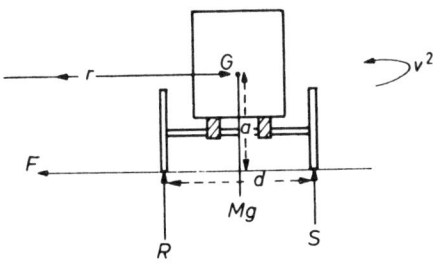

Figure 20.2

Figure 20.2 shows a vertical section through the centre of mass G of the car and the centre of the circle. R and S are the normal reactions of the ground on the wheels, while F is the total friction force at the wheels.

Applying $P = mf$ at G,

horizontally $F = Mv^2/r$ .... (i)

vertically $R + S - Mg = 0.$ .... (ii)

Taking moments about G, $L = I\alpha$. As there is no rotation $\alpha = 0$ and

$$Fa + (Rd/2) - (Sd/2) = 0$$

$\therefore$ $S - R = (2aF/d).$ .... (iii)

The force F, which causes the car to go round in a circle, is provided when the driver turns the wheels. F has an outward rotational effect about G which is balanced by the net inward rotational effect $d(S - R)/2$ [refer to equation (iii)] of the reactions. The difference between S and R produces this effect but $R + S$ is always equal to Mg [refer to equation (ii)]. As v is increased (or r decreased), the difference $S - R$ must increase to balance the increased value of $F(= Mv^2/r)$. When $S = Mg$ and $R = 0$, the difference $S - R$ cannot be increased and any further increase of speed (and hence of F) turns the car outward.

Subtracting equation (iii) from equation (ii)

$$2R = Mg - (2aF/d).$$

Substituting for F from equation (i)

$$2R = Mg - (2Mav^2/rd)$$

$\therefore$ $R = 0$ if $Mg - (2Mav^2/rd) = 0,$

i.e. $v^2 = (grd/2a).$

407

MOTION IN TWO DIMENSIONS

and hence if $v^2 \geq (grd/2a)$, the car will overturn, assuming that the coefficient of friction is large enough to prevent sideslip.

In order to overcome the tendency to overturn and/or sideslip the track can be banked so that no frictional force is necessary to keep the car moving in a circle. Suppose its inclination to the horizontal is θ (refer to *Figure 20.3*) and that there is no tendency to sideslip when the car is travelling at speed v in a circle of radius r.

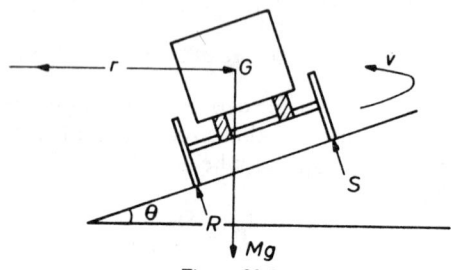

Figure 20.3

Since $P = Mf$,

horizontally $\qquad (R + S) \sin \theta = (Mv^2/r) \qquad \ldots$ (i)

vertically $\qquad (R + S) \cos \theta - Mg = 0$

$\therefore \qquad (R + S) \cos \theta = Mg. \qquad \ldots$ (ii)

Dividing equation (i) by equation (ii)

$$\tan \theta = (v^2/rg).$$

Example 3. A solid spherical ball, of mass m and radius a, is lying at the bottom of a fixed, hollow, perfectly rough sphere of radius b ($>a$). It is given a horizontal blow such that the initial speed of its centre is v. Find the ranges of values of v if the ball is not to leave the sphere.

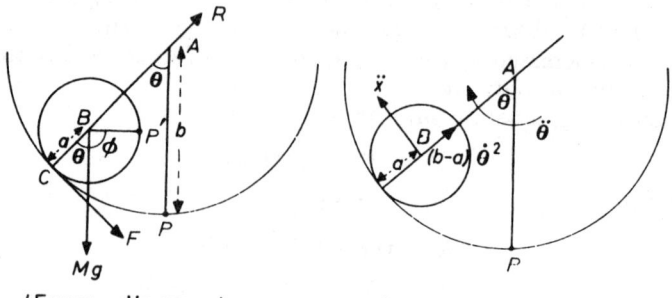

(Force diagram) $\qquad$ (Acceleration diagram)

Figure 20.4

408

GENERAL EQUATIONS OF MOTION

Referring to *Figure 20.4*, A and B are the centres of the two spheres: AP is the initial line of centres; P' is the point on the rolling sphere originally in contact with the fixed sphere at P. Let θ be the angle turned through by the line of centres. Let BP' make an angle ϕ with the vertical. This is the angle turned through by the rolling sphere about its own centre and thus its angular acceleration about B is $\ddot{\phi}$.

Since the sphere is perfectly rough, no slipping occurs and

$$\text{arc } CP' = \text{arc } CP \quad \text{(refer to Figure 20.4)}$$

$$\therefore \quad a(\theta + \phi) = b\theta$$

$$\therefore \quad a\phi = (b - a)\theta$$

$$\therefore \quad a\ddot{\phi} = (b - a)\ddot{\theta}. \quad \ldots \text{(i)}$$

Also $\quad \ddot{x} = AB\ddot{\theta}$

or $\quad \ddot{x} = (b - a)\ddot{\theta}. \quad \ldots \text{(ii)}$

For the motion of the centre of gravity of the rolling sphere we have $P = mf$. Considering components along and perpendicular to BA

$$R - mg\cos\theta = m(b - a)\dot{\theta}^2 \quad \ldots \text{(iii)}$$

$$F + mg\sin\theta = -m\ddot{x} \quad \ldots \text{(iv)}$$

Taking moments about B applying $(L = I\alpha)$,

$$Fa = m\tfrac{2}{5}a^2\ddot{\phi}. \quad \ldots \text{(v)}$$

From equations (i) and (v)

$$Fa = m\tfrac{2}{5}a^2\left(\frac{b-a}{a}\right)\ddot{\theta}.$$

$$\therefore \quad F = m\tfrac{2}{5}(b - a)\ddot{\theta}. \quad \ldots \text{(a)}$$

Substituting in equation (iv)

$$m\tfrac{2}{5}(b - a)\ddot{\theta} + mg\sin\theta = -m\ddot{x}$$
$$= -m(b - a)\ddot{\theta} \quad \text{[from equation (ii)]}.$$

$$\therefore \quad \ddot{\theta}(b - a)[(2/5) + 1] = -g\sin\theta.$$

Multiplying both sides of this equation by $2\dot{\theta}(= 2\,d\theta/dt)$, and integrating with respect to t,

$$(b - a)\tfrac{7}{5}\int 2\dot{\theta}\ddot{\theta}\,dt = -2g\int \sin\theta\frac{d\theta}{dt}dt.$$

$$\therefore \quad \tfrac{7}{5}(b - a)\dot{\theta}^2 = C + 2g\cos\theta \quad \ldots \text{(b)}$$

Now when $\theta = 0$, $\dot\theta = v/(b-a)$

$\therefore \quad \frac{7}{5}v^2/(b-a) = C + 2g.$ ....(c)

From equations (b) an (c)

$\frac{7}{5}(b-a)\dot\theta^2 - \frac{7}{5}v^2/(b-a) = 2g(\cos\theta - 1)$

$\therefore \quad (b-a)\dot\theta^2 = v^2/(b-a) + 10g(\cos\theta - 1)/7.$ ....(d)

There are two conditions for the ball to remain in contact with the sphere.

(a) If the centre of gravity of the ball never rises above the horizontal diameter of the fixed sphere: this limiting case is when $\dot\theta^2 = 0$ for $\theta = \pi/2$. Equation (d) becomes

$$0 \geqslant v^2/(b-a) + \tfrac{10}{7}g(0-1),$$

i.e. $\quad v \leqslant \sqrt{[10g(b-a)/7]}.$

(b) If the ball has sufficient speed to make complete revolutions around the inside of the sphere: in this case $R \geqslant 0$ for $\theta = \pi$, that is, from equation (iii)

$$m(b-a)\dot\theta^2 + mg\cos\theta \geqslant 0 \quad \text{for } \theta = \pi$$

i.e. $\quad (b-a)\dot\theta^2 \geqslant g$

substituting for $\dot\theta^2$ from equation (d) we have

$$v^2/(b-a) + 10g(\cos\theta - 1)/7 \geqslant g \quad \text{for } \theta = \pi$$

i.e. $\quad v^2/(b-a) + 10g(-2)/7 \geqslant g$

$$v^2/(b-a) \geqslant 27g/7$$

$$v \geqslant \sqrt{[27g(b-a)/7]}.$$

Exercises 20a

1. A uniform circular disc is rolling down a plane inclined at an angle α to the horizontal. Find the least value of μ, the coefficient of friction, which will prevent sliding, and the acceleration of the centre of the disc.

2. If in Question 1 the disc is replaced by a circular hoop and $\alpha = 30$ degrees, what is the value of μ and the acceleration of the centre of the hoop?

3. A light inextensible string is wound round a circular cylindrical reel of mass m and radius a. The end of the string is held fixed and the reel allowed to fall so that the thread unwinds. If the axis of the reel remains horizontal, find the acceleration of its centre and the tension in the thread.

4. A pendulum consists of a rod of mass m tied at one end to a light inextensible string of length l. The centre of mass of the rod moves in a horizontal circle of radius r with speed v. Find the angle the rod makes with the vertical.

5. A motor car is rounding a curve of 40 m radius on a level road. The distance between the lines of the wheels is 1·5 m and the centre of gravity of the car and passengers is 0·5 m from the ground and midway between the line of the wheels. Find the maximum speed at which the car can travel without overturning and the least value of μ, the coefficient of friction, which will stop sideslip at this maximum speed.

6. If to the car in Question 5 new tyres are fitted, and thus μ is increased to a value 1·6, can the car with the same load travel around the same curve at a higher speed?

7. Show that a cyclist travelling on level ground with a speed v in a circular path of radius r has to lean inwards through an angle θ from the vertical where $\tan \theta = v^2/gr$.

8. A set of rails in the form of a curve of radius 400 m is banked so that for a train travelling at 14 m/s round the curve there is no lateral thrust on the rails. The width between the rails is 1·34 m. Find how much the outer rail is raised above the level of the inner rail.

9. A reel wound with string is a cylinder, of mass $4m$ and radius a, and two discs, each of mass m and radius $b(>a)$. It is placed on a rough horizontal table and a horizontal force of magnitude $2mg$ is exerted on the free end of the string. The reel as it rolls winds up the string. If the string is perpendicular to the axis of the reel and no slipping occurs, find the acceleration of the reel.

10. A uniform rod AB, of mass m and length $2a$, has one end A attached by a smooth ring to a horizontal wire. The rod is held along the wire and allowed to fall. Find the speeds of its ends A and B when the rod is vertical.

20.2. ENERGY METHODS

In Section 16.4 it was shown that, for a system of particles

$$\text{Total K.E.} = \tfrac{1}{2}\bar{v}^2 \Sigma m + \tfrac{1}{2}\Sigma mv'^2,$$

where $\bar{v}$ is the velocity of the centre of mass and v' is the velocity relative to the centre of mass, i.e.

total K.E. = K.E. of centre of mass + K.E. relative to centre of mass

Therefore for the case of a rigid body moving in two dimensions

$$\text{K.E.} = \tfrac{1}{2}Mv^2 + \tfrac{1}{2}I\omega^2.$$

MOTION IN TWO DIMENSIONS

Example 1. A uniform rod, of mass m and length 2a, is perpendicular to a smooth horizontal plane with its lower end in contact with the plane. It is slightly disturbed from this position. Find its angular speed $\dot\theta$ when it has turned through an angle θ, and deduce the value of $\dot\theta$ when the rod is horizontal.

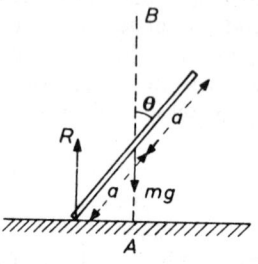

Figure 20.5

The forces acting on the rod are its weight mg acting vertically downwards through G, the centre of gravity, and the reaction of the ground R acting vertically. Therefore, G moves in a vertical line AB (refer to *Figure 20.5*). Thus if the height GA of G above the plane is x then the speed of G is

$$\dot x = d(a\cos\theta)/dt = -a\sin\theta\dot\theta. \qquad \ldots\text{(i)}$$

The angular velocity of the rod is $\dot\theta$ $\qquad\ldots$(ii)

The work done by R is zero because the point of the rod in contact with the horizontal plane moves along the plane, i.e. perpendicular to the direction of R. The work done by gravity as G descends from a height a to a height $a\cos\theta$ above the plane is $mg(a - a\cos\theta)$.

Work done = Final K.E. − Initial K.E. (both rotational and translational)

$\therefore \quad mga(1 - \cos\theta) = \tfrac{1}{2}m(\dot x^2 - 0) + \tfrac{1}{2}mk^2(\dot\theta^2 - 0)$

$\therefore \quad mga(1 - \cos\theta) = \tfrac{1}{2}ma^2\sin^2\theta\dot\theta^2 + \tfrac{1}{2}\dot\theta^2 ma^2/3$

[from equations (i) and (ii)]

$\therefore \qquad\qquad \dot\theta^2 = 6g(1 - \cos\theta)/a(3\sin^2\theta + 1).$

When the rod is horizontal, $\theta = \pi/2$.

$\therefore \qquad\qquad \dot\theta^2 = 6g/4a$

and $\qquad\qquad \dot\theta = \sqrt{(3g/2a)}.$

412

ENERGY METHODS

Example 2. A uniform sphere (mass m, radius a, centre B) is balanced on top of a fixed rough sphere (radius 2a, centre A). It is slightly disturbed from its position of unstable equilibrium and rolls down the surface of the fixed sphere. If at any time t while the sphere is still rolling AB makes an angle θ with the vertical, show that $21a\dot\theta^2 = 10g(1 - \cos\theta)$. If μ is the coefficient of friction between the two spheres, show that slipping will occur when $2\sin\theta = \mu(17\cos\theta - 10)$ (provided that $\cos\theta > \frac{10}{17}$).

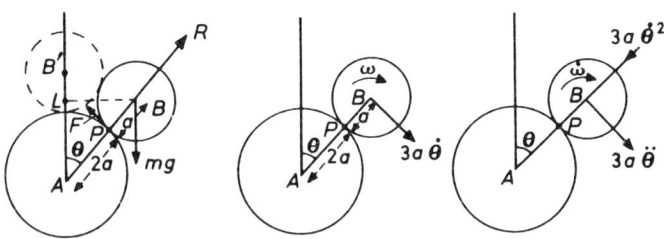

(Force diagram)   (Velocity diagram)   (Acceleration diagram)

Figure 20.6

Let P be the point of contact of the two spheres. P has two velocities both perpendicular to AB but in opposite directions, one is $3a\dot\theta$ due to the movement of centre B and the other $a\omega$ due to the movement about B. Since P is stationary,

$$3a\dot\theta = a\omega$$

i.e. $$3\dot\theta = \omega. \qquad \ldots\text{(i)}$$

Let B' be the original position of B (refer to *Figure 20.6*) and BL the perpendicular to AB'. The frictional force F and the reaction R do no work while the body is rolling, since the point P at which they act is stationary, thus

$$\text{Work done} = \text{Final K.E.} - \text{Initial K.E.}$$

becomes $$mgB'L = \tfrac{1}{2}mv^2 + \tfrac{1}{2}mk^2\omega^2$$

i.e. $$mg3a(1 - \cos\theta) = \tfrac{1}{2}m(3a\dot\theta)^2 + \tfrac{1}{2}m\cdot\tfrac{2}{5}a^2\omega^2.$$

Substituting for ω from equation (i),

$$3mga(1 - \cos\theta) = \tfrac{1}{2}m9a^2\dot\theta^2 + m\tfrac{1}{5}a^2 9\dot\theta^2$$

$\therefore$ $$ga(1 - \cos\theta) = (\tfrac{9}{2}a^2 + \tfrac{9}{5}a^2)\dot\theta^2$$

$\therefore$ $$21a\dot\theta^2 = 10g(1 - \cos\theta). \qquad \ldots\text{(ii)}$$

413

MOTION IN TWO DIMENSIONS

Differentiating this equation with respect to θ

$$42a\dot\theta\ddot\theta = 10g \sin \theta \dot\theta.$$

$$\therefore \qquad 21a\ddot\theta = 5g \sin \theta. \qquad \ldots\text{(iii)}$$

For the equation of motion of the sphere regarded as a particle of mass m at B,

$$P = mf$$

$$\therefore \qquad mg \sin \theta - F = m \cdot 3a\ddot\theta \qquad \ldots\text{(iv)}$$

and $\qquad mg \cos \theta - R = m \cdot 3a\dot\theta^2. \qquad \ldots\text{(v)}$

Taking moments about B,

$$Fa = m \cdot \tfrac{2}{5}a^2 \dot\omega$$

From equation (i) $\qquad F = 6ma\ddot\theta/5. \qquad \ldots\text{(vi)}$

Substituting for F from equation (vi) in equation (iv)

$$mg \sin \theta - 6ma\ddot\theta/5 = 3ma\ddot\theta,$$

$$\therefore \qquad \ddot\theta = 5g \sin \theta / 21a. \qquad \ldots\text{(a)}$$

From equations (a) and (vi)

$$F = 2mg \sin \theta / 7. \qquad \ldots\text{(b)}$$

From equations (ii) and (v)

$$R = \frac{mg}{7}[17 \cos \theta - 10]. \qquad \ldots\text{(c)}$$

After F reaches its maximum possible value μR the sphere will slip. When $F = \mu R$,

$$2mg \sin \theta / 7 = \mu(mg/7)[17 \cos \theta - 10]$$

i.e. $\qquad 2 \sin \theta = \mu(17 \cos \theta - 10)$

provided that the spheres are still in contact, i.e. provided $R > 0$, i.e. $\cos \theta > \tfrac{10}{17}$ [from equation (c)].

Exercises 20b

1. A uniform rod of length 2 m is held inclined at 60 degrees to the horizontal with one end on smooth ground. If the rod is released from rest, find its angular speed just before it hits the ground. What path does the centre of mass of the rod follow?

2. A pair of steps consists of two identical ladders hinged at O, each of length $2l$. The steps rest on smooth ground held at an angle of 120 degrees with one another by a light string. If the string breaks,

find the speed of O just before it hits the ground and the angular speed of one half of the step ladder.

3. A uniform circular cylinder rolls directly up a plane inclined at an angle α to the horizontal, its axis being at right angles to the line of greatest slope. If initially its centre of mass was moving with speed v, find how far up the plane it goes before coming instantaneously to rest.

4. A solid spherical ball of mass 100 g and radius 5 cm is rolling without slipping along a horizontal road at 6 m/s. Calculate the total kinetic energy stored in the ball. If it comes to a hill of inclination $\sin^{-1}(\frac{1}{10})$ find how far it will roll before coming instantaneously to rest.

5. A particle is fastened to the surface of a uniform sphere of radius a. The particle and sphere are of equal mass. The sphere is now placed on a rough surface with the particle at the highest point of the sphere. If when the system is released the sphere rolls without slipping, find the speed of the centre of the sphere when it has turned through an angle θ.

6. A uniform circular cylinder (mass m, radius b and centre B) is balanced on top of a fixed rough cylinder (radius a, centre A). It is slightly disturbed from its position of unstable equilibrium and rolls down the surface of the fixed cylinder. If at any time t, while the cylinder is still rolling, AB makes an angle θ with the vertical, show that $(a + b)\dot\theta^2 = 2g(1 - \cos\theta)$. If μ is the coefficient of friction between the two cylinders, show that slipping will occur when $\tan\theta = \mu/(1 + 2\mu)$.

7. A uniform circular disc (centre B, mass m and radius b) rolls on the inside of a fixed hollow cylinder (radius a) whose axis is horizontal. The plane of the disc is vertical and cuts the axis of the cylinder in A. If θ is the angle AB makes with the vertical, find an expression for $\dot\theta$ given that when the disc is in its lowest position $\dot\theta = \Omega$. In order that the disc may make complete revolutions show that $\Omega \geqslant \sqrt{[11g/3(a - b)]}$.

8. Rework Question 10 of Exercises 20(a) using the principle of work.

9. In Example 3, Section 20.1, obtain equation (d) by the principle of work.

20.3. IMPULSIVE MOTION

Regarding a rigid body as a collection of particles rigidly joined together, the result of Section 16.2.

$$\boldsymbol{P} = M\frac{d^2\bar{\boldsymbol{r}}}{dt^2} \qquad \ldots \text{(i)}$$

shows that a force P acting on a rigid body is equal to the product of the total mass and the acceleration of the centre of mass. Integrating equation (i) with respect to time we have

$$\int_{t_0}^{t_1} P \, dt = M(\bar{v}_1 - \bar{v}_0).$$

Thus, if P is an impulsive force, its impulse is measured by the change of momentum of the whole body, regarded as a point mass at the centre of gravity. Similarly from Section 16.4, for a single force

$$r \times P = \frac{d}{dt} \Sigma(r \times mv)$$

$$\therefore \quad r \times P = \frac{d}{dt} \text{ (angular momentum)}$$

$$= \frac{d}{dt}(A).$$

Integrating we have

$$\int_{t_0}^{t_1} (r \times P) \, dt = A_1 - A_0.$$

In the case of an impulsive force the interval of time is small and r will be approximately constant

$$\therefore \quad r \times \int_{t_0}^{t_1} P \, dt = A_1 - A_0$$

or the moment of the impulse is equal to the change of angular momentum.

Example 1. *A circular disc, of mass m and radius a, is spinning freely on a smooth horizontal table with angular speed ω about its centre O. The plane of the disc is horizontal. If a point A on its circumference is suddenly fixed, find its new angular speed about its centre. Also find the impulse applied at A.*

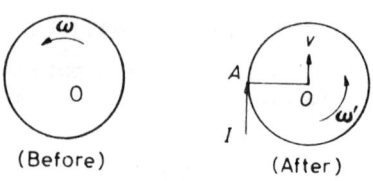

(Before) (After)

Figure 20.7

IMPULSIVE MOTION

Let I be the impulse acting when the point A is fixed. It will act tangentially. Let ω' be the new speed of rotation and v the speed of the centre O (refer to *Figure 20.7*). Since the impulse is at A, the angular momentum about A is conserved.

$$m\frac{a^2}{2}\omega + 0 = m\frac{a^2}{2}\omega' + mva.$$

But
$$v = a\omega' \qquad \ldots \text{(i)}$$

∴
$$ma^2\omega/2 = 3ma^2\omega'/2$$

∴
$$\omega' = \omega/3. \qquad \ldots \text{(ii)}$$

Also Impulse = change of momentum

∴ $\qquad I = mv$

$\qquad\qquad = ma\omega' \quad$ from equation (i)

∴ $\qquad I = ma\omega/3 \quad$ from equation (ii).

Example 2. Two equal uniform rods AB, BC, each of mass m, are freely jointed at B and lie in a straight line on a smooth horizontal table. The point A is given an impulse of magnitude I at right angles to AB. Find the initial speeds of the centres of the rods.

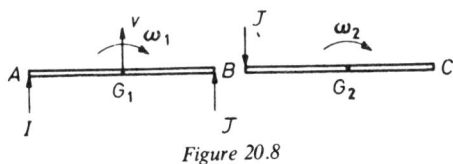

Figure 20.8

Let $2a$ be the length of each rod. Let G_1, G_2 be the centres of AB, BC. Let ω_1, ω_2 be the angular speeds of AB and BC respectively about G_1 and G_2, and v the speed of G_1 (refer to *Figure 20.8*) (all considered immediately after the impact). Let the internal impulsive reaction at B be of magnitude J on AB and BC. B will have a speed of $-a\omega_1$ relative to G_1

∴ $\qquad$ the speed of B is $v - a\omega_1$.

B has a speed of $a\omega_2$ relative to G_2, i.e. G_2 has a speed of $-a\omega_2$ relative to B

∴ $\qquad$ the speed of G_2 is $v - a\omega_1 - a\omega_2$. $\qquad \ldots \text{(i)}$

417

MOTION IN TWO DIMENSIONS

For each rod we shall apply

$$\text{Impulse} = \text{change of momentum at } G$$

and Moment of Impulse = change in angular momentum about G.

Rod AB
$$I + J = mv \qquad \ldots \text{(ii)}$$
$$Ia - Ja = m \cdot \tfrac{1}{3}a^2\omega_1$$
$$\therefore \quad I - J = ma\omega_1/3. \qquad \ldots \text{(iii)}$$

Rod BC $\qquad -J = m$ (speed of G_2)

$\therefore$ from equation (i)
$$-J = mv - ma\omega_1 - ma\omega_2 \qquad \ldots \text{(iv)}$$
$$Ja = -\tfrac{1}{3}ma^2\omega_2$$
$$\therefore \quad J = -ma\omega_2/3. \qquad \ldots \text{(v)}$$

Substituting for J from equation (v) in equations (ii), (iii) and (iv)
$$I - ma\omega_2/3 = mv \qquad \ldots \text{(a)}$$
$$I + ma\omega_2/3 = ma\omega_1/3 \qquad \ldots \text{(b)}$$
$$ma\omega_2/3 = mv - ma\omega_1 - ma\omega_2$$
$$\therefore \quad v = (4a\omega_2/3) + a\omega_1. \qquad \ldots \text{(c)}$$

Substituting for v from equation (c) in equation (a)
$$I - ma\omega_2/3 = 4ma\omega_2/3 + ma\omega_1$$
$$\therefore \quad I - 5ma\omega_2/3 = ma\omega_1. \qquad \ldots \text{(d)}$$

From equations (b) and (d)
$$I - 5ma\omega_2/3 = ma\omega_1 = 3I + ma\omega_2$$
$$\therefore \quad -2I = 8ma\omega_2/3.$$
$$\therefore \quad \omega_2 = -3I/4ma. \qquad \ldots \text{(e)}$$

Substituting for ω_2 from equation (e) in equation (d)
$$I - 5ma/3(-3I/4ma) = ma\omega_1$$
$$\therefore \quad \omega_1 = 9I/4ma \qquad \ldots \text{(f)}$$

and from equation (c)
$$v = 4a/3(-3I/4ma) + a(9I/4ma)$$
$$= 5I/4m \qquad \ldots \text{(g)}$$

which is the speed of G_1 and from equations (i), (e) (f) and (g)

the speed of $G_2 = v - a\omega_1 - a\omega_2$
$$= (5I/4m) - a \cdot (9I/4ma) - a(-3I/4ma)$$
$$= -I/4m.$$

Exercises 20c

1. Two particles A and B, each of mass m, are joined by a light rod of length $2a$. A is given an impulse of magnitude I at right angles to AB, describe the motion which follows.

2. Two particles A and B, of mass m and M respectively, are joined by a light rod of length $2a$. An impulse I is given at a point X on the rod and at right angles to the rod. Find an expression for the kinetic energy of the system after the blow. Find the position of X if the kinetic energy is to be a minimum.

3. A uniform rod AB, centre G, mass m and length $2a$ is lying on a smooth horizontal table. It is given a horizontal impulse of magnitude I perpendicular to the rod at a point X where $GX = x$. Find the initial speed of the point X and its position for this speed to be a minimum.

4. A uniform rod, of mass m and length $2a$, lies at rest on a smooth horizontal surface. It receives an impulse mu at right angles to its length and applied at a point distant $a/2$ from its centre. Show that after a time $\pi a/3u$, the rod will have turned through a right angle and find the distance travelled by its centre during this time.

5. A square lamina of mass m and side $2a$, is rotating freely in a horizontal plane with angular speed ω about a perpendicular axis through its centre of gravity. One corner is suddenly fixed. Find its new angular speed and the change in kinetic energy.

6. A circular disc (radius a) is rolling with speed V along a rough horizontal plane. It strikes, perpendicularly, a rough horizontal peg fixed at a height $2a/5$ above the plane. Find the angular speed with which the disc begins to rotate about the peg.

7. A uniform rod, of mass $4m$ and length $2a$, is lying at rest on a smooth horizontal table. A particle, of mass m moving with speed v along the table in a direction perpendicular to the rod, strikes it at one end and adheres to it. Find the loss of kinetic energy of the system due to the impact.

EXERCISES 20*

1. A uniform circular lamina of radius a is lying on a smooth horizontal table. A point on its circumference is given a tangential

* Exercises marked thus, †, have been metricized.

impulse which starts it moving with a tangential velocity u. Find the angular speed with which the lamina starts to move.

2. A uniform solid sphere, of radius a, rolls without slipping directly up a slope inclined at an angle α to the horizontal. If its initial angular speed is ω, find the time taken, the distance travelled and the number of revolutions made until it first comes to instantaneous rest.

3. A uniform rod AB of length $2a$ lies at rest on a smooth horizontal table. It is given an impulse at one end A at right angles to its length. It is noted that one point X in the rod remains stationary. Find the distance AX.

4. A circular cylinder, of radius a, rolls down a rough plane inclined at an angle α to the horizontal. The centre of mass of the cylinder is in its axis which is perpendicular to the line of greatest slope of the plane. Show that the acceleration of the cylinder is $a^2 g \sin \alpha/(a^2 + k^2)$, where k is the radius of gyration of the cylinder about its axis.

5. A sphere of radius a can roll without slipping on the outside of a larger fixed sphere. If it is released gently from the top of the larger sphere, find the angular distance the small sphere moves through before it leaves the surface of the larger sphere.

6. A uniform cube of side a is placed with one edge in contact with a horizontal plane. It is slightly disturbed from the position of unstable equilibrium. Find an expression for ω^2, where ω is the angular speed of the cube, when one face meets the plane. Consider the two cases when the plane is (a) smooth, (b) rough enough to prevent sliding.

7. A rigid body is rotating about an axis through its centre of gravity. A parallel axis distance h from the original axis suddenly becomes fixed and the body starts to rotate about this new axis. If ω was the original angular speed and k the radius of gyration about the old axis, find the new angular speed.

8. Two thin rods AB, BC, each of mass m and length $2a$, are freely hinged at the point B. They are placed on a smooth horizontal table with $\angle ABC = 90$ degrees. An impulse of magnitude I is applied at the end A of the first rod in the direction AB. Find the initial speeds of the centres of the two rods.

9. A thin uniform rod, of mass m and length $2a$, lies in a vertical plane with its ends in contact with two smooth planes, one horizontal the other vertical. The rod is released from rest inclined at an angle α to the horizontal. Assuming that the rod remains in contact with the vertical plane, find its angular speed of rotation when it is inclined at an angle θ to the horizontal.

EXERCISES

10. A uniform sphere, of radius b and centre B, is balanced on top of a fixed rough sphere of radius a, centre A. It is slightly disturbed from its position of unstable equilibrium and rolls down the surface of the fixed sphere. If at any time t while the sphere is still rolling, AB makes an angle θ with the vertical, show that $\dot\theta^2 = 10g(1 - \cos\theta)/7(a + b)$. Also show that the sphere will slip, if $\cos\theta > \frac{10}{17}$, when $\mu(17\cos\theta - 10) = 2\sin\theta$, where μ is the coefficient of friction.

11. A uniform rod AB, of mass M and length a, lies on a smooth horizontal table, with the end A smoothly jointed to a point of the table. A particle of mass m is at B in a smooth groove cut along the length of the rod. Motion is started in the plane of the table by giving the system an angular velocity ω about A and the particle a speed u towards A. Prove that the particle reaches A if
$$Mu^2 > a^2\omega^2(M + 3m).$$
(Oxford).

12. Two fixed equal cylinders rotate in opposite directions about their axes which are parallel and in the same horizontal plane. A plank is placed symmetrically across the cylinders perpendicular to their axes and is in equilibrium slipping on the cylinders. The plank is then pushed a short distance in the direction of its length and released. Show that it subsequently moves in Simple Harmonic Motion.

13. Two uniform rods AB and BC each of length $2a$ and of mass M are freely jointed at B and lie in a straight line on a smooth horizontal plane. An impulse I is applied in this plane at A perpendicular to AB. Show that the initial speeds of the centres of mass of the rods are in the ratio $5:1$ and find the ratio of the initial angular velocities of the rods. (London)

14. A uniform rod, of mass m and length $2a$, lies on a smooth horizontal table and is rotating freely in the plane of the table about one end O which is fixed. The rod strikes a particle of mass m at rest at a distance x from O, whereupon the particle acquires a speed u. Find the impulsive reaction at O and show that this reaction vanishes if $x = 4a/3$. If the particle adheres to the rod on impact and there is no impulsive reaction at O, find what fraction is lost of (a) the angular velocity of the rod, (b) the kinetic energy of the system. (London)

15. Prove that the moment of inertia of a uniform solid sphere, of radius a and mass m, about an axis through its centre is $\frac{2}{5}ma^2$. A hollow hemisphere, of radius $3a$, is fixed with its base horizontal and uppermost. A uniform solid sphere, of mass m and radius a, rolls without slipping on the inner surface of the hemisphere, the motion of the centre of the sphere being in a vertical plane. Prove that the motion is identical with that of the bob of a simple pendulum.

If, instead, the contact between the hemisphere and the sphere is perfectly smooth, prove that the motion of the centre of the sphere in a vertical plane is again that of the bob of a simple pendulum and that the ratio of the lengths of the pendulums in the first and second case respectively is 7:5. (London)

16. A uniform circular disc of mass M and radius a lies flat on a horizontal table with its centre at the origin of coordinates. A horizontal blow J is applied to the disc at the point $(a, 0)$ in a direction making an acute angle α with the positive x-axis. If the initial velocity of a point (x, y) of the disc has components u and v parallel to the axes of z and y respectively, show that

$$u = \frac{J}{Ma}(a\cos\alpha - 2y\sin\alpha) \qquad v = \frac{J\sin\alpha}{Ma}(a + 2x).$$

If the disc begins to rotate about a point on its circumference, show that $\alpha = \pm 30$ degrees and that all points which have an initial speed V lie on a circle of radius MVa/J. (London)

17. A uniform disc, of mass m and radius a, is projected in its own vertical plane along a rough horizontal table. The initial velocity of the centre of the disc is V and the velocity of the point of contact is nV in the same direction. The coefficient of friction between the disc and the table is μ. If the disc ceases to slip on returning to the point of projection, find the value of n and the greatest distance moved in the direction of projection.

Show also that the disc will *not* return to its starting point if $n < 3$. (London)

18. A uniform rod AB, of mass M and length $2a$, has a particle of mass M attached to the end B. The rod is allowed to fall freely without rotation and with its length horizontal. When the velocity of the rod is v vertically downwards, the end A engages with a smooth pivot about which the rod can rotate freely. Find the impulsive reaction at the pivot.

Find also the least value of v if the rod proceeds to make complete revolutions about the hinge. (London)

19. Find the moment of inertia of a uniform circular disc of mass m and radius a about the line through its centre perpendicular to its plane.

Two such discs are in the same vertical plane, and are free to rotate in the plane about their centres, which are fixed. A uniform rod, of mass M and length equal to the distance between the centres, is smoothly jointed to each disc at a point on its rim, so that, when it moves as the discs rotate, it remains parallel to the line joining their centres. Find the length of the simple equivalent pendulum as the system performs small oscillations under gravity. (Oxford)

20. A car is travelling without skidding at 36 km/h round a circular bend of radius 50 m on a horizontal surface. Show that the coefficient of friction between the wheels and the road is at least $\frac{10}{49}$.

Find the angle at which this road must be banked, if the coefficient of friction is $\frac{1}{3}$ and the maximum speed at which a car can travel round the bend without skidding is 72 km/h. (WJEC)†

21. A uniform circular cylinder, of radius a and mass M, rolls without slipping down the rough inclined face of a wedge. The axis of the cylinder is horizontal and perpendicular to the line of greatest slope of the inclined face of the wedge. The wedge rests on a smooth horizontal plane and is prevented from slipping by a force P. Find P and the angular acceleration of the cylinder in terms of M, a and α (the inclination of the face).

22. Two thin uniform rods AB, BC, each of mass m and length $2a$, are freely jointed at B. They lie at rest on a horizontal table with $\angle ABC = 90$ degrees. If a horizontal impulse of magnitude I is given to the system at the point B, show that, whatever the direction of I, B will begin to move with a speed of $4I/5m$ (in the direction of I).

23. Three uniform rods AB, BC, CD are freely jointed at B and C. They lie at rest on a horizontal table forming three sides of a square $ABCD$. A horizontal impulse of magnitude I is given to the system at the point A, perpendicular to AB and in the direction of AD. Find the initial angular speeds of the three rods and the reactions at the joints B and C, given that each rod is of mass m and length $2a$.

24. A uniform rectangular lamina, of mass m and with sides of lengths a and b, lies at rest on a smooth horizontal table. An impulse of magnitude I is given to one corner of the rectangle. Find the magnitude of the greatest angular speed with which the lamina begins to move.

25. A lamina, of mass m and radius of gyration k about a perpendicular axis through its centre of mass G, is moving in its own plane with velocity u and without rotation. It is suddenly pinned at a point P in its own plane such that the length of GP is x and the direction $\overline{GP}$ is perpendicular to u. Find the magnitude of the angular velocity immediately after pinning, and the loss of kinetic energy. Prove that the magnitude of the angular velocity immediately after pinning is a maximum when $x = k$. (Oxford)

26. A rough cone of semi-vertical angle α, with its axis vertical and its vertex downwards, is rotating with constant angular velocity ω about its axis. A heavy particle is on the inside surface of the

cone at a vertical height h above the vertex. The coefficient of friction between the particle and the cone is μ, and the particle is on the point of slipping *up* the cone. Prove that, if $g \cot^2 \alpha < \omega^2 h$ and $\mu < \tan \alpha$, then

$$\omega^2 h \tan \alpha (\tan \alpha - \mu) = g(\mu \tan \alpha + 1).$$

(Oxford)

21

EQUILIBRIUM

21.1. GENERAL CONDITIONS OF EQUILIBRIUM

A body is said to be in equilibrium when its centre of mass G is at rest or moving with uniform velocity *and* its angular momentum relative to G is constant (or zero). This includes the important and common case of a rigid body at rest.

Consider a body in equilibrium under the action of forces $P_1, P_2, P_3, \ldots$, whose lines of action pass through points whose position vectors are $r_1, r_2, r_3, \ldots$, respectively. Since its angular momentum is constant and its centre of mass has zero acceleration, we have, from the results obtained in Sections 16.2 and 16.3,

$$\Sigma(r \times P) = 0 \qquad \ldots \text{(i)}$$

$$\Sigma P = 0. \qquad \ldots \text{(ii)}$$

If each force P is resolved in the directions of $\hat{a}, \hat{b}, \hat{c}$ (three non-coplanar unit vectors), so that $P = P_1\hat{a} + P_2\hat{b} + P_3\hat{c}$, then equation (ii) becomes $\Sigma(P_1\hat{a} + P_2\hat{b} + P_3\hat{c}) = 0$ or $\Sigma P_1 = 0$, $\Sigma P_2 = 0$ and $\Sigma P_3 = 0$.

Thus our general conditions of equilibrium may be stated as follows:

If a body is in equilibrium then ... (i) the sum of the moments of the external forces about any point is zero and (ii) the sums of the resolved parts of the external forces in three non-coplanar directions are separately zero.

Conversely, if these two conditions are satisfied, the body is in equilibrium.

Usually the resolution is carried out in three directions mutually at right angles, and in the case of coplanar forces it is sufficient that the sums in *two* directions are zero.

These relations will be true whatever units of force are used. As in Chapter 11, it is convenient to use the weight of unit mass as a unit of force, i.e. 1 kg wt. 1 kg wt. $= g$ N, where g is the magnitude of the local gravitational constant.

EQUILIBRIUM

Example 1. *A thin uniform rod AB, of length 4 m and mass 7 kg, is freely hinged to a wall at A. To the other end B is attached a light elastic string BC of natural length 2 m, C being fixed at the same level as A and 5 m away from it. If, when released, the rod finally comes to rest with angle ABC = 90 degrees, find the modulus of the string. Find also the reaction at A.*

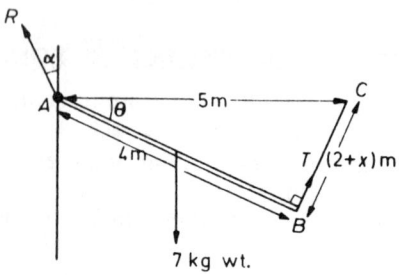

Figure 21.1

When a body is *freely* hinged there is no frictional couple at the hinge, while the normal reaction between the parts of the hinge may be in any direction. In this case let the reaction of the hinge at A on AB be as shown in *Figure 21.1*. Also let the extension of the string BC in the equilibrium position be x m.

Then in triangle ABC, $AC^2 = AB^2 + BC^2$ (Pythagoras), so that $BC = 3$ m and $x = 1$ m. And, if angle $CAB = \theta$, then $\sin \theta = \frac{3}{5}$, $\cos \theta = \frac{4}{5}$.

By Hooke's law $\qquad T = \lambda . x/l$

$\therefore \qquad\qquad\qquad T = \lambda . 1/2.$

Taking moments about A (anticlockwise positive),

$$T \times 4 - 7 \cos \theta \times 2 = 0. \qquad \ldots \text{(i)}$$

Resolving,

horizontally $\qquad R \sin \alpha - T \sin \theta = 0 \qquad \ldots \text{(ii)}$

vertically $\qquad R \cos \alpha + T \cos \theta - 7 = 0. \qquad \ldots \text{(iii)}$

Substituting for T and for $\cos \theta$ in (i) gives

$$\tfrac{1}{2}\lambda \times 4 - 7 \times \tfrac{4}{5} \times 2 = 0$$

426

GENERAL CONDITIONS OF EQUILIBRIUM

or
$$\lambda = 28/5 \text{ kg wt.}$$

Substituting for T and θ in (ii) and (iii) gives
$$R \sin \alpha = 42/25,$$
$$R \cos \alpha = 119/25.$$

Squaring and adding, $R^2 = (\frac{7}{25})^2(17^2 + 6^2)$, ∴ $R = 7\sqrt{13}/5$ kg wt.

Dividing, $\tan \alpha = 6/17$ and $\sin \alpha$ is positive

∴ $\alpha = 19°\ 26'$.

The elastic string has modulus $5\frac{3}{5}$ kg wt. and the reaction at A is a force of magnitude $\frac{7}{5}\sqrt{13}$ kg wt. acting at $19°\ 26'$ to the vertical.

Example 2. *A uniform plank AB, of mass 40 kg and length 6 m, rests horizontally on two supports C and D. AC = 2 m and CD = 3 m. If a man weighing 80 kg stands at the centre of the plank, find the reactions at C and D.*

How far can the man walk towards the end B of the plank without overturning it? Find the reactions at C and D when it is about to overturn.

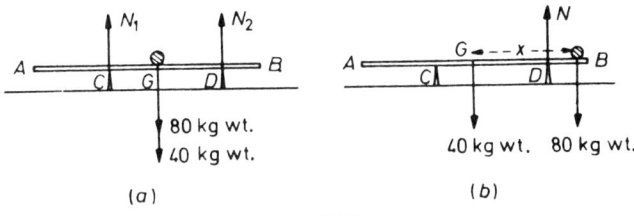

Figure 21.2

Since the weights of the man and the plank are vertical, the reactions at C and D must also be vertical.

Consider first the case when the man stands at the centre of the plank (refer to *Figure 21.2a*).

Taking moments about C
$$N_2 \times 3 - 120 \times 1 = 0.$$

Vertically
$$N_1 + N_2 - 120 = 0$$

giving $\quad N_2 = 40, \quad N_1 = 80$ kg wt.

Now let the man walk a distance x towards B until the plank begins to tilt. At this point the plank begins to lose contact with the

EQUILIBRIUM

support at C so there will be no reaction there (refer to *Figure 21.2b*). Taking moments about G,

$$N \times 2 - 80x = 0$$

vertically, $\quad N - 120 = 0.$

giving $\quad N = 120$ kg wt. and $x = 3$ m.

When the man stands at the centre the reactions at C and D are 80 and 40 kg wt. respectively. The plank will start to overturn just as the man reaches B and then the reactions at C and D will be zero and 120 kg wt. respectively.

Example 3. A uniform solid hemisphere, of weight $2W$, rests with its base inclined at an angle θ to the horizontal and with its curved surface in contact with a rough horizontal floor and a smooth vertical wall. The coefficient of friction between floor and hemisphere is 5/12. If the hemisphere just begins to slip when a particle of weight W is attached to its highest point, find θ.

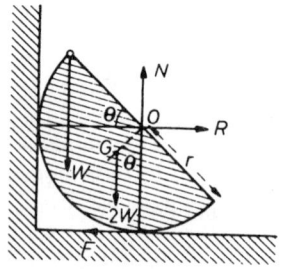

Figure 21.3

Let the hemisphere have radius r. Then its centre of gravity G is at a distance $\frac{3}{8}r$ from the base, i.e. $OG = \frac{3}{8}r$ (refer to *Figure 21.3*). Since the hemisphere is about to slip

$$F = \mu N$$

∴ $$F = \tfrac{5}{12}N.$$

Taking moments about O,

$$Wr \cos \theta + 2W\tfrac{3}{8}r \sin \theta - \tfrac{5}{12}Nr = 0. \quad \ldots \text{(i)}$$

Resolving,

vertically $\quad N - W - 2W = 0 \quad \ldots \text{(ii)}$

GENERAL CONDITIONS OF EQUILIBRIUM

[horizontally $\quad\quad R - \tfrac{5}{12}N = 0. \quad\quad\quad\quad \ldots (iii)]$

From (ii) $N = 3W$ and substituting this value in (i)

$$Wr\cos\theta + \tfrac{6}{8}Wr\sin\theta = \tfrac{5}{12}3Wr$$

$\therefore \quad\quad \cos\theta + \tfrac{3}{4}\sin\theta = \tfrac{5}{4}$

$\therefore \quad\quad \tfrac{4}{5}\cos\theta + \tfrac{3}{5}\sin\theta = 1$

$\therefore \quad\quad \cos(\theta - \alpha) = 1$

where $\cos\alpha = \tfrac{4}{5}$, $\sin\alpha = \tfrac{3}{5}$ and hence $\alpha = 36° 52'$.

$\therefore \quad\quad \theta - \alpha = 0 \quad \text{or } 2\pi \quad \text{or} \ldots$

But θ is acute,

$\therefore \quad\quad \theta = \alpha = 36° 52'.$

The base of the hemisphere is inclined at $36° 52'$ to the horizontal. [Note that equation (iii) was not needed in this particular example, but would have enabled us to find the reaction between the hemisphere and'the wall.]

Example 4. A solid body consists of a cube, of weight W and side 4a, surmounted by a pyramid of weight W and height 4a. The pyramid has a square base, also of side 4a, fixed to coincide with the top of the cube. The body is placed on a rough plane, inclined at an angle α to the horizontal, with an edge of its base along a line of greatest slope of the plane. A force P, acting up the plane and parallel to it, is applied to the vertex of the pyramid. Initially it is not large enough to disturb the equilibrium of the body. Find the force of friction, the normal reaction and the point on the base at which it acts.

If P now increases and the plane is rough enough to prevent slipping, indicate how the forces change and find the value of P which causes the body to overturn.

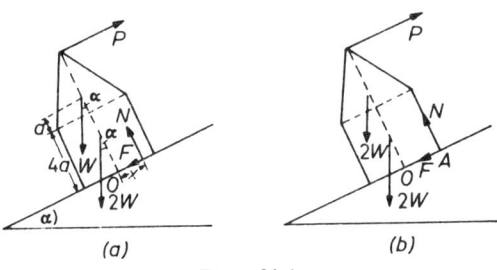

Figure 21.4

Initially the forces are as shown in *Figure 21.4a*. The centre of gravity of the pyramid part, being $\frac{1}{4}(4a)$ above its base, is distant $5a$ from the plane. Let the normal reaction N act at a point distant x from O the centre of the base of the body.

Taking moments about O,

$$W5a \sin \alpha + 2W2a \sin \alpha + Nx - P \cdot 8a = 0. \quad \ldots (i)$$

Resolving,

perpendicular to the plane $\quad N - 3W \cos \alpha = 0 \quad \ldots (ii)$

up the plane $\quad\quad\quad\quad P - F - 3W \sin \alpha = 0. \quad \ldots (iii)$

From (ii) $N = 3W \cos \alpha$ and substituting this value in equation (i) gives

$$9Wa \sin \alpha + 3W \cos \alpha x - 8Pa = 0$$

$$\therefore \quad\quad x = \frac{8Pa - 9Wa \sin \alpha}{3W \cos \alpha}.$$

From (iii) $\quad\quad F = P - 3W \sin \alpha$.

Initially then, $F = P - 3W \sin \alpha$, $N = 3W \cos \alpha$ and acts at a distance $(8P - 9W \sin \alpha)a/(3W \cos \alpha)$ from O.

As P increases, F will increase to prevent sliding, but the magnitude of N remains constant at $3W \cos \alpha$. To preserve the balance of moments the point of action of N moves away from O until finally N acts through the upper edge of the base at A (refer to *Figure 21.4b*). Then, if P increases still further, there will be a net clockwise moment and the body will overturn.

On the point of overturning $x = 2a$

$$\therefore \quad\quad (8P - 9W \sin \alpha)/(3W \cos \alpha) = 2$$
$$\therefore \quad\quad P = (9W \sin \alpha + 6W \cos \alpha)/8.$$

This is the least value of P which will cause the body to overturn.

It should be noted that, in examples like these and others, the point about which to take moments and the directions in which to resolve are arbitrary. A careful choice may reduce the amount of work in a problem considerably.

Exercises 21a

1. One end of a uniform rod, of mass 4 kg is freely hinged to a fixed point A. The rod is held inclined at 30 degrees to the downward vertical by a force F applied at the other end. F is at right angles to the rod. Find the magnitude of F and the components of the reaction at A along and perpendicular to the rod.

GENERAL CONDITIONS OF EQUILIBRIUM

2. A uniform plank AB, of mass 40 kg and length 3 m, rests horizontally on two supports C and D. It is placed so that $AC = 60$ cm and $BD = 120$ cm. Find the reactions at C and D.

3. A uniform rectangular door, 200 cm by 80 cm, weighs 16 kg. It is in a vertical plane with its longer edges vertical and can turn freely about two hinges attached to one of its longer edges. The hinges are 20 cm from either end. Find the horizontal components of the reactions at the hinges.

If the door is fitted badly so that the whole weight of the door is carried by the upper hinge A, find the resultant reaction at A.

4. A solid body weight W_1 consists of a hemisphere, radius a, and a right circular cone, radius a, and height $4a$ of different density joined by their plane faces. The centre of gravity of the body lies on the axis of symmetry distant $a/8$ inside the hemisphere. A weight W_2, whose size may be neglected, is attached to the apex of the cone. Show that, if $W_1 = 32W_2$, the solid will rest in equilibrium when placed on a horizontal plane whichever part of the hemispherical base is in contact with the plane. Find also the reaction of the plane on the body in terms of W_1.

5. A rectangular packing case, $90 \times 90 \times 150$ cm and of mass 100 kg, rests with one of its square ends on a horizontal floor. A horizontal force F is applied at right angles to an upper edge. If the floor is rough enough to prevent sliding, find the least value of F which will cause this case to overturn.

6. A uniform ladder, weight W, rests with one end against a smooth vertical wall and the other end on a smooth horizontal floor. It is prevented from slipping by a horizontal string attached to its lower end. If θ is the inclination of the ladder to the vertical, find the tension in the string and the reactions of the wall and the floor on the ladder.

7. A uniform solid hemisphere rests with its base inclined at an angle θ to the horizontal. Its curved surface is in contact with a rough horizontal floor and with a smooth vertical wall. If the hemisphere is about to slip, find the value of θ in terms of μ the coefficient of friction between hemisphere and floor.

8. A light rod AB of length $2a$ is smoothly jointed to a vertical wall at the point A. It is maintained in a horizontal position by a light inextensible string of length $4a$ attached to the end B and to a point C in the wall above A. Weights W and $2W$ are attached to the mid-point and the end B of the rod respectively. Find the tension in the string and the reaction at the hinge A.

9. A uniform ladder has one end resting against a rough vertical wall and the other end on a rough horizontal floor. The coefficients

EQUILIBRIUM

of friction at the wall and the floor being $\frac{1}{4}$ and $\frac{4}{9}$ respectively. Find the greatest possible inclination of the ladder to the horizontal.

10. A cube, of mass 75 kg and side 50 cm, rests with one face on a rough horizontal floor. A vertical plane through the centre of the cube perpendicular to two opposite faces intersects the cube in a square $ABCD$, CD being in contact with the ground. A force of 25 kg wt. is applied at B vertically upwards, and a force of 20 kg wt. is applied at A in the direction of AB. If the cube is on the point of slipping, find μ the coefficient of friction between the cube and the ground.

Find also the point at which the reaction of the ground on the cube acts.

11. A uniform ladder rests with one end against a smooth vertical wall and the other end on a rough horizontal floor, coefficient of friction $5\sqrt{3}/18$. The inclination of the ladder to the wall is 30 degrees. Show that a man whose weight is equal to twice the weight of the ladder can just ascend to the top without the ladder slipping.

12. A uniform sphere, of weight W, rests on a rough plane which is inclined at an angle α to the horizontal. It is prevented from moving by a light inextensible string which is tangential to the sphere and makes an angle θ with the plane. Find the tension in the string, the frictional force and the normal reaction between the sphere and the plane.

13. A lamina, of weight W, is in the form of an equilateral triangle ABC of side $2a$. It rests in a vertical plane with one vertex A against a smooth vertical wall and another vertex B on a rough horizontal floor. AB is inclined at an angle of 60 degrees to the wall. Find the normal and frictional reactions at B.

Deduce that, if the lamina is not to slip and μ is the coefficient of friction between floor and lamina, μ must be greater than or equal to $1/\sqrt{3}$.

14. A man weighing 72 kg stands on a uniform plank AB of length 3 m and mass 24 kg. The plank is supported in a horizontal position on two trestles. The points of support, C and D, are such that $AC = 20$ cm and $DB = 30$ cm. Find the reactions at C and D when the man stands at a distance a m from A.

Deduce the region of the plank over which the man can move without it tipping over.

15. A uniform ladder, of weight W, rests with one end against a smooth vertical wall and the other end on a smooth horizontal floor. Its lower end is attached by means of a light inextensible string to the junction of the wall and the floor. A man of weight W climbs the ladder. Show that as the man moves from a point one-sixth of

TWO-FORCE PROBLEMS

the way up the ladder to a point five-sixths of the way up, the tension in the string doubles.

21.2. TWO-FORCE PROBLEMS

When a body is in equilibrium under the action of two forces only, the forces must be equal and opposite (so that their vector sum is zero) and in the same straight line (so that their total moment is zero).

Example. A uniform solid oblique circular cone has its axis inclined at an angle θ to its base. The radius of the base is r and the length of its axis is $4l$. If the cone is suspended from the point on the base nearest to its axis, find the inclination of the base to the vertical.

If the cone is placed on its base on a horizontal surface and is released, find the range of values of l for which it does not topple.

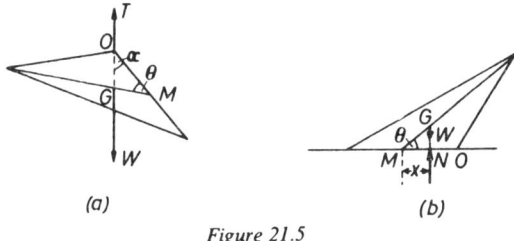

Figure 21.5

The centre of gravity of the cone lies one quarter of the way up its axis so that $MG = l$.

When the cone is suspended from O in equilibrium, the tension and weight must be in the same line. Hence OG must be vertical and the angle $MOG = \alpha$ is the inclination of the base to the vertical. (Refer to *Figure 21.5a*.)

Applying the sine rule to triangle OMG,

$$\frac{l}{\sin \alpha} = \frac{r}{\sin(\alpha + \theta)}$$

$\therefore \quad l(\sin \alpha \cos \theta + \cos \alpha \sin \theta) = r \sin \alpha$

$\therefore \quad \sin \alpha (r - l \cos \theta) = l \cos \alpha \sin \theta$

$\therefore \quad \tan \alpha = \frac{l \sin \theta}{r - l \cos \theta}.$

The base of the cone makes an angle $\tan^{-1}(l \sin \theta/(r - l \cos \theta))$ with the vertical.

EQUILIBRIUM

When the cone stands on the surface the weight and normal reaction N must be in the same straight line, i.e. the point in the base at which N acts must be directly beneath G (refer to *Figure 21.5b*). However, for sufficiently large values of l the line of action of W may lie outside the base while N acts at O. Then there is a clockwise moment and the cone overturns.

Hence for no overturning

$$x < MO$$

i.e. $$l \cos \theta < r$$

$$\therefore \quad l < r \sec \theta.$$

The cone will not topple provided $l < r \sec \theta$ (indeed if $l < r$ the cone cannot overturn whatever the value of θ).

Exercises 21b

1. A hemisphere, of radius a, is suspended by a light inextensible string attached to a point on the circumference of its plane face. Find the inclination of the plane face to the vertical when (a) it is solid and homogeneous (b) it is hollow.

2. A uniform lamina is in the form of a triangle ABC, right-angled at B, in which $AB = 12$ cm and $BC = 9$ cm. It is suspended from A. Find the inclination of AC to the vertical.

3. A uniform solid oblique circular cylinder has base radius a and its axis is inclined at 60 degrees to its base. If the cylinder does not topple when placed on a horizontal surface, find its greatest possible height.

4. A uniform thin hollow right circular cone (without a base) has base radius a and height h. It is placed with its base in contact with a rough plane. If the plane is gradually tilted and the friction is sufficient to prevent slipping, find the inclination of the plane when the cone topples.

5. A uniform right prism has a cross-section in the form of a triangle ABC. With the usual notation for such a triangle, show that the prism can stand on the face containing BC provided $\cos c < -a/b$.

21.3. THREE-FORCE PROBLEMS

Problems in which a body is in equilibrium under the action of three forces can be solved by the general method of taking moments and resolving as described in Section 21.1. However, in the three-force case, there are alternative methods available due to two important properties.

THREE-FORCE PROBLEMS

(a) If a body is in equilibrium under the action of three coplanar* forces, they meet in a point or are parallel.

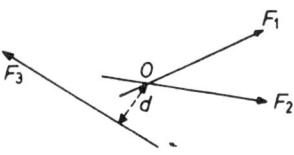

Figure 21.6

Referring to *Figure 21.6*, let two of the forces intersect at O and the third F_3 lie at a distance d from O. Then taking moments about O,

$$F_3 \times d = 0$$

$$\therefore \quad d = 0$$

and F_3 must pass through O as well. Hence if two of the forces intersect they all meet in a point. Alternatively, if none intersect, the three forces are parallel.

(b) When a body is in equilibrium, the vector sum of the external forces acting on it is zero. Hence they can be represented by the sides of a closed polygon taken in order. When there are three forces this polygon becomes a triangle which is particularly suitable for trigonometrical calculation.

So, instead of taking moments and then resolving, we may if convenient (a) obtain a trigonometric relation from the space diagram having made the three forces meet in a point, and (b) obtain two more equations from the trigonometry of the vector triangle.

Example 1. A sphere, of mass 5 kg and radius 5 cm, has a light inextensible string 5 cm long attached to its surface. The other end of the string is fastened to a smooth vertical wall and the sphere hangs in equilibrium resting against the wall. Find the inclination of the string to the wall. Find also the tension in the string and the reaction between the wall and the sphere.

* It can be shown that if a body is in equilibrium under three forces they must be coplanar. Hence the property applies in *any* three-force equilibrium problem.

EQUILIBRIUM

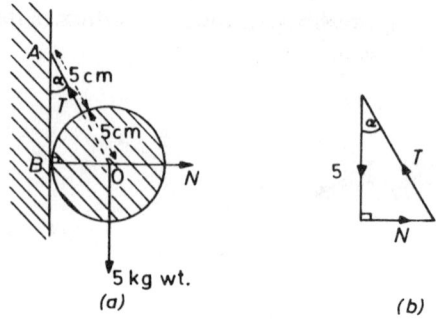

Figure 21.7

The normal reaction of the wall and the weight of the sphere act through its centre O (refer to *Figure 21.7a*). Since it is under the action of three forces, the sphere will turn until the line of the string also passes through O. Then from triangle ABO, $\sin \alpha = OB/OA = \frac{1}{2}$, and since α is clearly acute this gives $\alpha = 30$ degrees.

From the vector triangle (*Figure 21.7b*),

$$N = 5 \tan \alpha \quad \text{and} \quad T = 5 \sec \alpha$$

$\therefore \qquad N = 5/\sqrt{3} \text{ kg wt.} \quad \text{and} \quad T = 10/\sqrt{3} \text{ kg wt.}$

The string is inclined at 30 degrees o the downward vertical, the tension in the string is $10/\sqrt{3}$ kg wt. and the reaction of the wall is normal to the wall and of magnitude $5/\sqrt{3}$ kg wt.

A useful formula in these problems is the cotangent rule for a triangle.*

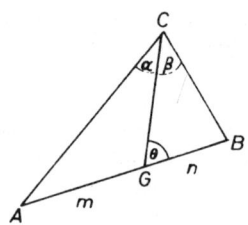

Figure 21.8

---

* Proofs of this formula may be found in standard mathematical texts including *Pure Mathematics for Advanced Level* by Bunday and Mulholland.

THREE-FORCE PROBLEMS

If, in the triangle shown in *Figure 21.8*, G divides AB in the ratio $m:n$, then

$$(m + n) \cot \theta = m \cot \alpha - n \cot \beta$$

or
$$(m + n) \cot \theta = n \cot A - m \cot B.$$

An especially useful case is when G is the mid-point of AB so that $m = n = 1$. Then

$$2 \cot \theta = \cot \alpha - \cot \beta = \cot A - \cot B.$$

Example 2. A uniform ladder AB, of weight W, leans against a smooth vertical wall at A and rests on rough horizontal ground at B. λ is the angle of friction between ground and ladder and the ladder is inclined at an angle θ to the wall. Show that $\tan \theta \leq 2 \tan \lambda$, and find the reactions at A and B when the ladder is about to slip.

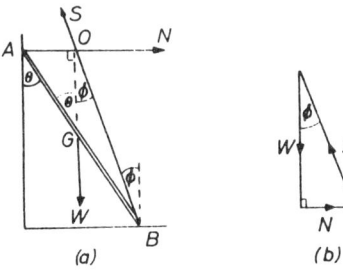

Figure 21.9

With the ladder in equilibrium the lines of action of the normal reaction N at A, the weight W and the total reaction S at B must meet at O (refer to *Figure 21.9a*). Let ϕ be the angle between S and the vertical.

Applying the cotangent rule to triangle AOB,

$$2 \cot \theta = \cot \phi - \cot 90°$$

$$\therefore \quad \tan \theta = 2 \tan \phi. \quad \ldots \text{(i)}$$

From the vector triangle (refer to *Figure 21.9b*) which is right angled,

$$N = W \tan \phi \quad \ldots \text{(ii)}$$

$$S = W \sec \phi. \quad \ldots \text{(iii)}$$

EQUILIBRIUM

Now for no slipping at B, $\phi \leq \lambda$ the angle of friction, or since ϕ, λ are acute angles

$$\tan \phi \leq \tan \lambda$$

$\therefore$ $\qquad 2 \tan \phi \leq 2 \tan \lambda$

and from (i) $\qquad \tan \theta \leq 2 \tan \lambda.$

When the ladder is about to slip, $\phi = \lambda$ and substituting for ϕ in (ii) and (iii) gives

$$N = W \tan \lambda, \qquad S = W \sec \lambda.$$

Example 3. A smooth hemispherical bowl, of internal radius r, is fixed with its rim horizontal. A thin uniform rod of length l ($l < 4a$) rests with one end inside the bowl and the other projecting over the rim. If the rod is inclined at an angle θ to the horizontal, show that $4r \cos 2\theta = l \cos \theta$.

Find also the reactions between rod and bowl in terms of θ and W (the weight of the rod) only.

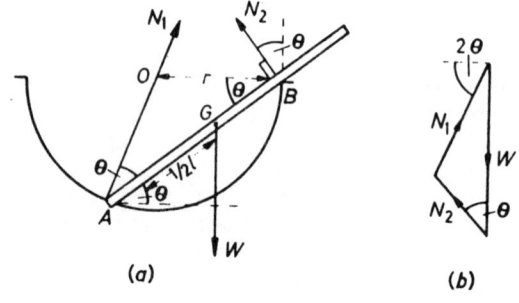

Figure 21.10

Refer to *Figure 21.10*. The reactions at A and B are normal to the common tangents of the surfaces in contact. Hence N_1 passes through the centre O of the hemisphere and N_2 is perpendicular to the rod. Since in triangle AOB, $AO = OB = r$, therefore $\angle OAB = \angle OBA = \theta$. Thus N_1 is inclined at an angle 2θ to the horizontal.

Taking moments about A,

$$N_2 \times AB - W \times \tfrac{1}{2}l \cos \theta = 0$$

$\therefore \qquad N_2 2r \cos \theta = W \tfrac{1}{2} l \cos \theta.$ $\qquad \ldots$ (i)

THREE-FORCE PROBLEMS

From the vector triangle, applying the sine rule

$$\frac{N_1}{\sin\theta} = \frac{N_2}{\sin(90° - 2\theta)} = \frac{W}{\sin(90° + \theta)}$$

$$\frac{N_1}{\sin\theta} = \frac{N_2}{\cos 2\theta} = \frac{W}{\cos\theta}.$$

Giving $\qquad N_1 = W\tan\theta \qquad$ .... (ii)

and $\qquad N_2 = W\cos 2\theta/\cos\theta.\qquad$ .... (iii)

Substituting this value of N_2 in equation (i),

$$W\frac{\cos 2\theta}{\cos\theta} \cdot 2r\cos\theta = W\tfrac{1}{2}l\cos\theta$$

or $\qquad 4r\cos 2\theta = l\cos\theta.$

(Note that this last relation can also be obtained from the geometry of the figure after drawing N_1, N_2 and W to meet in a point. However, taking moments seems easier in this case.)

Exercises 21c

1. A spherical steel ball, of mass 200 kg, is suspended by a chain fastened to it by a smooth joint. It is pulled sideways by a force of 50 kg wt. Find the force exerted by the chain on the ball.

2. A uniform square plate $ABCD$, of side $2a$, has a string attached to the mid-point of BC. The other end of the string is fastened to a point O in a smooth vertical wall. The plate rests, in a vertical plane, with the vertex C against the wall and CA horizontal. Find the length of the string.
Also, if the plate has weight W, find the tension in the string and the reaction at the wall.

3. A circular hoop, of weight W, hangs over a rough horizontal peg. An equal weight W is attached by a string to the rim of the hoop. If the string is tangential to the hoop which is just about to slip, find the angle of friction between the peg and the hoop.

4. A rod AB, of length $5a$, rests inside a smooth sphere, radius $5a$. $AG = 2a$, where G is the centre of gravity of the rod. Find the inclination of the rod to the horizontal.

5. A uniform semi-circular plate rests with one point of its curved surface against a smooth vertical wall, and another point on a rough horizontal floor. The plane of the plate is vertical and perpendicular to the wall. If the bounding diameter makes an

angle θ with the horizontal, show that the total reaction of the floor on the plate is inclined at $\tan^{-1}(4\sin\theta/3\pi)$ to the vertical.

If μ is the coefficient of friction between floor and plate, deduce the range of values μ may take for equilibrium to be possible.

6. A sphere rests on the curved surface of a fixed smooth hemisphere radius $2a$. Equilibrium is maintained by a light inextensible string, length a, attached to the sphere and the highest point of the hemisphere. Find the tension in the string and the reaction between hemisphere and sphere. (Radius a, weight W).

7. A uniform ladder rests with one end against a rough vertical wall and the other on a rough horizontal floor. The coefficients of friction at the wall and floor are respectively μ_1 and μ_2. Find the inclination of the ladder to the floor when the equilibrium is limiting.

8. A solid homogeneous sphere, of radius a, rests in contact with a smooth plane inclined at an angle $\sin^{-1}(3/5)$ to the horizontal. It is maintained in equilibrium by a light inextensible string of length $18a/7$, one end of which is attached to the sphere and the other end to the plane. Find the tension in the string and the reaction of the plane on the sphere in terms of W the weight of the sphere.

9. A uniform rod AB, of length $2a$ and weight W, rests on a rough peg P with its lower end A under a smooth peg Q at a lower level than P. The rod is inclined at an angle θ to the horizontal and $PQ = c$ ($<a$). Find the coefficient of friction at the peg P, if the rod is just about to slip.

10. A uniform rod rests inclined at an angle θ to the horizontal with its ends on two planes, each inclined at an angle α to the horizontal. The line of intersection of the planes is horizontal. The angle of friction at both points of contact is λ. Show that, if the rod is about to slip,

$$\tan\theta = (\sin 2\lambda)/(\cos 2\alpha + \cos 2\lambda).$$

[It is suggested that students also pick out those problems in Exercises 21a for which the above methods are suitable and rework them.]

21.4. SEVERAL BODIES IN CONTACT

By Newton's third law, the reactions between bodies in contact are equal and opposite. Hence, when considering the equilibrium of several bodies as a whole, the internal forces may be ignored. In addition, we may consider the equilibrium of each body separately.

Take the case of two bodies in equilibrium under the action of coplanar forces. Taking moments and resolving for both bodies

SEVERAL BODIES IN CONTACT

together and for each separately, gives nine equations. Only six of them, however, are independent, and perhaps not all of these are required in a particular problem. It is important to try to select those equations which lead most simply to the required solution.

Example. A uniform thin smooth rod, of weight W and length $4a$, is freely hinged to a rough horizontal surface. The rod is lifted and a rough cube, of side $2a$ and weight W, is pushed under it until the rod is inclined at 45 degrees to the horizontal. Show that, when released, the system will remain at rest provided $7\mu \geqslant 2\sqrt{2} - 1$, μ being the coefficient of friction between the cube and the surface.

Find the horizontal and vertical components of the reaction at the hinge in this position.

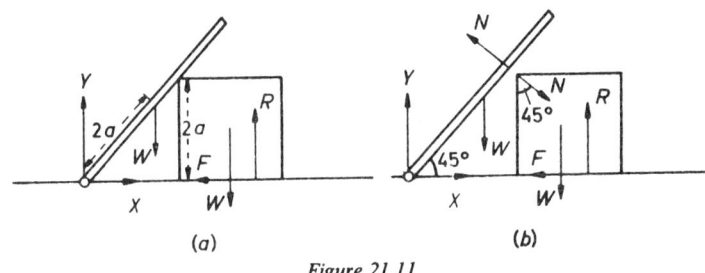

Figure 21.11

Suppose that no slipping takes place and the rod and cube remain in equilibrium. *Figure 21.11a* shows the forces on the bodies considered together: *Figure 21.11b* shows the forces on the bodies separately.

Consider first the rod and cube together. Resolving,

horizontally $\qquad X - F = 0 \qquad$ ....(i)

vertically $\qquad Y + R - 2W = 0. \qquad$ ....(ii)

Consider the rod alone and take moments about the hinge.

$$N \cdot 2a \operatorname{cosec} 45° - W \cdot 2a \cos 45° = 0. \qquad \text{....(iii)}$$

Consider the cube alone. Resolving,

horizontally $\qquad N \sin 45° - F = 0 \qquad$ ....(iv)

vertically $\qquad R - N \cos 45° - W = 0. \qquad$ ....(v)

EQUILIBRIUM

From (iii) $N = \frac{1}{2}W$ and substituting this value in (iv) and (v) we obtain

$$F = \tfrac{1}{4}W\sqrt{2}, \qquad R = W\left(1 + \frac{\sqrt{2}}{4}\right).$$

But for no slipping $F \leqslant \mu R$.

$$\therefore \qquad \frac{\sqrt{2}}{4} \leqslant \mu\left(1 + \frac{\sqrt{2}}{4}\right)$$

$$\therefore \qquad \mu \geqslant \frac{\sqrt{2}}{4 + \sqrt{2}}$$

$$\therefore \qquad 7\mu \geqslant 2\sqrt{2} - 1.$$

Also, substituting for F and R in (i) and (ii),

$$X = \tfrac{1}{4}W\sqrt{2}, \qquad Y = W\left(1 - \frac{\sqrt{2}}{4}\right),$$

and these are the horizontal and vertical components of the reaction of the hinge on the rod.

21.5. JOINTED RODS

The methods outlined in the previous section can be applied to frameworks consisting of heavy rods freely hinged (pin-jointed) together. However, to determine the reactions between the rods the equilibrium of the pins has to be considered.

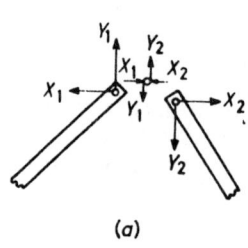
(a)

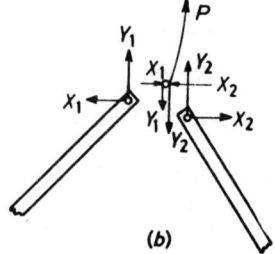
(b)

Figure 21.12

JOINTED RODS

When there is no external force acting on the pin, the forces acting will be as shown in *Figure 21.12a* (the pin is assumed to be of negligible weight). Since the pin is in equilibrium, $Y_1 = Y_2$ and $X_1 = X_2$. Hence the reactions on the rods are equal and opposite so that they may be treated as if the pin were not there and the rods were acting directly on one another (*see* Example 1 following).

When there is an external force P acting on the pin (refer to *Figure 21.12b*), then these reactions will no longer be equal and opposite and the equilibrium of the pin must be considered in each case. However, in particular problems, it is usually possible to omit consideration of the reactions at such a joint. This is done either by taking moments about the joint or by considering the adjacent rods and the pin as one unit (*see* Example 2 following).

Example 1. Three identical rods, each of weight W, are freely jointed together to form a triangle. The framework is suspended from the mid-point of one of the rods. Find the tension in the string with which it is suspended and the reactions at the joints.

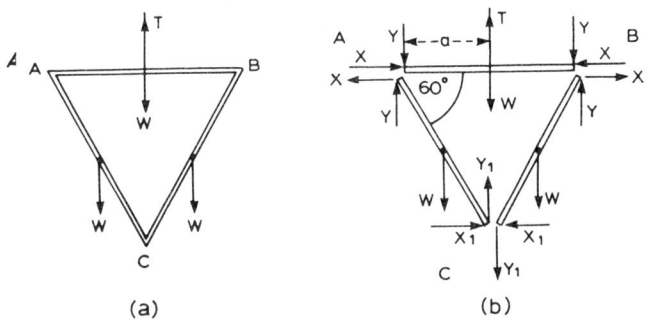

Figure 21.13a, b

Referring to *Figure 21.13a*, consider the framework as a whole.

Vertically $\qquad T - 3W = 0.$ ....(i)

The framework and the external forces acting on it are, in this case, symmetrical about a vertical line. Hence the reactions at the joints, which arise in response to the external forces, are also symmetrical. These reactions are shown in *Figure 21.13b*.

Consider the rod AB. Resolving,

vertically $\qquad T - 2Y - W = 0.$ ....(ii)

EQUILIBRIUM

Consider the rod AC. Resolving,

vertically $\qquad Y + Y_1 - W = 0 \qquad \ldots$ (iii)

and horizontally $\qquad X_1 - X = 0. \qquad \ldots$ (iv)

Taking moments about A

$$Y_1 . 2a \cos 60° + X_1 2a \sin 60° - W . a \cos 60° = 0. \quad \ldots (v)$$

Now solving successively equations (i), (ii), (iii), (v) and (iv) we obtain

$$T = 3W, \quad Y = W, \quad Y_1 = 0, \quad X_1 = X = \frac{\sqrt{3}}{6}W.$$

Thus the components of the reaction on the rod AC at A are as shown in Figure 21.13c.

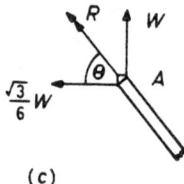

(c)

Figure 21.13c

Hence the resultant reaction R on AC at A is given by

$$R^2 = W^2 + \left(\frac{\sqrt{3}}{6}W\right)^2$$

$\therefore \qquad R = W\sqrt{39}/6$

also $\qquad \tan \theta = 6/\sqrt{3} = 2\sqrt{3}.$

The tension in the string is $3W$, the reaction at C is horizontal and of magnitude $\sqrt{3}W/6$, and the reaction at A is of magnitude $\sqrt{39}W/6$ inclined at $\tan^{-1} 2\sqrt{3}$ to the horizontal.

(Note that the fact that $Y_1 = 0$ could also have been deduced from the symmetry of the figure.)

Example 2. A framework ABCD consists of four equal rods, each of weight W, freely jointed together. The framework is suspended from A and prevented from collapsing by a light inextensible string, of the same length as the rods, joining A to C. Find the reaction at B and the tension in the string AC.

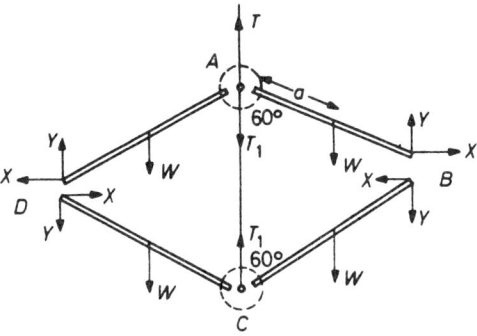

Figure 21.14

The framework and the external forces acting on it are symmetrical about the line AC. Wherever, because of this symmetry, it is clear that forces are equal, they have been marked as such on the diagram (Figure 21.14).

Consider the rod AB. Taking moments about A,

$$X \cdot 2a \cos 60° + Y \cdot 2a \sin 60° - Wa \sin 60° = 0. \quad \ldots (i)$$

Consider the rod BC. Taking moments about C,

$$X \cdot 2a \cos 60° - Y 2a \sin 60° - Wa \sin 60° = 0. \quad \ldots (ii)$$

Consider the rods DC and CB together. Resolving vertically,

$$T_1 - 2Y - 2W = 0. \quad \ldots (iii)$$

Adding (i) and (ii)

$$X \cdot 4a \cos 60° = 2Wa \sin 60°.$$

$$\therefore \quad X = \frac{\sqrt{3}}{2} W.$$

Subtracting (ii) from (i)

$$Y \cdot 4a \sin 60° = 0$$

$$\therefore \quad Y = 0.$$

Substituting for Y in (iii)

$$T_1 = 2W.$$

EQUILIBRIUM

The reaction at B is horizontal and of magnitude $\sqrt{3}W/2$; the tension in the string AC is of magnitude $2W$.

21.6. LIGHT FRAMEWORKS

Consider a light rod AB, in equilibrium, whose weight is negligible compared with resultant forces R_1, R_2 acting on its ends (refer to Figure 21.15). Taking moments about A, $R_2 d = 0$. Hence $d = 0$ and R_2 acts along BA. Similarly R_1 acts along the rod. Thus the forces always act along a light rod.

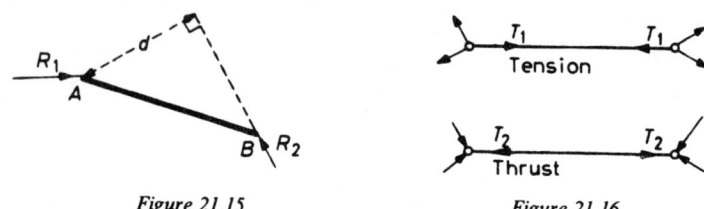

Figure 21.15   *Figure 21.16*

In a framework consisting of *light* rods, the rods transmit the forces between the joints. *Figure 21.16* shows the forces T_1, T_2 exerted by two such rods on the joints at their ends. When the rod pulls on the joints (like a string) the force in it (T_1) is called a *tension*; when it pushes on the joints the force (T_2) is called a *thrust*.

When considering the equilibrium of such a framework we can (*a*) consider the framework as a whole and (*b*) consider the equilibrium of each *joint*.

Example. A framework consists of five light rods, four of length a and one of length $a\sqrt{3}$, joined together as shown in Figure 21.17. The framework is in a vertical plane hinged freely to a smooth wall at A and with a load W at C. Find the horizontal and vertical components of the reaction at A and the stresses in the rods.

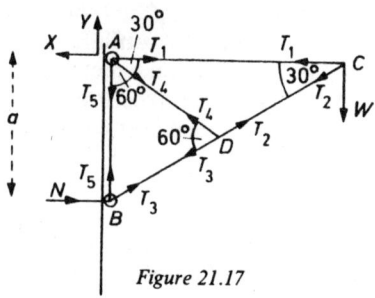

Figure 21.17

JOINTED RODS

Consider the whole framework, refer to *Figure 21.17*. Taking moments about A,
$$Na - Wa\sqrt{3} = 0 \quad \ldots \text{(i)}$$

Resolving,

vertically $\quad\quad\quad\quad\quad Y - W = 0 \quad\quad\quad \ldots \text{(ii)}$

horizontally $\quad\quad\quad\quad X - N = 0. \quad\quad\quad \ldots \text{(iii)}$

Consider the joint at C. Resolving,

vertically $\quad\quad\quad\quad T_2 \sin 30° + W = 0 \quad \ldots \text{(iv)}$

horizontally $\quad\quad\quad T_1 + T_2 \cos 30° = 0. \quad \ldots \text{(v)}$

Consider the joint at D. Resolving,

perpendicular to BC $\quad\quad T_4 \sin 60° = 0 \quad\quad \ldots \text{(vi)}$

along BC $\quad\quad\quad T_2 - T_3 - T_4 \cos 60° = 0. \quad \ldots \text{(vii)}$

Consider the joint at B. Resolving,

vertically $\quad\quad\quad\quad T_5 + T_3 \cos 60° = 0. \quad \ldots \text{(viii)}$

Equations (i), (ii) and (iii) give $Y = W$ and $X = N = W\sqrt{3}$.

From (iv) and (v), $T_2 = -2W$ and $T_1 = W\sqrt{3}$

From (vi) and (vii), $T_4 = 0$ and $T_3 = -2W$

From (viii), $\quad\quad\quad T_5 = W.$

Thus the horizontal and vertical components of the reactions at A are $W\sqrt{3}$ and W respectively. The forces in AC and AB are tensions of magnitude $W\sqrt{3}$ and W respectively, and the forces in BD and DC are thrusts each of magnitude $2W$. There is no stress in the rod AD, indeed it is unnecessary for the framework.

Engineering structures of this type (bridges, cranes etc.), often have many members and there is a large number of joints to consider. Methods have been developed to "streamline" the process of determining the stresses in the rods. One such method involves constructing a "force-diagram" for the whole structure by super-imposing the vector diagrams for each joint. However, we shall not pursue these matters further here.

Exercises 21d

1. A uniform rod, of length $6a$ and weight W, is smoothly hinged at one end to a rough horizontal floor. The rod rests on the curved

EQUILIBRIUM

surface of a hemisphere inclined at 45 degrees to the horizontal. The hemisphere, of weight W and radius a, has its base in contact with the floor. Find the reaction between the rod and the hemisphere and show that the coefficient of friction between the floor and the hemisphere is greater than or equal to $\frac{3}{5}$.

2. Two smooth spheres, of weights W and W' respectively, hang suspended from the same point by light inextensible strings. The strings are such that the spheres hang with their line of centres horizontal. If the strings suspending W and W' make angles of θ and ϕ respectively with the vertical, show that

$$W:W' = \tan\phi : \tan\theta.$$

3. A smooth sphere, of weight W, rests on the face of a prism, weight $4W$, the face being inclined at an angle θ to the horizontal. The sphere is in contact with a vertical wall and the prism rests on a rough horizontal floor, coefficient of friction μ. Show that $\mu \geqslant (\tan\theta)/5$.

4. Three rough cylinders, each of radius a and weight W are in equilibrium with their axes horizontal. Two are in contact with a rough horizontal plane and the distance apart of their axes is $12a/5$. The third cylinder rests symmetrically on top of the first two. If the equilibrium is limiting at all lines of contact, find the coefficients of friction.

5. Two identical rods AB and BC, each weight W, are smoothly jointed at B. They rest in a vertical plane with A and C on a rough horizontal surface; angle $ABC = 2\theta$. The coefficient of friction at both A and C is μ. Show that $2\mu \geqslant \tan\theta$ and find the reaction at B.

6. A framework $ABCD$ consists of four equal rods, of equal length and each of weight W, freely jointed together. The framework is suspended from the mid-point of AB so that it hangs in a vertical plane. Find the reactions at B and D.

7. Three identical rods, each of weight W, are freely jointed together to form a triangle ABC. The framework is suspended from A. Find the horizontal and vertical components of the reactions at the joints.

8. A rhombus $ABCD$ is formed of four uniform rods, each of length $2a$ and weight W, freely jointed at the corners. A light inelastic string, of length $2a$, connects A and C. The system hangs in equilibrium by two vertical strings attached to A and B so that the rod AB is horizontal. Find the tension in each of the three strings.

Find the reaction between the rods at the joint D and show that no horizontal force acts on the rod BC at C. (WJEC)

9. A light rod AB is freely hinged to a vertical wall at A and the

ALTERNATIVE CONDITIONS FOR EQUILIBRIUM

end B is connected to a point C on the wall above A by another light rod which is horizontal. A weight W hangs from B. If $\angle CAB = \theta$, find the stresses in the rods.

10. A framework consists of three light rods AB, BC and CA freely jointed at A, B and C. They form a triangle in which angle ABC is a right angle and angle $CAB = 30$ degrees. The frame is in a vertical plane, supported at A and C, with AC horizontal and B above AC. A weight W is hung from B. Find the stresses in the rods.

21.7. ALTERNATIVE CONDITIONS FOR EQUILIBRIUM

Our general method, for solving problems of equilibrium under the action of coplanar forces, has been to obtain three equations by taking moments about a point and resolving in two directions at right angles. We shall show that alternative methods are:

(a) to take moments about two points A and B, and to resolve in a direction *not* at right angles to AB, or

(b) to take moments about three points *not* in the same straight line.

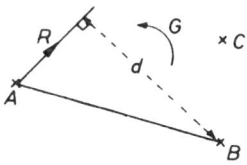

Figure 21.18

Let A, B, C be three non-collinear points in the plane of the system of forces. Then the system can be reduced to a single force R acting through A together with a couple of moment G (see *Figure 21.18* and Chapter 17).

If the total moment about A is zero, then $G = 0$. If, in addition, the total moment about B is zero, then $Rd = 0$ and either $R = 0$ or R lies along AB. That is, the system is in equilibrium ($R = G = 0$) or R lies along AB.

But if R is not zero and lies along AB, it must have a component in a direction not perpendicular to AB, and it must have a moment about C. Hence if either of these is zero, R must be zero and the system is in equilibrium.

Example. A uniform rod AB, of length 4 m and mass 10 kg, has a particle of mass 5 kg attached at B. It is freely hinged at A to a vertical wall and held inclined at 60 degrees to the downward vertical by a string attached to B and to a point on the wall 4 m, above A. Find the tension in the string and the horizontal and vertical components of the reaction at the hinge.

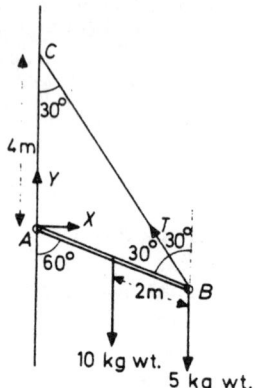

Figure 21.19

Refer to *Figure 21.19*. Since $AC = AB = 4$ m and the exterior angle of triangle $ABC = 60$ degrees, $\angle ACB = \angle CBA = 30$ degrees. Taking moments about A,

$$T \times 4 \sin 30° - 10 \times 2 \sin 60° - 5 \times 4 \sin 60° = 0. \quad \ldots \text{(i)}$$

Taking moments about C,

$$X \times 4 - 10 \times 2 \sin 60° - 5 \times 4 \sin 60° = 0. \quad \ldots \text{(ii)}$$

Resolving vertically,

$$Y + T \cos 30° - 10 - 5 = 0. \quad \ldots \text{(iii)}$$

From (i) $T = 10\sqrt{3}$ kg wt.

From (ii) $X = 5\sqrt{3}$ kg wt.

From (iii) $Y = 0$.

The tension in the string is $10\sqrt{3}$ kg wt. and the reaction at A is horizontal and of magnitude $5\sqrt{3}$ kg wt. (Note that instead of (iii), a third equation could have been obtained by taking moments about some point not on CA, say B for example.)

MISCELLANEOUS EXAMPLES

21.8. MISCELLANEOUS EXAMPLES

Example 1. A cone of radius r and height h rests on a rough horizontal surface, the coefficient of friction between cone and surface being μ. A gradually increasing horizontal force P is applied to the vertex of the cone. Find the range of values of μ for which the cone will slide before it topples.

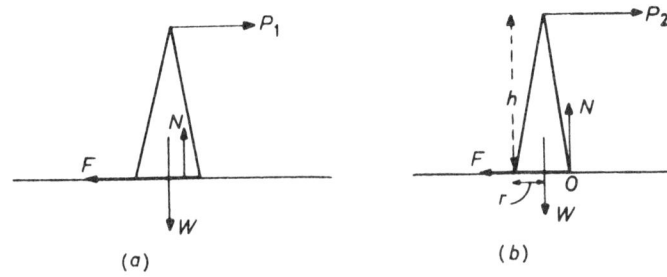

Figure 21.20

Figure 21.20a shows the forces acting when the cone is about to slip and $F = \mu N$. In this case resolving

vertically $\qquad N - W = 0,$

horizontally $\qquad P_1 - \mu N = 0.$

This gives
$$P_1 = \mu W. \qquad \ldots (i)$$

Figure 21.20b shows the forces when the cone is about to topple about O. In this case, taking moments about O, $P_2 h - Wr = 0$.

$\therefore \qquad P_2 = Wr/h. \qquad \ldots (ii)$

The cone will slide before it topples if P reaches the value P_1 before it reaches the value P_2, i.e.

if $\qquad \mu W < Wr/h$

or $\qquad \mu < r/h.$

Note that in this kind of problem we investigate the least values of the *varying* quantity which cause toppling and sliding. In this problem the varying quantity was the force P, in another it might be the gradually increasing inclination of a plane.

EQUILIBRIUM

Example 2. A step-ladder consists of two equal arms BA, AC freely hinged at A. The weight of AB is three times that of AC. Show that if the angle between BA and AC is steadily increased then slipping first takes place at C.

If BA, AC are to be placed so that $\angle BAC = 90$ degrees, find the least value of μ to prevent slipping.

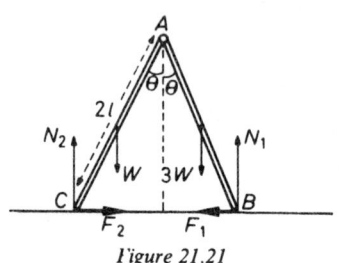

Figure 21.21

Consider CA and AB together (refer to *Figure 21.21*). Taking moments about C,

$$N_1 . 4l \sin \theta - 3W . 3l \sin \theta - Wl \sin \theta = 0 \qquad \ldots \text{(i)}$$

Resolving,

vertically $\qquad\qquad N_1 + N_2 - 4W = 0 \qquad \ldots \text{(ii)}$

horizontally $\qquad\qquad F_1 - F_2 = 0. \qquad \ldots \text{(iii)}$

From (i) $N_1 = 5W/2$ and substituting this value in equation (ii) gives $N_2 = 3W/2$. Also from equation (iii) $F_1 = F_2 = F$ (say).

Consider AB alone. Taking moments about A,

$$N_1 . 2l \sin \theta - 3W . l \sin \theta - F . 2l \cos \theta = 0,$$

and on substituting for N_1 this gives

$$F = W \tan \theta. \qquad \ldots \text{(iv)}$$

Hence as θ increases, F will increase and, since it will reach the value $\mu.\frac{3}{2}W$ before it reaches the value $\mu.\frac{5}{2}W$, slipping will take place at C.

When $\theta = 45$ degrees, $F = W$ from (iv) and for no slipping.

$$F \leqslant \mu N_2$$

i.e. $\qquad\qquad W \leqslant \mu 3W/2$

i.e. $\qquad\qquad \mu \geqslant 2/3.$

The least value of μ to prevent slipping, when $\theta = 45$ degrees, is 2/3.

MISCELLANEOUS EXAMPLES

Example 3. A sphere of weight W is hung on a string from a point on the vertical line of intersection of two perpendicular walls, and hangs in smooth contact with them. If the length of the string and the radius of the sphere are both a, find the tension in the string and the reactions due to the walls.

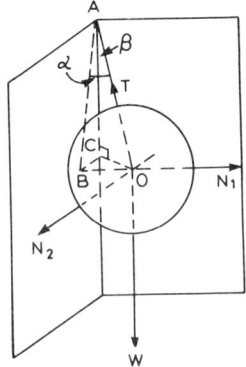

Figure 21.22

Refer to *Figure 21.22*. Since the normal reactions and weight pass through the centre O of the sphere, then so must the line of action of the tension.

Let A be the point where the string meets the walls: B the point of contact on one wall: C the point on the line of intersection of the walls at the same level as B. Then α, the inclination of the string to the wall is given by

$$\sin \alpha = OB/OA = a/(2a) = \tfrac{1}{2}$$

$$\therefore \quad \alpha = 30°.$$

Similarly β, the inclination of the string to the vertical is given by

$$\sin \beta = OC/OA = a\sqrt{2}/(2a) = 1/\sqrt{2}$$

$$\therefore \quad \beta = 45°.$$

Resolving,

vertically	$T \cos 45° - W = 0$	....(i)
normal to wall at B	$N_1 - T \sin 30° = 0$	....(ii)
normal to other wall	$N_2 - T \sin 30° = 0.$	....(iii)

From equation (i) $T = \sqrt{2}W$ and substituting this value in (ii) and (iii) gives $N_1 = N_2 = W/\sqrt{2}$.

EQUILIBRIUM

The tension in the string is of magnitude $W\sqrt{2}$ and the reactions at the walls are normal and each of magnitude $W/\sqrt{2}$.

Example 4. A table has an equilateral triangular top, of side 2a, supported by three vertical legs at the mid-points of the sides of the triangle. The whole table has a weight W. Show that the greatest weight that can be placed at one of the vertices of the triangle without overturning it is $\frac{1}{3}W$.

If a particle of weight $\frac{1}{6}W$ is placed at a corner of the table, find the vertical reactions on the legs.

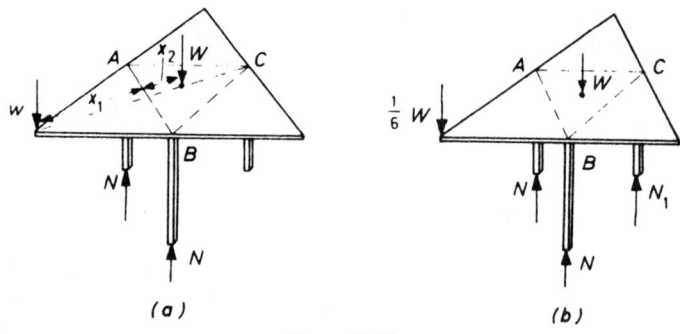

Figure 21.23

Consider the table when a weight w is placed at a vertex and it is about to topple (refer to *Figure 21.23a*). There will be no reaction on the leg at C and the reactions at A and B will be the same by symmetry.

Taking moments about the *line AB*,

$$wx_1 - Wx_2 = 0.$$

But, since the triangle is equilateral and of side $2a$,

$$x_1 = \tfrac{1}{2}a\sqrt{3} \quad \text{and} \quad x_2 = \tfrac{1}{6}a\sqrt{3}.$$

$$\therefore \quad w \cdot \tfrac{1}{2}a\sqrt{3} - W \cdot \tfrac{1}{6}a\sqrt{3} = 0,$$

giving
$$w = \tfrac{1}{3}W.$$

This is the greatest weight that can be placed at a vertex without overturning the table.

Hence if a weight $\frac{1}{6}W$ is placed there the table will not topple and the reaction at C will not be zero (refer to *Figure 21.23b*).

MISCELLANEOUS EXAMPLES

Taking moments about the line AB in this case,

$$\tfrac{1}{6}W \cdot \tfrac{1}{2}a\sqrt{3} - W \cdot \tfrac{1}{6}a\sqrt{3} + N_1 \cdot \tfrac{1}{2}a\sqrt{3} = 0. \quad \ldots \text{(i)}$$

Resolving vertically,

$$2N + N_1 - W - \tfrac{1}{6}W = 0. \quad \ldots \text{(ii)}$$

From equation (i) $N_1 = \tfrac{1}{6}W$ and substituting this value in (ii) gives $N = \tfrac{1}{2}W$.

Thus the reactions at A and B are of magnitude $\tfrac{1}{2}W$ and that at C of magnitude $\tfrac{1}{6}W$.

Exercises 21e

1. A uniform rod, of length 40 cm and weight 20 kg, is freely hinged at one end A to a vertical wall. The other end B is tied to a point C, vertically above A, by a string of length 30 cm. The string is at right angles to the rod. By taking moments about two points and resolving in a suitable direction, find the tension in the string and the horizontal and vertical components of the reaction at A.

2. A uniform right circular cylinder, of radius a, weight W and height h, is placed on a rough plane inclined at an angle α to the horizontal. A horizontal force P is applied at the highest point of the cylinder directed towards the plane. The plane is rough enough to prevent slipping and the magnitude of P is such that the cylinder is about to overturn. By taking moments about three suitable points, determine directly the magnitude of P and of the normal reaction and friction.

3. A homogeneous right circular cylinder, of height h and base radius a, is placed with a plane face in contact with a rough plane, the coefficient of friction being μ. The plane is gradually tilted. Show that the cylinder will topple before it slides if $\mu > 2a/h$.

4. A uniform solid cube is placed on a plane inclined at $\sin^{-1} \tfrac{3}{5}$ to the horizontal, one of the edges of the cube being horizontal. A gradually increasing force is applied, parallel to and directly up the plane, to the mid-point of the upper edge of the cube. The cube eventually topples over without sliding. Find the range of possible values of μ, and the coefficient of friction between cube and plane.

5. Two equal uniform rods AB and BC, of weights W and $2W$, respectively, are freely jointed at B. They rest in a vertical plane with A and C on rough horizontal ground. The coefficient of friction at both A and C is μ. The angle ABC is gradually increased and slipping occurs when angle ABC is 120 degrees. Find μ.

6. A pair of steps consists of two uniform ladders AB and BC, each of weight W and length $2a$, freely hinged at B. They are placed

on rough horizontal ground with angle ABC a right angle. A man of weight $2W$ climbs the steps on one side. Calculate the reactions at the floor when his distance from the top is x. If the coefficient of friction is $1/3$ at both A and C, find the distance the man can climb before the steps collapse.

7. A sphere, of weight W, rests in the corner formed by two vertical walls and a horizontal floor. It is held in position by a horizontal force, of magnitude $2W$, directed through the centre of the sphere and at 45 degrees to each wall (the walls being at right-angles to each other). Find the reactions at the three points of contact.

8. In the situation described in Question 7, the force of magnitude $2W$ is replaced by a force of magnitude $W\sqrt{3}$ directed through the centre of the sphere directly into the corner. Find the reactions in this case.

9. A square table, of weight W, is supported by four vertical legs one at the mid-point of each side. What is the greatest weight that can be placed at a corner without upsetting the table?

10. A table, in the form of an equilateral triangle ABC, is attached to a vertical wall by a smooth hinge along the side AB. It is supported in a horizontal position by two strings attached to points D, E in AC, BC respectively. D divides AC in the ratio $2:1$ and E divides BC in the same ratio. The strings are parallel, perpendicular to AB and inclined (towards the wall) at an angle α to the table. If the table has a weight W and two particles, each of weight W, are placed at the mid-points of AC and BC, find the tensions in the strings. Find also the horizontal and vertical components of the reaction on the hinge.

EXERCISES 21*

1. Two smooth planes inclined at angles α and β to the horizontal face each other their line of intersection horizontal. A sphere of weight W rests between them, find the reaction of each plane on the sphere.

2. A circular hoop, of radius a weight W, has a light inextensible string of length $2a$ attached to its perimeter. The other end of the string is attached to a smooth vertical wall and the hoop hangs in equilibrium resting against the wall with its plane perpendicular to the wall. Find the tension in the string and the reaction between the hoop and the wall.

3. A uniform rod, of weight W, rests on two equally rough pegs A and B at different levels in the same vertical plane. The rod is inclined at an angle θ to the horizontal and the coefficient of friction

* Exercises marked thus, †, have been metricized.

EXERCISES

between a peg and a rod is μ. Show that, so long as the centre of the rod lies between A and B equilibrium is possible if $\mu \geq \tan \theta$.

4. A plane lamina in the form of a sector of a circle centre O, angle 60 degrees is freely suspended by a string attached to one end A of its bounding circular arc. Find the inclination of OA to the downward vertical.

5. A rod is freely hinged to a horizontal floor and rests, inclined, across a circular cylinder. The cylinder also rests on the floor and its axis is perpendicular to the rod. Show that equilibrium is impossible if either the rod, the floor or the cylinder are smooth.

6. A uniform solid right circular cone, of height h and radius r, is placed with its base on a rough plane. The coefficient of friction between cone and plane is μ. The inclination of the plane is gradually increased and the cone begins to slide and topple simultaneously. Show that $\mu = 4r/h$.

7. A uniform rod AB, of length $2a$, rests with one end A against a rough vertical wall. It is supported, at an angle α to the horizontal, by a light smooth ring R threaded on it and tied to a point P on the wall vertically above A. Deduce that angle PRA is a right angle.

If $PA = d$ ($d \sin \alpha < a$) and μ is the coefficient of friction, show that when the rod is on the point of slipping down the wall

$$d \sec \alpha = a(\cot \alpha + \mu).$$

8. Two equal uniform rods AB and BC, each of weight W, are freely jointed at B and A is attached to a smooth fixed pivot. A horizontal force, of magnitude W, acts at the end C. Show that, in equilibrium, the rods are inclined at $\tan^{-1}(2/3)$ and $\tan^{-1}(2)$ to the vertical respectively. Find the resultant reaction at A.

9. A uniform ladder, of length $4a$ and weight W, rests with one end against a smooth wall and the other end on a rough horizontal floor. A light elastic string, of modulus W and natural length a, is attached to the mid-point of the ladder and to the junction of the wall and the floor. Show that, whatever the inclination of the ladder, the stretched length of the string is constant. Find the frictional force at the foot of the ladder when it is inclined at an angle θ to the vertical.

10. A uniform rod AB, of length $2a$, has a ring fastened to the end A which is threaded on a fixed rough horizontal wire. The end B is firmly attached by a light inextensible string, of length $2a$, to a fixed point C on the wire. If μ is the coefficient of friction at A, and the rod is about to slip when $\angle ABC = 2\theta$, show that $3\mu = \tan \theta$.

11. Three uniform rods AB, BC and CD, each of length $2a$ and weight W, are freely jointed at B and C. They rest, symmetrically in

a vertical plane, with A and D in contact with a rough horizontal surface. If $\angle BAD = 60$ degrees and $\angle ABC = 120$ degrees, find the frictional force at A. Find also the reaction at B.

12. A uniform rod AB, of length $2a$ and weight W, rests with one end A on a rough horizontal floor, coefficient of friction μ. Another point C of the rod, where $AC = c(< 2a)$ touches a smooth peg. If the rod is inclined at an angle θ to the vertical, show that

$$\mu \geq \frac{a \sin \theta \cos \theta}{c - a \sin^2 \theta}.$$

13. A piece of uniform thin wire is bent and joined to form a semicircle together with its bounding diameter AB. The wire is hung with AB in contact with a rough horizontal peg. The coefficient of friction between wire and peg is μ. If equilibrium is possible with any point of AB in contact with the peg, find the range of possible values of μ.

14. A uniform thin plank, of length $8a$ and weight W_1, rests in a horizontal position on a fixed rough circular cylinder of radius a. A weight W_2 is attached to one end and the plank rotates, without slipping, on the cylinder to a new position of equilibrium inclined at an angle θ to the horizontal. If the plank is just about to slip, show that $\lambda = 4W_2/(W_1 + W_2)$ where λ is the angle of friction.

15. A uniform ladder rests against a smooth vertical wall with its lower end on a rough floor, coefficient of friction μ. The floor slopes away from the wall at an angle of inclination of α to the horizontal. The ladder is inclined to the wall at an angle θ and is just about to slip, find μ.

16. Three uniform rods AB, BC and CA, whose weights are proportional to their lengths are freely jointed to form a triangular framework. $AB = 30$ cm, $BC = 40$ cm and $CA = 50$ cm. The framework is hung with CA horizontal and over a peg P. Show that $AP = 24$ cm.

If 5 g/cm is the mass per unit length of the rods, find the horizontal and vertical components of the reactions at A and C.

17. A uniform cube lies on a rough horizontal floor, coefficient of friction μ. A gradually increasing force is applied to the mid-point of and perpendicular to a top edge acting in an upward direction at an angle θ ($< 90°$) to the top force. Show that the block will rotate about a bottom edge, without sliding first, if $\tan \theta > (1 - 2\mu)/\mu$.

18. Three spheres of equal radii but of weights W_1, W_2, W_3, rest in order, with their centres on a horizontal line. They hang on three light inextensible strings from the same point which is vertically above their line of centres. If θ_1, θ_2, θ_3 are the inclinations of the

EXERCISES

strings, respectively, to the vertical (measured in the same sense), show that $\Sigma(W \tan \theta) = 0$.

19. A step ladder consists of two equal arms AB and BC, each of weight W, freely hinged at B. It rests with A and C on a smooth horizontal floor and with the arms each inclined at 30 degrees to the vertical. A light inextensible string connects C to the mid-point of AB. Find the tension in the string and the reaction at B.

20. Two planes, each inclined at an angle α to the horizontal, face each other. Their line of intersection is horizontal. Three identical uniform cylinders are placed symmetrically in the groove between the planes, with their axes horizontal. The two lower cylinders are in contact with one another and each with a plane. The third cylinder rests on these two. If all points of contact are smooth and the lower cylinders are about to separate, show that $\tan \alpha = 1/3\sqrt{3}$.

21. The axles of a four-wheeled railway truck of weight W are of equal length and the distance between them is a. The horizontal line which bisects each of them perpendicularly is intersected by the vertical line through the centre of mass of the truck at a distance b in front of the rear axle ($b \leq \frac{1}{2}a$). If, when the truck is resting on level rails, one of the front wheels, of negligible weight, is removed, find the reactions on the other three wheels. (Oxford)

22. A smooth sphere, of weight $2W$, and a uniform rod, of weight W, are suspended from the same point B by two light inextensible strings BA, BC respectively. A is a point on the sphere and C an end of the rod. They rest in equilibrium with the mid-point of the rod touching the sphere. If θ and ϕ are the inclinations of BA and BC to the vertical respectively, show that

$$\sin \phi = 2 \sin \theta / \cos (\theta + \phi).$$

23. A uniform solid right circular cone, of semi-vertical angle $\tan^{-1}(1/2)$ and weight W, has a light inextensible string fastened to its vertex and to a point on the rim of its base. When the string is hung over a smooth peg, the axis of the cone is horizontal. Find the length of the string in terms of the radius of the base of the cone, also the tension in the string.

24. A smooth circular cylinder, centre O and radius a, is fixed with its axis horizontal and parallel to a smooth vertical wall. The distance from the axis of the cylinder to the wall is d. A uniform rod length $6a$ rests on the cylinder with one end B touching the wall (B lower than O). The rod is in a vertical plane at right angles to the axis of the cylinder. If the rod rests in equilibrium at an angle of 45 degrees with the horizontal, find d in terms of a.

EQUILIBRIUM

25. A solid uniform hemisphere, of radius 13 cm, rests in equilibrium with its curved surface touching a rough horizontal plane and a smooth peg at a height of 8 cm above the plane. Show that the reaction at the peg is inclined at $\sin^{-1}(5/13)$ to the horizontal and that $\mu \geqslant 4/9$.

26. Two equal uniform rods AB, BC, of weight W and length $2a$, are rigidly jointed at B with $\angle ABC = 90$ degrees. AB rests with a point P in contact with a rough horizontal peg, coefficient of friction $\mu = 3/4$. Show that $9a/8 \leqslant AP \leqslant 15a/8$.

27. Two uniform rods AB, BC, each of length $2a$ and of weights $3W$ and $4W$, respectively, are smoothly jointed at B. The end A is smoothly jointed to a fixed point. The rods are in equilibrium with ABC horizontal and BC supported at a point D. Find the distance AD.

28. A framework $ABCD$ consists of four equal rods, each of weight W length $2a$, freely jointed together. It is suspended from A and prevented from collapsing by a light rod of length $2a$ joining B and D. Find the reaction at C and the stress in the rod.

29. Two smooth uniform cylinders, each of weight W and radius a, rest, touching each other, on a smooth horizontal plane. A third equal cylinder is placed symmetrically on top of the other two. Equilibrium is maintained by an endless elastic band passing round all three cylinders. If the two lower cylinders are on the point of separating, find the tension in the band and the reaction between an upper and lower cylinder.

30. A smooth hemispherical bowl of internal radius a is fixed with its rim horizontal. A thin uniform rod of length l weight $2W$ rests with one end A in contact with the curved surface of the bowl and the other end B projecting over the rim. A particle of weight W is attached to the rod at B. By taking moments about the point of contact of the rod and the rim of the bowl, show that for equilibrium to be possible the maximum value of l is $3a$. When the rod is in equilibrium inclined at an angle θ to the horizontal, show that $3a \cos 2\theta = l \cos \theta$.

31. A uniform rod AB, of length $2a$, rests inside a rough fixed sphere of internal radius $3a$. The rod lies in a vertical plane passing through the centre of the sphere. Prove that, in the limiting position, the rod is inclined to the horizontal at an angle $\tan^{-1}[9\mu/(8 - \mu^2)]$, μ being the coefficient of friction.

32. A mast, of length $3l$ and weight W, is hinged to a point on level ground at O and held upright by three ropes attached to its top P. The ropes are fastened to points A, B and C on the ground such that $AO = BO = 4l$ and $CO = 3l$. Also angle AOB is a right angle and

EXERCISES

CO produced bisects angle AOB. Show that the tensions in the ropes AP and PC are in the ratio $5:8$.

If the tension in PC is $\frac{1}{5}W$, find the reaction at O.

33. A step-ladder consists of two equal legs, hinged at the top, the centre of gravity of each leg being at its mid-point. It stands on smooth level ground with a taut cord joining the points one-quarter of the way up each leg. Show that, if a man whose weight is twice the weight of the step-ladder stands three-quarters of the way up one side, the tension of the cord is quadrupled. (London)

34. Five equal light rods are freely jointed together to form a rhombus $ABCD$ and its diagonal BD. This framework hangs freely from A weights of 4 and 2 kg wt. hang from B and C respectively. Prove that BD is inclined to the horizontal at an angle $\cot^{-1}(2\sqrt{3})$. Show also that the tension in DC is three times that in BC. (London)†

35. A heavy uniform sphere of radius a has a light inextensible string attached to a point on its surface. The other end of the string is fixed to a point on a rough vertical wall. The sphere rests in equilibrium touching the wall at a point distant h below the fixed point. If the point of the sphere in contact with the wall is about to slip downwards and the coefficient of friction between the sphere and the wall is μ, find the inclination of the string to the vertical.

If $\mu = h/2a$ and the weight of the sphere is W, show that the tension in the string is $(W/2\mu)(1 + \mu^2)^{1/2}$. (London)

36. Two uniform cylinders, each of same radius but of unequal weights W_1 and W_2, rest with their axes horizontal on a rough inclined plane. The two cylinders are in contact with their central sections in the same plane, and the cylinder of weight W_1 is uppermost. Show that all the frictional forces on the two cylinders are equal and hence that $W_1 > W_2$.

If μ is the coefficient of friction at both points of contact with the plane and if the inclination of the plane to the horizontal is less than 45 degrees, show that

$$\mu \geqslant (W_1 + W_2)/(W_1 - W_2).$$

37. A uniform rod of weight $4W$ and length $2a$ is maintained in a horizontal position by two light inextensible strings each of length a attached to the ends of the rod. The other ends of the strings are attached to small rings each of weight W which can slide on a fixed rough horizontal bar with which the coefficients of friction are each $\frac{1}{2}$. Show that in equilibrium the distance between the bar and the rod cannot be less than $4a/5$, and find the greatest and least possible distances apart of the rings. (London)

38. An oblique cone of uniform density has a circular base of radius a, the perpendicular distance of the vertex from the plane of the base is b and the inclination of the axis to the plane of the base is 45 degrees. If the cone is placed with its base on a horizontal table, find the condition that it will remain at rest.

If the cone is placed with its curved surface in contact with the table, show that it will not tilt over on to its base if

$$4a^2 + 2b^2 > 5ab.$$ (London)

39. Two parallel rails l_1 and l_2 lie fixed in the same horizontal plane, the distance between them being $5a$. A uniform rod, of length $16a$ and weight W, is placed on the rails. It is at right angles to the rails with its centre distant $2a$ from l_1. A force kW is applied horizontally and at right-angles to the rod at the end nearest l_2. If k is steadily increased from zero and the coefficient of friction μ is the same at both points of contact, find at which point slipping first takes place.

If the rod remains in equilibrium when $k = \frac{1}{10}$, find the range of possible values of μ.

40. Two uniform rods AB, BC, each of weight W and length $2a$, are smoothly hinged at B and rest in one horizontal line on supports at P and Q, where $PB = x$ and $BQ = y$. Prove that

$$\frac{1}{x} + \frac{1}{y} = \frac{2}{a},$$

and that $y \geqslant 2a/3$.

If the support at P cannot withstand a load greater than $\frac{3}{4}W$, prove that $y \leqslant 4a/5$. (JMB)

41. Two smooth fixed planes, each inclined at an angle α to the horizontal, intersect in a horizontal line which is at the bottom of the planes. Two identical uniform heavy rods AB and BC, freely hinged at B, rest symmetrically with the ends A and C one on each of the inclined planes. The plane of the rods is vertical and perpendicular to the line of intersection of the inclined planes. If each rod makes an angle θ with the downward vertical, show that $\tan \theta = 2 \tan \alpha$.

If, instead, $\theta = 60$ degrees and $\alpha = 30$ degrees and equilibrium is maintained by a light inextensible cord joining the mid-points of the rods, show that the magnitude of the action at the hinge is twice that of the tension in the cord. (London)

42. One end A of a uniform rod AB, of length $2a$ and weight W, is smoothly hinged to a point of a vertical wall. The rod is kept in a horizontal position by a weightless inextensible string fixed to the

EXERCISES

rod at B and to a point C of the wall which is vertically above A. A weight w is suspended from the rod at B. If the reaction at A is perpendicular to BC, prove that

$$AC = 2a\sqrt{\{1 + 2w/W\}}$$

and find the reaction at A in terms of W and w. (WJEC, part)

43. A uniform wire is bent to form a circular arc subtending an angle 2α at the centre. Prove that the distance of the centre of mass from the centre of the circle is $(r \sin \alpha)/\alpha$, where r is the radius.

A uniform wire in the form of a semicircle of radius r rests against a smooth vertical wall and is in a vertical plane perpendicular to the wall with its middle point in contact with the wall. It is kept in equilibrium in this position by means of a light string of appropriate length joining a point B which divides the length of the wire in the ratio $1:3$ to a point A on the wall. Explain how to fix the direction of the string and prove that it makes an angle θ with the horizontal where

$$\tan \theta = \frac{\pi}{\pi - 2\sqrt{2}}.$$

If the weight of the wire is W, find the reaction at the point of contact with the wall. (JMB)

44. Show that the centre of gravity of a uniform triangular lamina coincides with that of three equal particles, one at each vertex.

A uniform lamina of weight W is in the shape of a convex quadrilateral $ABCD$ in which ABC is equilateral, $AD = CD$ and the angle $ADC = 90$ degrees. The lamina rests in a vertical plane with DC in contact with a horizontal table. A particle of weight W' is attached at B. If the lamina is about to topple, show that

$$W' = \tfrac{1}{6}(3\sqrt{3} + 1)W.$$ (London)

45. A uniform circular cylinder of weight W, whose axis is horizontal, rests on a fixed plane inclined to the horizontal at an angle α. A uniform rod, also of weight W, rests in a horizontal position with one end on the highest generator of the cylinder and the other smoothly hinged to the plane. The rod lies in a vertical plane through the centre of gravity of the cylinder and perpendicular to its axis. The system is in equilibrium and the coefficient of friction between the rod and cylinder and also between the cylinder and inclined plane is μ. Show that the normal reaction of the rod on the cylinder has magnitude $\tfrac{1}{2}W$. Show also that the frictional forces acting on the cylinder are equal in magnitude.

Hence show that $\mu \geqslant 3 \tan \tfrac{1}{2}\alpha$. (JMB)

46. A uniform lamina of mass 3 lb, in the shape of an equilateral triangle of side 3 in, is suspended from a point by three light inextensible strings of length 5 in, 4 in, 4 in, attached to the vertices of the lamina. Find the tensions in the strings in the position of equilibrium, and the angle which the lamina makes with the horizontal in this position. (Oxford)

[A "lb" is a unit of mass and an "in" is a unit of length. Give your answer in "lb wt."]

47. Three weightless rods AB, BC, and CD of respective lengths l, $2a$ and l are smoothly jointed at B and C and carry equal weights W attached to A and D. They hang in equilibrium symmetrically over a fixed smooth circular cylinder, whose axis is horizontal and whose radius is $a\sqrt{3}$, so that they are in a vertical plane with BC horizontal and AB, CD in contact with the cylinder. Prove that, if BC is also in contact with the cylinder, then $l < 4a$ and find the reaction of the cylinder on BC.

If $l > 4a$, show that the angle θ made with the horizontal by AB is given by

$$l \cos^3 \theta = a(\sqrt{3} \sin \theta - 1).$$ (WJEC)

48. Show that the centre of gravity of a uniform thin hemispherical bowl of radius a is at a distance $a/2$ from the plane of the rim.

The bowl is placed with its curved surface in contact with a horizontal table. A uniform smooth rod whose weight is half that of the bowl rests with one end in contact with the smooth inner surface of the bowl and with a point of the rod in contact with the rim. In equilibrium the rod and the plane of the rim each make an angle θ with the horizontal. Show that $\theta = \pi/8$ and find the length of the portion of the rod protruding from the bowl. (London)

49. The ends of a light elastic string of natural length $2b$ and modulus of elasticity W are attached to the ends of a uniform rod of length $2b$ and weight W. The string passes over a fixed smooth circular cylinder of radius $a(< b)$, the axis of the cylinder being horizontal. The rod rests horizontally below the cylinder and the acute angle between the rod and each straight portion of the string is α.

(a) If the rod is not in contact with the cylinder, show that $\tan \alpha/2 > a/b$ and that α satisfies the equation

$$b(\operatorname{cosec} \alpha - 2 \sec \alpha + 2) = 2a(\alpha - \tan \alpha).$$

(b) If the rod touches the cylinder, find the reaction at the contact in terms of W, a and b. (London)

EXERCISES

50. A uniform rod AB, of length $\sqrt{a^2 + b^2}$, rests with one end A on a rough horizontal floor and the other end B on a rough vertical wall. A is distant a from the foot of the wall and the plane of the rod and its projection on the wall makes an angle θ with the floor. If the ground is rough enough to prevent slipping at A and the coefficient of friction at B is μ, show that

$$\mu \geq b \cot \theta / a.$$

51. A uniform sphere, of radius a and weight W, is supported by three equal fixed rough rods joined at their ends to form a horizontal equilateral triangle of side b. Show that the centre of the sphere is at a height

$$\sqrt{\left(a^2 - \frac{b^2}{12}\right)}$$

above the level of the rods.

A couple is applied to the sphere about a vertical axis through the centre of the sphere. If the sphere is on the point of slipping, show that the moment of the couple about this vertical axis is

$$\frac{\mu Wab}{\sqrt{(12a^2 - b^2)}},$$

where μ is the coefficient of friction between each rod and the sphere. (AEB)

52. A uniform rod AB of weight W rests horizontally with A attached to a light inextensible string, which passes over a smooth pulley fixed above the rod and supports a heavy particle of weight W' hanging freely, whilst B is in contact with a rough plane of inclination α to the horizontal. The string, rod and the line of greatest slope through B are in the same vertical plane. If λ is the angle of friction and B is about to slip downwards, draw a figure indicating clearly all the forces acting on the rod when $\lambda < \alpha$.

Show that $W' = \tfrac{1}{2} W \sec(\alpha - \lambda)$.

Draw another figure for the case $\lambda > \alpha$.

Show that, if $\lambda + \alpha \geq \tfrac{1}{2}\pi$, B cannot slip upwards whatever the weight of the particle. (WJEC)

ANSWERS*

EXERCISES 1

1. kg m/s
2. m²
3. m³, kg/m³
4. N s
5. kg m²
6. 175 N
7. 300 000 kg, 5 m/s, 50 000 N, 510 000 W

Exercises 2a

1. Vectors: b, e, j. Scalars: a, c, d, f, g, h
2. a and d are equal and opposite
   a and e represent two or more vectors
3. (a) 105°   (b) 30°   (c) 105°
   (d) F_1 and v_2; v_1 and v_2
   F_2 and d; F_3 and v_2
   (e) F_1 and F_3, F_3 and v_1
5. 12·5 km at an angle $\tan^{-1}(\tfrac{3}{4})$ E. of N.

Exercises 2b

1. 7 N at an angle of 38° 13′ to the 8 N force.
2. 174·6 N at an angle of 9° 54′ to the 120 N force
3. 17 units at an angle of 28° 4′ to the 15 unit force
4. 8 knots N. 36° 52′ E.
5. $3\sqrt{43}$ km/h at an angle 7° 35′ with the forward motion of the ship
6. 97° 11′
7. 72° 33′; 3·15 min
8. 3 N, 2 N
9. $|\mathbf{Q}| = 12$
10. 6 N, 12 N

Exercises 2c

1. (a) $-5, 0$ (b) $-15, 15\sqrt{3}$
   (c) $6\sqrt{2}, -6\sqrt{2}$ (d) $-4\sqrt{3}, -4$
   (e) $0, -25$ (f) $4\sqrt{3}, 4$
2. $50, 50\sqrt{3}$ cm/s

* These answers, including those given to questions from past examination papers of the various examining boards, are entirely the responsibility of the authors.

EXERCISES

3. Vertically $R\cos\alpha + \mu R\sin\alpha - mg$
   Horizontally $\pm(R\sin\alpha - \mu R\cos\alpha)$
4. $F\cos(45° - \theta + \alpha)$ and $F\cos(45° + \theta - \alpha)$
   or $F\cos(45° - \theta - \alpha)$ and $F\cos(45° + \theta + \alpha)$
5. Horizontally $F\sin\theta$; vertically $F\cos\theta$; yes

Exercises 2d
1. 1·2 km/h, N. 3·5° W.　　2. 98·4 N, N. 85° 50′ E.
3. 28·1 m/s² at 7° 8′ to the upward vertical
4. $\cos^{-1}(-\frac{1}{7}) = 98° 13′$
5. $\sqrt{3}$ units at 30° to the third force
6. 14·2 units, N. 169° 11′ W.

EXERCISES 2

1. 23·43 at 26° 20′ to the force of magnitude 15
2. $7\sqrt{2}$　　　　3. 9·60 km/h, N. 57° 47′ W.
4. 287·3 km/h, N. 157° 37′ W.　　5. 11·8 km, N. 98° 5′ W.
6. 354·6 km/h, 62·5 km/h
7. $2\sqrt{7}$ N at $\tan^{-1} 3\sqrt{3}$ to BC on the opposite side to A
8. No　　　　　　　　　　9. 106° 16′
10. 23·09 m　　　　　　　11. 5
14. 20 km/h
17. $\sqrt{6}F$ along the diagonal of the cube from the corner P
18. (a) $P\sqrt{3}$　　　　(b) 150°　　　　(c) 120°, 60°

Exercises 3a
2. (a) 2　　　　　(b) 5　　　　　(c) 4
   (d) 4·64　　　　(e) 2·95　　　　(f) 5
3. Yes
4. **a** and **b** parallel and in the same sense
6. 120°　　　　　　　　　7. 60°
9. $-\mathbf{a}, \mathbf{b}, -(\mathbf{a}+\mathbf{b})$　　　　10. $\mathbf{b} - \mathbf{a}$

Exercises 3b
1. $\lambda = \pm|\mathbf{a}|/|\mathbf{b}|$　　　　　2. $m = n = 0$
6. (a) $3\mathbf{a}$　　　　(b) $-\mathbf{a}$　　　　(c) **0**
7. (a) $\mathbf{x} = \mathbf{b} - \mathbf{a}$
   (b) $\mathbf{x} = \frac{2}{3}\mathbf{a} + \frac{2}{9}\mathbf{b} - \frac{1}{9}\mathbf{c}$
8. $2\overrightarrow{AD}$ newton where D is the mid-point of BC
9. $2\overrightarrow{AB}, 2\overrightarrow{AC}$
10. $FA\ \mathbf{p} - \mathbf{q}$, $CD\ \mathbf{q} - \mathbf{p}$, $DE\ -\mathbf{p}$, $EF\ -\mathbf{q}$

Exercises 3c

1. (a) $10i + 10j$ (b) $-10i - 10j$
   (c) $-5\sqrt{3}i/2 + \frac{5}{2}j$ (d) $8i - 8j$
   (e) $-10\sqrt{3}i - 10j$
2. (a) $3i + 2j$ (b) $3i + j$ (c) $10i/3 + 2j$
3. (a) $-4i + 4j - 10k$ (b) $\sqrt{20}$
   (c) $i - k$ (d) $\sqrt{58}$
4. $3, (\frac{2}{3}, -\frac{1}{3}, \frac{2}{3}); 2i/3 - j/3 + 2k/3$
5. $2\sqrt{3}(i + j + k)$
6. $\overrightarrow{AB}\ 3i + j, \overrightarrow{BC}\ i - 7j, \overrightarrow{CD} - 11i + 5j, \overrightarrow{DA}\ 7i + j$
7. $\overrightarrow{AB}\ 4i + 6j, \overrightarrow{BC}\ 3i - 2j, \overrightarrow{CA}\ -7i - 4j$;
   $\sqrt{52}, \sqrt{13}, \sqrt{65}$
8. 16·6 m, N. 77° 34′ W.
9. $8\cos\phi\cos\theta i + 8\cos\phi\sin\theta j + 8\sin\phi k$
10. $\overrightarrow{AB}\ i - 2j + 3k, \overrightarrow{BC}\ i + 2j + k, \overrightarrow{CA}\ -2i - 4k$;
    $\sqrt{14}, \sqrt{6}, \sqrt{20}$

Exercises 3d

1. (a) $2ta$ (b) $-a/t^2$ (c) 0
   (d) 0 (e) $6ta$
2. $(3t^2 + t + 3)i + (1 + 4t - t^2 - t^3)j$; $6ti - j, i - (2t + 3t^2)j$,
   $5j$
3. $r = t^2a/2 + c_1 t + c_2$ where c_1 and c_2 are any two constant vectors including 0
4. $r = t^3a/6 + t^2b/2$ 5. $3a\sin\theta\cos\theta$
6. (a) $10\cos 5ti - 15\sin 5tj, -50\sin 5ti - 75\cos 5tj$
   (b) 15, 50
7. $i/2 - j/2$ 8. $-\sin\theta i - \cos\theta j$
9. $r = (1 - t^2)i + 2tj$

Exercises 3e

1. $-1, -6, 0$ 2. $-1, -32, 0$
3. $-\frac{9}{5}, -\frac{9}{\sqrt{14}}$ 5. 135°
6. $\cos^{-1}(-\frac{1}{3})$ 7. $2\cos u - u\sin u$
8. $6ti - j, i - (2t + 3t^2)j, 4t^3 - 2t + 9t^2$
9. $5t^2 - 6t^5, 10t - 30t^4, i - j$
10. (a) $a^2 + a \cdot b + a \cdot c$ (b) $a^2 - b^2$ (c) $a^2 + b^2 + 2a \cdot b$
11. either $b = c$ or a is perpendicular to $b - c$
13. $-17i + 23j - 7k$

EXERCISES

Exercises 3f
1. (a) $a \times b$ (b) $b \times a$ (c) 0
   (d) $2b \times a$
2. (a) $-3k$ (b) $-3i + 6j$
3. They are all vectors
   (a) 12 units vertically up   (b) 16 units vertically down
   (c) $12\sqrt{3}$ units vertically up
4. $a = b + kx$ (k a constant)
5. (a) $i - 13j - 9k$ (b) $-7i - 25j + 5k$
7. $6(i - j + k)$ kg m²/s
8. $12\hat{n}$ Nm, where $\hat{n}$ is upwards out of the paper if ABC is anticlockwise
9. $5a \times b, 6a \times b$ 10. 60/7
11. (a) $2Pa$ (b) $2Pa$
    (c) $4Pa$ (d) 0

Exercises 3g
1. $-i + 9j$ 2. $1:2, 3|3b - 8a|/5$
3. $a + c - b$ 6. $4b/3 - a/3, 3a - 2b$
8. $(a + b + c)/3, (a + b + c + d)/4$

Exercises 3h
2. $3a$
4. (a) $r = t(-i + k)$
   (b) $r = (3i - j + k) + s(-i + k)$
   (c) $r = (3i - j + k) + u(3i - 3j - k)$
   or $r = 2\hat{j} + 2\hat{k} + t(3i - 3j - \hat{k})$
5. $3i + 5j - 7k$ 6. $p = -2, 4(i - j + k)$
7. $\cos^{-1}(\frac{33}{65})$
8. $(x - a_1)/b_1 = (y - a_2)/b_2 = (z - a_3)/b_3 = t$
9. $2a + b, 3a - 2b, a + b$

EXERCISES 3

1. $\overrightarrow{DE} = -p/2; \overrightarrow{EF} = (p - q)/2; \overrightarrow{FD} = q/2$
4. (a) $a = 2b$ (b) $a = -b$
   (c) $a \times b = O$ or $a = kb$ (d) $F_1 + F_2 = ka$
   (e) $x \cdot y = 0$ (f) $P + Q = -R - S$
   (g) $x \times y = z, y \times z = x, z \times x = y$
   (h) $a \cdot b = 0, b \cdot c = 0, c \cdot a = 0$
5. No, the reciprocal of a vector has not been defined
6. $P = -3Q - R$ 7. $\lambda = \frac{3}{2}$ or $-\frac{1}{3}$
8. $(i - 2j)/\sqrt{5}, -1/\sqrt{5}$ 9. 22 units

469

ANSWERS

12. (a) $(i - 2j + 3k)/\sqrt{14}$, $(4i - 3j - k)/\sqrt{26}$,
    $(5i - 5j + 2k)/\sqrt{54}$
    (b) 7   (c) $7/\sqrt{26}$ (d) $\cos^{-1}(7/\sqrt{364})$
13. $4:1$ 14. $\cos^{-1}(\frac{1}{3}) = 70°\,32'$
16. $\cos^{-1}(-1/5\sqrt{13})$ 17. $|a| = \sqrt{2}|b|$
19. 7 units, 7 units   20. (a) $4:1$, $\lambda = 6$ or -24
21. $|v| = 3a \sin t \cos t$
23. (a) $2i + j + k$ (b) $3k$ (c) 1
24. $a(2\sqrt{3} - 1)$ 25. $-8j + 16k$, $\frac{32}{5}$
26. $3\sqrt{3}Pa$; Pa
27. $r = 4i - 2j + 2k + t(4i - j + 2k)$, $9i - 7j$, $12i - 4j + 6k$
28. $a = 4$, $b = -2$, $c = +1$ 29. 1
30. (a) $\lambda(a + b)$, $(1 - \mu)b + a/2$ (b) 2, $4i + j - 2k$, $\frac{2}{5}$, $\frac{\sqrt{21}}{2}$
32. $2i + 3j$, $6i + 4j$, $-i + 6j$
33. $P(\hat{a} + \hat{b} - \hat{c})$, $-\hat{a}.\hat{b} + 3(\hat{a}.\hat{c} + \hat{b}.\hat{c}) = 1$;
    L_3 is $r = (\hat{a} + \hat{b} - \hat{c}) + u(\hat{b} \times \hat{a} - 3\hat{c} \times \hat{a} - 3\hat{c} \times \hat{b})$
34. (a) $a(dr/dt).(d^2r/dt^2)$ (b) $(a.b)dr/dt$
    (c) $(a.dr/dt)b$ (d) $2r.dr/dt - 2dr/dt . 1/r^3$
    (e) $r \times d^2r/dt^2$ (f) $(dr/dt)^2 + r.d^2r/dt^2$
36. $BP:PC = 2:3$, $AQ:QC = 1:3$, $k = \frac{20}{3}$
38. $10\overrightarrow{OE}$
40. (a) $i + 2k + t[2i + 2j + k]/3$
    (b) $\frac{2}{3}ti + (\frac{2}{3}t - 1)j + (\frac{1}{3}t + 1)k$, $\cos\theta = (t - 6)/\sqrt{(18t^2 - 12t + 36)}$
41. $(i + j + k)/\sqrt{3}$, $\alpha = \frac{1}{3}$, $\beta = \frac{2}{3}$, $\gamma = 0$;
    $\cos\angle POQ = \frac{1}{3}$; $c = i - j$; $\omega.(c \times \omega) = 0$
42. $i + 3j + 2k$; (i) $\sqrt{10}$, $r = (i + 3j + 2k) + t(3i + j)$;
    (ii) $r.(i - 3j - 7k) = -22$; (iii) $-2i + 6j - 8k$
43. (ii) $\overrightarrow{PX} = \frac{1}{3}\overrightarrow{PQ} + \frac{2}{3}\overrightarrow{PS}$; $\overrightarrow{QX} = -\frac{2}{3}\overrightarrow{PQ} + \frac{2}{3}\overrightarrow{PS}$;
    $\overrightarrow{RX} + \overrightarrow{TX} = \frac{1}{3}(\overrightarrow{PS} - \overrightarrow{PQ})$
44. $r = (i - 2j + k) + t(3i + 4k)$;
    (i) 50 units; (ii) $(i + 5j - 7k)/3$

Exercises 4a

1. 27, 40, 36, 27 km/h   2. 30, 28, 30, 22 km/h
3. 27·2 and 27·5 respectively   4. 3rd one
5. 30π km/h

Exercises 4b

2. 12 m/s, 24 m/s   3. 12 km/h, $10\frac{10}{13}$ km/h
4. (a) 48 m/s   (b) 44 m/s
   (c) 96 m/s   (d) 80 m/s and 160 m/s

EXERCISES

5. $13\frac{1}{3}$ km/h
6. $20\sqrt{3}$ m/s, $-20\sqrt{3}$ m/s, 10 m
7. 5·1, 5·01, 5·001, 5 cm/s
8. (a) 50 m from A after 5 s (b) 30 m from A after $\sqrt{15}$ s
9. 59 min
10. 16·20 h

Exercises 4c

1. $\sqrt{65}$ cm/s, S. $\tan^{-1} 8$ E
2. $|S| = 1$, $|v| = 2$, a circle
3. $12\sqrt{10}$ m/s, S. $\tan^{-1} 3$E.
4. (12, 2)
5. $(7, -3)$, $5\sqrt{2}$
6. $(17, 0)$, $(68, -11)$, $[5 + 4t, 3 - 2t]$
7. 13 m/s
8. 1 m, $(4 + \cos^2 2t)$ m/s
9. $y^2 = 4ax$
10. $|s| = 1$, $|v| = \sqrt{(4 + 9\cos^2 2t)}$

Exercises 4d

1. $\sqrt{34}$, $\sqrt{37}$, 5
2. 24·4 km/h, S. 35° W.
3. $7i + 2j$, $-8i + 7j$, $12i - 2j$
4. $3\sqrt{5}$ km/h at an angle $\tan^{-1} 2$ with the direction of the first man
5. 10·6 km/h
6. 7·37 km/h from S. 61° 18′ W.
7. $d/[u\sqrt{(1 - \lambda^2)}]$
8. 15·0 km/h, West
9. $30\sqrt{5}$ km/h from S. $\tan^{-1}\frac{1}{2}$ E.
10. After 1 h, 4 km

Exercises 4e

1. $20\pi/3$, $15\pi/2$, $\frac{5}{2}$
2. 12, 30π, $\pi/30$ m/s
3. 100π, 4π m/s
4. $\pi/43\,200$ rad/s, $4\,000\pi/27$ m/s
5. 25 m/s, 25 m/s
6. $21\frac{9}{11}$ min past 4 o'clock
7. 8 rev/min
8. $\tan \theta = t/2$

EXERCISES 4

1. 30, 45, $32\frac{1}{7}$, 30, 60, $38\frac{1}{3}$ km/h
2. $12·22\frac{1}{2}$, $12·35\frac{5}{9}$, $12·45\frac{1}{7}$, $13·19\frac{1}{9}$ h
4. 101·2 km, 21·1 km/h
5. 11·3 h
6. $t = 6$, $s = 4$, $t = 5·05$ s
7. $10\sqrt{2}$, N. 45° W.
8. $i + 9j$, No
9. 16 min
10. $-2i - 6j$ km/h from N. $\tan^{-1}(\frac{1}{3})$ E.
11. N. 36° 52′ E., 989 km
12. 22·24 m, 19 min 12 s
13. 10·6 km, 0·536 h
15. 13 m/s, N 112° 37′ E., $76\frac{12}{13}$ m
16. $2\sqrt{5}i + \sqrt{5}j$, $-\sqrt{5}i + 3\sqrt{5}j$, $4i + j$, 21 min
17. 10 knots, S. 60° W.
18. 3:4
19. 0·5 h, 5·70 km
20. After 1 h, $5i + 9j$

ANSWERS

21. 1st and 3rd ships at $2 - 20$, 2nd ship arrives at $2 - 30\cdot77$
22. $a\sqrt{(4u^2 - 2v^2)/(u^2 - v^2)}\,h$ 23. $28°\,41'$
24. $61\cdot48$ km/h towards N. $42°\,20'$ E.; due W.
25. (a) S. $40°\,32'$ W., $5\cdot26$ min   (b) 250 km/h
    (c) $V_2 = 500$ km/h
26. $4\cdot5$ km/h $180°$; $6\cdot1$ km/h $305°$
28. $2\cdot51$ h
29. (a) $(x_2 - x_1), (y_2 - y_1)$ (b) $d(x_2 - x_1)/dt, d(y_2 - y_1)/dt$
    A_2 relative to A_1 $(d/\sqrt{2} - u_1 t)$
    $(d/\sqrt{2} + u_2 t)$
    A_3 relative to A_1 $(\sqrt{2}\,d - u_1 t)$
    $(\sqrt{2}\,d - u_3 t)$
31. 8 m/s, 4 m/s   32. $\tan^{-1}(0\cdot5)$
33. $(10\mathbf{i} - 2\mathbf{j})/3$; $t = 14/13$
34. $V = \sqrt{86}$ km/h, $\theta = 35\cdot4°$; Direction of current is S$35\cdot4°$E

Exercises 5a

1. (a) 27 m/s, 90 s   (b) 3 m/s^2, 6 s
   (c) 21 m/s, 4 s   (d) $1\,224$ m, 42 m/s
2. 4 m/s^2 3. 50 s, 50 m
4. $s_1 = 1\,800$ m, $s_2 = 1\,200$ m   5. $0\cdot10$ m/s, $0\cdot24$ m/s^2
6. $12\cdot5$ m   7. 12 m/s, 2 m/s^2
8. After 8 s, 80 m from A 9. 146 s
10. $f = (b - 2a)/n^2$; $u = (4a - b)/2n$

Exercises 5b

1. It accelerates uniformly at 5 m/s^2 for 10 s. It then decelerates uniformly at $2\cdot5$ m/s^2 for 40 s. Initial velocity 50 m/s; maximum velocity 100 m/s; Distance covered $2\,750$ m
2. 200 m   4. $2\,520$ m, 19 m/s, $-0\cdot2$ m/s^2
5. 376 m, $51\tfrac{2}{3}$ m, No

Exercises 5c

1. $(2t^3 + 9t^2)/6$ m/s, $(t^4 + 6t^3)/12$ m, 126 m/s; 216 m
2. $4(1 - e^{180})/(1 - 4e^{180})$ m/s   3. $k = 7g/16V^2$
4. $t = \log_e s/15 + s/75 + C$ 5. $20\sqrt{2}, 100$
6. $4\,000(\pi + 2)$

Exercises 5d

1. (a) $-0\cdot789$ m/s$^2 \pm 1\cdot51$ m/s
   (b) $1\cdot89$ m/s^2, $0\cdot628$ m/s
2. (a) $\sqrt{2\pi}$ seconds   (b) π seconds   (c) $\pi/10$ seconds

EXERCISES

3. π seconds, 5 m, 20 m/s²
4. 3 m, $8\sqrt{2}$ m/s
5. $6\sqrt{2}$ m, $\sqrt{2\pi}$ seconds
6. 15/2 m/s, 4·5 m/s
8. 5 units, 5 units
9. $6\sqrt{2}$ cm, 16 s
10. π, 5π seconds, 7π seconds

EXERCISES 5

1. 1 m/s
2. Velocity $-2 \sin 2t\mathbf{i} + 2 \cos 2t\mathbf{j}$
   Acceleration $-4 \cos 2t\mathbf{i} - 4 \sin 2t\mathbf{j}$
   (a) $t = (2n+1)\pi/2 + \pi/6$; (b) $t = (2n+1)\pi/2 - \pi/6$
3. 4 cm/s, 6 cm/s²
4. S.H.M. about $x = C/n^2$ as centre
5. 33 m/s, 11(3 log 3 − 2) s
6. 13 min, 11·04 km
7. $C = -1/2k^2$
9. 100 m
10. 239 m, 48 m/s
11. 40 s after it starts to decelerate; 90 cm/s
12. $2\sqrt{(g/a)}$
13. $7\sqrt{7}/2$ m/s, 36·75 m/s²
14. 8π cm/s, $4\pi^2$ cm/s²
15. $\dfrac{(t_2^2\theta_1 + 2t_1t_2\theta_1 - t_1^2\theta_2)}{t_1t_2(t_1 + t_2)}$

   $\dfrac{(t_2^2\theta_1 + 2t_1t_2\theta_1 - t_1^2\theta_2)}{2(t_1\theta_2 - t_2\theta_1)}$
16. (a) $\dot\omega = (n_2 - n_1)/900$ rev/s²
   (b) $(n_1 - 3n_2)^2/8(n_1 - n_2)$

18. 5·86 s
19. 57·6 km/h, 8·55 km h⁻¹ min⁻¹
20. 0·8 m/s², 20 m
21. 60 km/h
22. $t = \sqrt{(a/2g)}$, $9a/4$
23. 14 s, $66\tfrac{2}{3}$ m, 3·2, 4·4, 11·6 s
24. $\sqrt{[2s(f_1 + f_2)/f_1 f_2]}$, $s/v + v(f_1 + f_2)/f_1 f_2$
25. 6·9 m/s, $(13 + 4\sqrt{10})/3$ m/s
27. 7·7 s
28. $3u$, $3·5u$, $4u^2/8$
29. (a) 4·55, (b) 7·24
30. 79/22, 35/22, $2\pi\sqrt{3/22}$, $11\tfrac{2}{3}$
32. (i) $5/(3\pi)$ m; (ii) $(5\sqrt{3})/2$ m/s; (iii) $1/(2\pi)$ m; (iv) 1/18 s
33. Acceleration, $-\omega^2 x$; force, $m\omega^2 x$; periodic time, $2\pi/\omega$;
    Next simultaneously at centres after 12 s;
    Next simultaneously at centres, moving
    in the same direction, after 24 s

EXERCISES 6

1. 1 600 N
2. 4 m/s²
3. 160 kg

473

ANSWERS

4. 0.64 m/s^2, S.H.M. about O of period $2\pi/\sqrt{8}$ seconds
5. 5 m/s^2, $\sqrt{(\frac{6}{5})}$ s   6. 2.3 m/s^2
7. At rest *or* moving with uniform velocity
8. $\frac{5}{3} \text{ m/s}^2$ parallel, $1/\sqrt{3} \text{ m/s}^2$ perpendicular
9. 4 m/s^2 between 416 and 320 N forces at $\tan^{-1}(5\sqrt{3}/11)$ to 416 N force
10. $8i + 16j \text{ m/s}^2$ 11. $2mta$
12. $-map^2[\sin pt \, i + \cos pt(2j + k)]$
13. 8N
14. (i) $\frac{3}{2}(1 + \pi^2 \cos^2 \pi t + 4t^2) + 2\left[\frac{\pi^2}{16}\cos^2\left(\frac{\pi t}{4}\right) + 9t^4 + 9\right]$;
    (ii) $3\sqrt{(\pi^4 \sin^2 \pi t + 4)}$; (ii) $2/\sqrt{(5 + \pi^2)}$; $i + 6k$

Exercises 7a

1. Tension (T), weight $3g$ newton also $T = 6(\frac{2}{4})$ newton
2. Normal reaction, weight $5g$ newton, horizontal force 13 newton
3. Normal reaction through centre weight $3g$ newton
4. Normal reaction (N), weight $2g$ newton, horizontal force (Q), frictional force (F) and $F = N/2$
5. Total reaction at ϕ to the normal, weight $8g$ newton and $\phi = 30$ degrees
6. Tension (T), weight $2g$ newton, and $T = 980(\frac{2}{50})$ newton
7. Tension, weight mg, normal reaction outwards through the centre
8. Q, weight mg, normal reaction N, friction up the plane (F) and $F = \mu N$
9. On the 3 g mass: weight $0.003g$ newton, normal reaction, tension, friction
   On the 2 g mass: weight $0.002g$ newton, tension
10. On particle: weight $3g$ newton, normal reaction
    On wedge: weight $7g$ newton, normal reaction with particle, normal reaction with plane

Exercises 7b

1. $1/\sqrt{5} \text{ m/s}$ 2. remains at rest
3. 11 m/s 4. $2s \text{ m/s}^2$, $\pi\sqrt{2}$ seconds
5. 5.27 s 6. $(t^2 + 1)/2 \text{ m/s}^2$, 6 m/s
7. (i) $t_1 = 2/k$, $t_3 = 5/k$; (ii) $k = 0.7$
   (iii) 67.5 N, 50 N, 43 N

Exercises 8a

1. 80 m 2. 14.9 cm, 198 cm/s

EXERCISES

3. 5·15 s
4. 50 m
5. 12 000 N
7. $1/\sqrt{7}$ s, $\sqrt{7}$ m/s
8. $\frac{19}{28}$
9. $\frac{1}{7}$ s, $\frac{3}{7}$ s, $\pm 1\cdot 4$ m/s
11. $(u^2 - v^2)/2g(\sin\alpha + \mu\cos\alpha)$
12. 2·18 m/s², 1·09 m/s
14. $21/\sqrt{2}$ m/s, N.W.

Exercises 8b

1. 1 000 m
2. 3·83 m/s
3. 0·515 N
4. $4\sqrt{10/7}$ s
5. $u^2/2\mu g$
6. 0·687 s
7. 156 m
8. 1 800 m

Exercises 8c

1. 0·7 m/s
2. 23·7 s
3. 2·95 m/s
4. 4·9 m/s² downwards (9 kg)
   2·45 m/s² downwards (3 kg)
   12·25 m/s² upwards (1 kg)
5. 4·94 m/s²
6. $-\frac{5}{7}g$, 1 s

EXERCISES 8

1. 6 m from O in the direction of I
2. $4i + 8j$ m/s; $8i + 16j$ m
3. 4·9 m/s
4. 3 s
6. $mg(4m + M)/(3m + M)$; $2mg(2m + M)/(3m + M)$
7. East, 2 m
8. 1·08 N.
9. 9·375 m
10. 540 N, 21·6 N; 440 N, 17·6 N
11. $2(2 + \sqrt{3})m$
12. $6i + 3j + 5k$
14. $0\cdot 2g$ m/s², 7·2 m, $14\sqrt{3/5}$ m/s
15. $60mg(1 + \mu)/49$
17. $g\sqrt{3/5}$; $\sqrt{(5h/g)}$
19. 0 m/s, $i + 15j$ m
20. 1·5 mg
21. 400 N/t, 75 s; 1 in 73·5
22. $37g/166$, $7g/166$, $11g/83$
23. $g/49$, 0·2 m
24. vertically downwards 780 cm/s, $\sqrt{6}/12$ s
25. $20/7$ m, $7u/54$
26. $-g/17$, $48mg/17$, $24mg/17$
27. 40 N, 380 N, 1 800 m
28. $22u/g$
31. 0·426 m/s², 37·5 N
32. $x = 2$
33. (i) $6u/g$; (iii) $u\sqrt{5}$; (iv) $3(5 - \sqrt{5})u/g$

Exercises 9a

1. 2·5 m/s
2. $2\pi\sqrt{(l/g)}$, $210/\pi$

ANSWERS

3. $\pi/3$ s
4. $\pi/20$ s, 20 cm
5. 7·65 cm/s
6. $\pi^2/10$ N
7. mg/x
8. $l, 0·25\pi\sqrt{(2l/g)}; 0·5\sqrt{(2gl)}$

Exercises 9b

1. $(\sqrt{2}/10)(1 + \pi/2)$ s
3. $2\pi\sqrt{(a/27g)}; 0·5\sqrt{(3ga)}$
5. $(4 + 3\pi)/16$ s
6. 15 cm above the table; 17·5 cm

EXERCISES 9

1. $d\sqrt{k/m}$
2. $\pi^2/100$ N, Zero
3. $2\pi\sqrt{(R/g)}$
4. $l + u\sqrt{(ml/\lambda)}; 0·5\pi\sqrt{(ml/\lambda)}$
6. $2\pi\sqrt{am/2T}$
7. $2\pi/7$ s, 1·4 m/s
8. $(m + M)gl/a$, $Mg(a + b)/a$, Mg, zero
9. $\pi/7$ s, 5 cm, $14\sqrt{21}$ cm/s
10. $(10 + 5\sqrt{2})$ m from A on the same side as B, $5\sqrt{2}$ m/s towards A,
11. $\pi\sqrt{(l/18g)}$
12. $\sqrt{(\lambda a/3m)}$
13. $v/n, 1/n \sin^{-1}(hn/v)$
15. At O where $AO = 3a$, $2\pi\sqrt{(a/5g)}$, $\sqrt{(95ag)}$, zero
17. $2\pi\sqrt{(l/g)}$, $U\sqrt{(l/g)}$
21. $(4\pi/21 + 2\sqrt{3/7})$
22. $\sqrt{(15\lambda ag)}$
23. Point of trisection nearer M, $0·5l$, $2\pi\sqrt{(l/3g)}$
25. $\ddot{x} = -2gx/3a$; $\pi\sqrt{(3a/2g)}$, $2mg$, $4mg/3$
26. $2\pi\sqrt{(4a/5g)}$
28. $2\pi\sqrt{(2l/g)}, 2\pi\sqrt{(2l/9g)}$
29. $\lambda b(P + \lambda)$
30. 12·6 joules, 84 N; (i) 6 m/s; (ii) $(\pi/40 + \frac{1}{12})$ s

EXERCISES 10

1. 12 m/s
2. $\sqrt{(u^2 + ks^2/m)}$
5. $96\frac{2}{3}$ m, 9 m/s
6. $\sqrt{6}$ m/s
7. (a) $u - 0·5k_1t^2$, $ut - k_1t^3/6$
   (b) ue^{-k_2t}, $u(1 - e^{-k_2t})/k_2$
   (c) $u\cos(t\sqrt{k_3})$, $(u/\sqrt{k_3})\sin(t\sqrt{k_3})$; period $2\pi/\sqrt{k_3}$
8. 0·5
9. $x = 0, \pm 2\pi/p$
13. $S = ae^{(v^2 - u^2)/2k}$
14. $ktu^2/(1 + ktu)$, $ut - \frac{1}{k}\log_e(1 + ktu)$

EXERCISES

16. $\dfrac{1}{2b}\log_e(a-bu_1^2)/(a-bu_2^2)$

17. $\tfrac{3}{8}$
18. $x=g(1-\cos\omega t)/3\omega^2$
19. $a(1+\log_e 2)$
20. 40 cm

21. $\dfrac{1}{2k}\log_e(1+ku^2/g)$; $\tan^{-1}(u\sqrt{k/g})/\sqrt{kg}$

22. $\dfrac{1}{k}\log_e(2+\sqrt{3})$
25. $(u+a/n)\pi$

28. $\dfrac{(M+m)V^2}{2(M-m)g}\log_e 4/3$

30. $\dfrac{1}{2kg}\log_e(1+ku^2/\mu)$; $\dfrac{1}{g\sqrt{k\mu}}\tan^{-1}(u\sqrt{k/\mu})$

31. $[m\log_e(1+u^2)]/2k$

32. (a) $\mathbf{u}=(\sin t+t\cos t)\mathbf{i}+(\cos t-t\sin t)\mathbf{j}$
    $\mathbf{a}=(2\cos t-t\sin t)\mathbf{i}-(2\sin t+t\cos t)\mathbf{j}$

33. (i) $2uT$; (ii) $T\log_e 3$; (iii) $4mu^2$

34. F/M; $\dfrac{d(\tfrac{1}{2}Mv^2)}{dt}=K$

35. (i) $\dfrac{1}{v^2}-\dfrac{1}{u^2}=2kt$; (ii) $\dfrac{1}{v}-\dfrac{1}{u}=ks$; 1·5 s

36. 56 m

37. $T=\displaystyle\int_{a/4}^{a}\sqrt{\left(\dfrac{ax}{a-x}\right)}\,dx$

Exercises 11a

1. 30°, $6\sqrt{3}$ kg wt.
3. 50 g wt.
4. $4W/3$, W.
5. 6 kg wt., 3 kg wt.
6. $F=70/\sqrt{3}$ kg wt., $R=140/\sqrt{3}$ kg wt.
7. 35 kg wt.
8. 5 kg wt., 12 kg wt.
9. 0·845 W
11. 124 g wt.
13. $W\sin(\alpha+\lambda)\sec\lambda$
15. $W\sqrt{7}/2$

Exercises 11b

1. $2W/\sqrt{3}$, $W/\sqrt{3}$, $2W/\sqrt{3}$
2. $W\sin(\alpha+\lambda)$ when $\theta=\lambda$
3. $W\sqrt{(\tfrac{52}{5})}$
4. $\mu=8-13/\sqrt{3}$
5. 9° 24′ 5·28

ANSWERS
EXERCISES 11

1. $i - j$
2. $W \sin \theta$, $W \cos \theta$
3. $W \sin \lambda$
4. $2W \cos \theta / (2 \cos \theta - 1)$
5. 60 g wt., 80 g wt.
6. $F = 4$ kg wt., $T = 5$ kg wt.
7. $\mu = 0.89$
8. 67° 58′, 44° 4′
11. 52° 38′, 3·86 W. or 7° 22′, 7·96 W.
12. $\mu = \sqrt{3}/4$
13. $(5 + 3\sqrt{3})$ W.
15. $W = 10$ kg wt., $T_{AB} = 10\sqrt{3}$ kg wt., $T_{BC} = 10$ kg wt., $T_{CD} = 10$ kg wt.
18. The particles move up the plane
19. W, $W\sqrt{3}$, 60°, $2W$
20. $T = W \sec \beta$, $R = 2W$, $F = W \tan \beta$
21. $9W/4$, $15W/4$, $W(5\sqrt{5} - 9)/4$, $7W/2$
22. $\tan^{-1}(\frac{5}{3})$, 45°, $\tan^{-1}(\frac{1}{3})$
23. $2\mu m$
24. $g/\sqrt{2}$, $g\sqrt{6}/2$, $g/\sqrt{2}$, $g/\sqrt{2}$;
26. (i) $5mg$, $4\sqrt{2}mg$; (ii) $\tan^{-1}(4/3)$, 45°

Exercises 12a

1. $k(\mathbf{a} \cdot \mathbf{b} + 2b^2)$, $k(a^2 - 4b^2)$, $-k(a^2 + \mathbf{a} \cdot \mathbf{b} - 2b^2)$
2. 20 J, 45 J, 65 J
3. 94·0 J
4. 784 kJ
5. 500 kJ
6. 24·5 kJ, 34·5 kJ
7. 1·029 J
8. 9·53 N
9. 38·89 J
10. 70·9 kJ
11. 3 920 J
12. $175\sqrt{3}$ J
13. -2.5 J, 2·5 J, 7·5 J
14. 0·144 J

Exercises 12b

1. 200 kN
2. 21 kW
3. 210 kW
4. 600 gW
5. $5RMv/18$
6. $v_1 v_2/(2v_2 - v_1)$
7. $202\frac{2}{3}$ Force de Cheval
8. $\sin^{-1}(\frac{5}{98})$
9. 368 W, 613 W

Exercises 12c

1. 0·75 m/s²
2. 16 kW
3. 7/200 m/s²
5. 14 N
6. 0·256 m/s²
7. $\sin^{-1}(\frac{2}{49})$
8. 40 N/t, $\frac{3}{40}$ m/s²
9. 99 s

Exercises 12d

1. $5\pi^2/18$, 2 s
2. 2·5 m

EXERCISES

3. Zero, $(2 + \sqrt{3})$ m
4. $v = 2\sqrt{ga}$
5. l
6. $7\sqrt{5}$ m/s
7. 50 m
8. $20\sqrt{14}$ cm/s
9. 17·5 cm
10. $\left\{\dfrac{2gd}{(m_1 + m_2)}[m_2 \sin \beta \sim m_1 \sin \alpha]\right\}^{\frac{1}{2}}$
11. $\left\{\dfrac{2ga}{(m_1 + m_2)}[m_2(\sin \beta - \mu \cos \beta) - m_1(\sin \alpha - \mu \cos \alpha)]\right\}^{\frac{1}{2}}$
12. 6·86 cm/s
14. $v^2 = 2ga(M + m_1 - 2m_2)/(M + m_1 + 4m_2)$
15. 11·2 cm/s, Yes

EXERCISES 12

1. $(2t - 4)\boldsymbol{i} - 2t^2\boldsymbol{j} + (t^2 + 3)\boldsymbol{k}$, $10t^3 + 10t - 8$
2. 23 800 J, 28 800 J
3. 2·94 m/s
4. 0·56 J, 0·441 N
5. 12 N, 0·98 m/s^2
6. (a) 46·7 J (b) $-2·55$ J (c) $-44·1$ J
7. 306 kW
9. $a = \frac{50}{3}$, $b = \frac{5}{6}$
11. $n^2(a - t)W/2$
12. 15 kW
13. $v = 12$
14. 39·4 m
15. $44\frac{4}{9}$ s, $42\frac{6}{7}$ s
16. $32 \log_e (\frac{16}{11})$
17. $k^2(t^5 - 25t^4 + 150t^3)/3m$
18. $34g/3$ J, 4·71 m/s
21. 155 kW
22. $12u/5$; $\frac{7}{3}$ kg
23. 80 N/t, $\frac{4}{245}$
25. 5 kN, $(50 - 1·5t)$ kN, 12 m/s, 240 kW
26. 0·047 m/s^2, 6·36 m/s
27. $MV^2 \log_e (\frac{4}{3})/2H$
28. $M(2v^3 + V^3)/6PV$
29. $\sqrt[3]{H/R}$ m/s
30. $\sqrt{3gl}$, $\sqrt{2gl}$
32. 15·52 m/s, 28·41 s
34. (i) $8(t^2 - 2t + 2)$; (ii) 0;
 (iii) $16(-\sin 2t\boldsymbol{i} - \cos 2t\boldsymbol{j} + \frac{1}{2}\boldsymbol{k})$; (iv) 1 s

Exercises 13a

1. 15 N s
2. $54\boldsymbol{i} + 300\boldsymbol{j}$ N s
3. 7 500 N s
4. 12 m/s in the opposite direction
5. 2 N s
6. 10 m/s
7. 4 N s

ANSWERS

8. $5000\sqrt{93}$ N s at $141°\ 3'$ with the original direction of motion
9. 50 N s
10. $6\sqrt{6}$ N s

Exercises 13b

1. 2 m/s
2. 14 000 kg m/s, 5·6 m/s
4. 5 m/s, 5·1 J
5. $45g$ kN
6. $-mu/(M + m)$, $Mu/(M + m)$

Exercises 13c

1. $mMv/(m + M)$
2. 30 N s, 6 m/s
3. 2·4 m/s, 120 N s, No
4. 3·5 m/s, 7 N s
5. $I/(m + M)$
6. $0·6\sqrt{2gh}$
7. 5 m/s, $5\sqrt{5}$ m/s, $5\sqrt{29}$ m/s, $I = 10$ N s
8. $I\sqrt{2}/6$, $I\sqrt{2}/3$
9. $\tan^{-1}(\tan\theta/(1 + 6\tan^2\theta))$ with the bisector, produced, of the angle BAC
10. $\frac{2}{7}$ s

Exercises 13d

1. Both balls continue in the same direction with speeds $3\frac{5}{8}$ m/s, $5\frac{5}{8}$ m/s
2. 1st ball, $10\frac{10}{11}$ m/s in reverse direction; 2nd ball, $\frac{10}{11}$ m/s in the same direction
3. zero and 4 cm/s; $\frac{2}{3}$
4. $\frac{9}{16}$
6. $(1 - e)u/2, (1 - e^2)u/4, (1 + e)^2u/4$, all in the original direction in which A was projected
8. $3mu^2/8$
9. $(2e - 1)(e - 2)(e + 1)$
10. $e = \frac{1}{3}$

Exercises 13e

1. $A = 8\sqrt{3}$ m/s, $B = 4$ m/s
2. $\tan^{-1}(0·5)$
5. 19·6 cm/s, 24·0 cm/s
6. $\frac{1}{3}$
7. $mu^2/16$
8. $2·25\sqrt{3} - \sqrt{2} = 2·48$ m/s, $0·75\sqrt{3} - 3\sqrt{2} = -2·94$ m/s; 221 J
9. $(17ga/177)^{1/2}$
10. $\sqrt{13}$ m/s, 2 m/s, $\tan^{-1}(2\sqrt{3})$, 10 N s

EXERCISES 13

1. 7·2 N s
2. 1000π N
3. $\tan^{-1}(1·003)$
4. 8°

EXERCISES

5. 67·9 kW, 4·52 kN
6. 0·8 mV
8. 72 N s
9. $\frac{9}{5}$ m/s, $3\sqrt{73}/5$ m/s
10. u, $4d/9$, $12mu/5$
11. $2\frac{5}{9}$ m/s, $2\frac{8}{9}$ m/s, $\frac{16}{9}$ N s, $\frac{8}{297}$
12. distance $e^2u^2/g(1 - e^2)$; time $2eu/g(1 - e)$
13. $\frac{1}{3}$, $u/\sqrt{37}$
14. $4\sqrt{2}$ m/s, $2\frac{2}{3}$ m/s
15. $3\frac{1}{3}$ m/s, $8\frac{1}{3}$ m/s
16. $-30i$ cm/s, $6i$ cm/s, 0·0864 J
17. $1/2m$, $1/2\sqrt{2}m$, $1/2\sqrt{2}m$, zero
19. $24mv^2$
20. 3·5 m, 4·17, $11\frac{61}{81}$ s
21. $e = \frac{2}{3}$
22. $u_1 = a$, $u_2 = a/2$
23. $u(1 - ne)/(1 + n)$; $u(1 + e)/(1 + n)$; $u(1 + ne^2)/(1 + n)$; $u(1 - e^2)/(1 + n)$ all in the original direction of projection
26. $3u/4$
27. $u(1 - e)/2$; $u(1 - e^2)/4$; $u(1 + e)^2/4$
29. A $u/6$ away from B; B at rest; C $u/6$ away from B
30. $V/16$, $3V/16$
31. $(7i + 3j)u/8$
34. $4:3$
35. $a(1 - e^{2n})/ue^{2n-1}(1 - e)\sin\theta$
37. $u\sqrt{(13k^2 + 20k + 16)}/4(k + 1)$; $k = \frac{1}{2}$
40. $mu\sqrt{0·75(1 + e)}$
43. $u = V$ or $V/3$, velocity vector is $Vi + Vj$ or $5Vi/3 + Vj$
45. $v_A = u$, $v_B = 2u$
46. $4(u - v)/5$, $2mu^2/5$, $e = \frac{1}{8}$
47. (i) $27y^2 = x^3$; (ii) $12i + 12j$; $r = 12(1 + \lambda)i + (8 + 12\lambda)j$; $20\sqrt{2}$, 40
48. K.E. = 625 000 joules, P.E. = 10 000 joules, power = 72 500 watts, height attained = $31\frac{1}{4}$ m

Exercises 14a

1. Horizontal 24·5 m/s, vertical 0·97 m/s
2. Horizontal $5\sqrt{3}/2$ m, vertical 0
3. 30 m
4. 7·5 m, $10\sqrt{3}/7$ s, $10\sqrt{3}$ m
5. $10\,000\sqrt{3}$ m, $10\,000\sqrt{2}$ m, 10 000 m
6. 32 m, 24 m
7. $\sqrt{(2·5ag)}$
8. (a) 12 km, (b) 2·25 km, (c) $140\sqrt{5}$ m/s at $\tan^{-1} 0·5$ to the horizontal
9. $12\sqrt{10}$ m/s at $\tan^{-1} 3$ to the horizontal
10. 11·25 m, $15\sqrt{2}/7$ s [3·03 s]

Exercises 14b

2. $u = 35\sqrt{2}/4$ m, $\theta = \tan^{-1}(\frac{4}{3})$

ANSWERS

3. $R_{max} = 45$ m, $15°$ or $75°$
4. $(\sqrt{3} - 1)ut/\sqrt{2}$
6. 17 930 m, 36·6 s
7. 33·3 m/s

Exercises 14c

1. 0·2
2. 45°

EXERCISES 14

1. $20° 54·5'$, $69° 5·5'$
2. $120\sqrt{21}$ m
5. $\tan^{-1}(\frac{3}{2})$
6. $\sin^{-1}(\frac{13}{40})(18° 58')$, 4·54 km
7. $u^4 \geq R^2 g^2$
8. 4·5 m
9. $R/T = u/[\sqrt{2}(\cos \alpha/2 + \sin \alpha/2)]$, yes
11. $\tan^{-1}(\frac{1}{3})$
15. 17·5 m/s at $\tan^{-1}(\frac{4}{3})$ to the horizontal, $2\frac{6}{7}$ s
18. 600 m, $17\frac{1}{7}$ s, $70\sqrt{12\,301}$ cm/s $\simeq$ 77·6 cm/s at $\tan^{-1}(1·98)$ to the horizontal
19. $2 \sin \alpha V_2 [V_2 \cos \alpha - V_1]/g$
20. $\cos^{-1}(0·25)$
24. $V = 50$ m/s
27. $R = 2pq/g$
28. $d \tan \theta/(1 + e)$
29. 1·325 m
30. 5·95 m/s, $(3 + \sqrt{2})m$
31. $V^2 > gk/[2n \sin \alpha(n \sin \alpha - \sin \beta)]$ which is equivalent to $V^2 > gh/[2n \sin \alpha(n \cos \alpha + \cos \beta)]$
32. $v^2 = g[d^2 + (1 + e)^2 h^2]/2eh$
33. $31V/8g$
36. 45°
37. $PQ = 16\sqrt{5}V^2/125g$
38. $40\sqrt{5}$ m
39. (a) $V^2/2g$ (b) $3V^2/2g$
40. $e = (\cot \theta \cot \alpha - 1)$
42. $\tan \theta = 1$ or 7, least speed $= \sqrt{(g/a)}$
43. Time $= \sqrt{(5a/2g)}$
44. $v = 20\sqrt{5}$, $\sqrt{10}/2$ s

Exercises 15a

1. 13 m/s² at $\tan^{-1} \frac{5}{12}$ to the radius
2. $\frac{25}{32}$ m/s² towards the centre
3. 20 cm/s
4. 50 m/s² towards the centre
5. $5\pi^2$ m/s² towards the axle
6. 10·74 m/s² at $\tan^{-1} \frac{20}{9}$ to the radius
8. $k\sqrt{7}/2$

EXERCISES

Exercises 15b

1. 675 N towards the centre
2. Apply a force mv^2/r always towards O
3. 256 N
4. 96 N
5. $45/\pi^2$ cm, $0.392\pi^2$ N
7. $60°$, $0.7\sqrt{6}$ m/s
8. 20 N, 23·2 N
9. 163
10. $\sqrt{\mu rg}$
11. $2.05 \times 10^{21} t$

Exercises 15c

1. 1·4 m/s, 1·47 N
2. 75g mN, 525g mN (both tensions)
3. $m(g - u^2/8r)$
4. $\cos^{-1} \frac{2}{3}$
5. $7\sqrt{2}$ m/s at $\cos^{-1} \frac{1}{6}$ with the upward vertical
6. 5·71 m/s, 5g N, no
7. a, zero
8. $\sqrt{[0.5gr(4\cos\theta - 3\cos 2\theta - 1)]}$
   $0.5mg(3 - 4\cos\theta + 3\cos 2\theta)$

EXERCISES 15

1. $e/m = v/Hr$
2. $\tan^{-1} 0.5$ to the horizontal
3. $h = g/\omega^2$, no
4. 2 m/s
6. 63·1 m
7. $7\sqrt{7}/5$ m/s
8. 5 cm
9. $60°$
10. $0.5\sqrt{ga}$, $1.5\sqrt{ga}$
11. $60°$
12. 30 N
13. 1·4 m/s, $1.4\sqrt{2}$ m/s, 7 rad/s
14. $0.2\sqrt{3}$ m
15. $19mg/3$, $mg/3$, $10a/3$
16. 415 rev/min
17. 4
18. $\sqrt{(g/3a)}$
19. $3mg/\sqrt{2}$
20. $5a/3$, $7a/3$, $5a/2$
22. 1·45 N s, 2·77 m/s, 25·1 N
28. $2bmg/(a - b) + 3mg\cos\theta$
29. $ga(2\sqrt{2} - 1)$, $mg(\sqrt{2} - 1)$
30. $\sqrt{(5ga)}$, $140a/81$
31. $R = (ga^2/\omega^2)^{1/3}$
32. $N_1 = mg/9$, $N_2 = 2mg/3$
33. (a) $[(\lambda - 3)mg\sin\theta]/(\lambda + 1)$; (b) $(2mg\lambda\cos\theta)/(\lambda + 1)$

Exercises 16a

1. $(4\mathbf{a} + 5\mathbf{b})/6$
2. $(6\mathbf{i} - 4\mathbf{j} + 10\mathbf{k})/6$
3. $(a, 2a)$, none
5. $\Sigma m \dfrac{dx}{dt} \Big/ \Sigma m$, $53\mathbf{i} + 46\mathbf{j}$ units

Exercises 16b

1. $i + j$ m/s^2
2. 200 m/s^2 at $\tan^{-1} \frac{3}{4}$ with AB, unaltered
3. kg, 0·5 kg in the direction of the force. Acceleration of A—yes, C of M—no
5. $-5\frac{1}{8}k$ m/s, $-5\frac{1}{8}k$ m/s

Exercises 16c

1. $-i - j - 2k$ N m
2. $a(2\sqrt{3} - 1)$
3. $12i - 60j - 36k$ kg m^2/s
4. $5ma^2\omega$, sum of the moments is zero
5. Velocities $t^2i + tj$, $2ti + t^2j - \frac{1}{2}t^2k$;
   displacements $t^3i/3 + t^2j/2$, $t^2i + t^3j/3 - t^3k/6$;
   angular momentum $t^4(j + k)/3$
7. $2FaT$

EXERCISES 16

1. On the diagonal through 0 $8\sqrt{3}a/7$ from 0.
3. $0·6t\hat{a} + 0·8t^2\hat{b}$, $3\hat{a} + 16\hat{b}$ N
4. $\lambda m(2n + 1)/3$
5. $0·75i - 0·5j$ m/s, $3i - 2j$ kg m/s
6. Only if friction is limiting and the particle moves
7. $2i + 3j$, zero, no
8. (a) $4(i + j)$ m/s (b) $-192k$ kg m^2/s
   (c) $4i$ N (d) $-80k$ N m
10. 1 kg m^2/s
11. Speed is reduced. The further from the axis the lower the angular speed
13. $2i + 5j$ cm/s, $5i + 7j$ cm/s, $-5i + j$ cm/s, $-8j$ cm/s
14. (a) $1·5mv^2 + ma^2\omega^2$
    (b) $2ma^2\omega\hat{a}$ where $\hat{a}$ is perpendicular to the plane of rotation
    The vector sum is zero, the sum of moments is zero; "dynamic" equilibrium.
15. Top slows and its axis moves in the direction of the force
17. $H = mr \times v$
18. (ii) Sum = 0

Exercises 17a

1. $P = i$
2. 3 units along $x = 13$
3. $4i + k$, $2i + 2j$
4. $-2F, F\sqrt{3}, -4F$
5. $6\sqrt{2}$ N at 45° to BA 24 N m
6. $16\sqrt{3}$ N along FD
7. $2F\sqrt{29}$, $5y = 2x + 48$

EXERCISES

8. $3a\sqrt{3}$ in the sense ABC
9. 0·24 N m in the sense $ABCD$
10. 20 N, $10y = 9$

Exercises 17b

1. $\sqrt{2}$ N along CA 2. Mid-point of AC
3. $2\sqrt{6}$ kN at 45° to CB on opposite side to A, mid-point of BC
4. Divides AC in the ratio $1:2$, AB in the ratio $2:1$, CB in the ratio $4:-1$.
5. Forces are in equilibrium
6. $r = (6i - 2j - 6k) + u(4i + 2j)$

EXERCISES 17

1. $\sqrt{2}P$ parallel to $\overrightarrow{DB}$ $2a$ from D on CD produced
2. $\sqrt{31}$ N at 8° 57' to BC on the side of A, with $3\sqrt{3}$ N m in the sense ACB
3. $5i - 5j + 5k$ N, $2i + 9j + 3k$ N m
4. $5P, 4P, 3P$
5. External division in the same ratio
6. $\sqrt{205}$ N
7. $3i + j - k$ at P_1 $-3i - j + k$ at P_2
8. $2k \times$ area, $2\Delta/3l$
9. 64 kN m
12. $P\sqrt{5}$, $y = 2x - a$ 13. $P\sqrt{10}$ parallel to $\overrightarrow{DB}$
14. $p = -2, r = 2i + 2j + 3k + u(2i + 3j)$
15. $2\overrightarrow{AC}$
16. $3\sqrt{7}$ N m, 6 N parallel to $\overrightarrow{BA}$ along a line $\sqrt{7/2}$ from C
17. $xY - yX = G$ 18. Mid-point of BC, $2G/3$
19. $X' = 3P(\cos \omega t + \sin \omega t)$, $Y' = P(2 \sin \omega t + \cos \omega t)$, $G = -4aP(\sin \omega t + \cos \omega t)$, $xY' - yX' = G$, $(0, 4a/3)$
21. 10 N, CD 2 m from D, AD produced 1·5 m from D, AB produced 2 m from B
22. $-2i, 4j$
24. (a) $5, 1, 4\sqrt{2}$ (b) $2·5AB, 1·25AD$
25. $2\sqrt{2}$ N parallel to RP cutting BA produced at X where $AX = 9·5$ cm. Magnitude unaltered, moment $= 0·35$ N m
26. $P\sqrt{5}, 2P\sqrt{5}$ meeting BC at Y where $BY = 0·75a$. Couple of moment $4Pa/\sqrt{5}$
27. $R = \lambda(3a + b + 2c), G = \lambda(a \times c)$. a is parallel to c.
28. (a) $P\sqrt{5}/3, 2P/3$ (b) $10\overrightarrow{OE}$

ANSWERS

29. AD is parallel to BC, $q = 1, r = 1$
30. $F_1 = (8, 2)$, $F_2 = (-4, -1)$
32. $R = 19i - 5j + 8k$, $G = 30(2i - 10j + 11k)$
    $r = (37\cdot 5i + 7\cdot 5j) + t(19i - 5j + 8k)$ or its equivalent.
33. (i) $\pi/3$; (ii) $\pm(i - j + k)/\sqrt{3}$; (iii) $R = 3i + 3j$, $G = -i - 4j$;
    (iv) $3\sqrt{2}$; (v) $(3c - 1)i - (4 + 3c)j + (3b - 3a)k$

Exercises 18a

2. $r \sin \alpha/\alpha$, r/π 3. $4r/3\pi$ from the centre
4. $\frac{1}{4}$ of the way up the line from the centre of the base to the vertex
5. $\dfrac{2\rho l + 4\lambda l^3}{2\rho + \frac{8}{3}\lambda l^2}$ from the end where $x = 0$
6. $\frac{1}{3}$ up the axis from the base   8. $3(2r - h)^2/4(3r - h)$

Exercises 18b

1. $2a/3$ from the centre rod along its perpendicular bisector
2. $(3\sqrt{2} - 4)m$ from the hypotenuse along its perpendicular bisector
3. $4\frac{15}{19}$ cm from the edge of square opposite the triangle along the axis of symmetry
4. $9a/5$ from the base along the axis, $15a/7$
6. $\frac{4}{15}$ cm from the centre on the line of centres
9. $11a/12$, $5a/12$ 10. $\frac{43}{45}$ m, $\frac{17}{9}$ m

EXERCISES 18

1. $(11i + 4j + 6k)/12$ 2. $3a, 7a; 3a, 6a$
3. 9 cm, 4 cm   7. $14a/3$ from AD, $\frac{86}{3}$ from CD
8. $2a/(\pi + 2)$ from the diameter on its perpendicular bisector
9. 7 cm from BC, $6\frac{2}{3}$ from DC
10. $a(2 - \sqrt{3})/2$ from AB on the side of D; $5a/2$ from BD
12. $(R^4 - r^4)/(R^3 - r^3)$
13. $(a^2 - h^2/3)/(2a - h)$ from CD
14. $(\frac{1}{2}, \frac{1}{10})$ 15. $49a\sqrt{2}/48$
16. $23a/72$ 17. at the point $(2a/3, 0)$
18. $(h_1^2\rho_1 - h_2^2\rho_2)/4(h_1\rho_1 + h_2\rho_2)$, $h_2^2 : h_1^2$
19. $(5\sqrt{3} - 4\pi)a/(4\pi - 3\sqrt{3})$ 20. At the apex of the cone
22. $\sin^{-1}\frac{1}{3}$
23. $(a^2 + ac + c^2)/3(a + c)$, $b(a + 2c)/3(a + c)$
24. $11r/20\sqrt{3}$ along the axis from the base
25. $3\sqrt{3}a/8$ from the base on the axis of symmetry

26. $\frac{14}{15}$ cm from the larger section
27. $(11h^2 - 16h_1^2)/8(7h - 8h_1)$ 28. $a/2, 5a/16$

Exercises 19a

1. $\sqrt{10}$ m
2. $Ma^2 \sin^2 \theta/3$
3. (a) $4M(a^2 + b^2)$
   (b) $4Ma^2$
   (c) $4Mb^2$
   (d) $8Ma^2b^2/(a^2 + b^2)$
4. $Ma^2/3, Mb^2/3$
5. (a) Ma^2 (b) $Ma^2/2$ (c) $2Ma^2/5$
   (d) $M(a^2 + b^2)/2$ (e) $Ma^2/6$
6. 0.0116 kg m^2
7. $2Ma^2/3$

Exercises 19b

1. $5Ma^2/4$
2. $Ma^2/2$
3. $2Ma^2/3, 8Ma^2/3$
4. $2Ma^2b^2/3(a^2 + b^2)$
7. $(2h^2 + 3a^2)/20$
8. $3Ma^2/10$
9. 0.94 kg m^2
10. $156Ma^2/625, 303Ma^2/625$

Exercises 19c

1. $\pi/10$ N m
2. 20π rad/s, 48π rad/s^2
3. $t = 13\frac{1}{3}$ s
4. 20π rad/s^2, 56.25
5. 128π N m
6. gt^2/k
7. $\frac{1}{8}$ kg m^2
8. $4, 0.4$ kg m^2
9. 9.10 s

Exercises 19d

1. $2\pi\sqrt{(5a/g)}$
2. Along the rod $(73 mg \cos \theta)/5$, perpendicular to the rod. $(mg \sin \theta)/5$
3. $1.5a$
4. 0.891 m
5. $\sqrt{3a}$
6. 1.68 s
8. (a) $4\pi\sqrt{(2a/g)}/3$
   (b) Along the axis $\dfrac{mg(43 \cos \theta + 27)}{8}$
   Perpendicular to the axis $-0.25mg \sin \theta$. (θ measured to the downward vertical)
9. $\pi\sqrt{(a/kg)}$
11. $28a/27$

Exercises 19e

1. $\sqrt{(5g/4a)}$
2. 50π N m
3. $\sqrt{(g/a)}$
4. $2 \sin(\theta/2)\sqrt{(g/3a)}$
6. $\sqrt{(5g/3a)}$
7. $0.37\sqrt{(g/a)}$

ANSWERS

Exercises 19f

1. 14 J
2. $Mv/2$
4. $R = I(hx/(k^2 + h^2) - 1)$
5. $x = a/\sqrt{3}$
6. $1 \cdot 5I^2/m$
7. $\omega = \sqrt{(3g/2a)}, I = m\sqrt{(6ga)}$
8. $m\sqrt{(2ga)}$

EXERCISES 19

1. $2Ma^2/5, 13Ma^2/20$
2. 5 s, 60 rad
3. $\frac{2}{7}(g - G/am)$
7. $7a/3$
8. $2\pi\sqrt{(a/8kg)}$
9. $mg/3$
10. The acceleration is uniform of magnitude $2g/7$
13. $I/J = 2$
15. Acceleration is uniform of magnitude $g/8$
17. $Mh^2/2, Mh^2/6$
18. $3Ma^2/10, M(0 \cdot 6h^2 + 0 \cdot 15a^2)$
21. $Mh^2(a + 3h)/6(a + b)$
22. $\theta^2 = 3(\sqrt{2} - 1)g/4a$
24. $\mu = (9\sqrt{2} - 12)/(11\sqrt{2} - 12)$
26. $3Ma^2/2, Mg \sin 2\theta, Mg(4 - 3 \cos 2\theta)/3$
27. $I(\omega_2 - \omega_1)/N, I(\omega_2^2 - \omega_1^2)/2N$
31. $M\sqrt{(5ga/2)}$
32. $10Ma^2/9, 2\pi\sqrt{(5\sqrt{2}a/6g)}$
33. $(m_1 - m_2)/(M/2 + m_1 + m_2)$ the "amount" is $M/4$
34. $12Ma^2, 2\pi\sqrt{(12a/5g)}$
35. $2\pi\sqrt{(149a/21g)}$
36. $2\sqrt{dmg/(M + 2m)}, Mmg/(M + 2m)$
37. $\omega^2 = 3\sqrt{2g/10a}$
38. $2\sqrt{(15ga/7)}$

Exercises 20a

1. $u \geqslant \frac{1}{3} \tan \alpha, \frac{2}{3}g \sin \alpha$
2. $u \geqslant 1/2\sqrt{3}, g/4$
3. $2g/3, mg/3$
4. $\tan^{-1} (v^2/rg)$
5. $24 \cdot 2$ m/s, $u = 1 \cdot 5$
6. No
8. $6 \cdot 7$ cm
9. $2gb(b \pm a)/(7b^2 + 2a^2) +$ or $-$ according as the string is above or below the axle
10. $\sqrt{6ga}$

Exercises 20b

1. $(3\sqrt{3}g/4)^{1/2}$ rad/s. The path is a vertical straight line
2. $\sqrt{3gl}, \sqrt{3g/4l}$
3. $3v^2/4g \sin \alpha$

EXERCISES

4. 2·52 J, 25·7 m
5. $\sqrt{[0\cdot 8ga(1-\cos\theta)]}$
7. $\dot\theta^2 = \Omega^2 - 4g(1-\cos\theta)/3(a-b)$

Exercises 20c

1. A and B rotate about the C of M with angular speed $I/2ma$. The C of M has a speed $I/2m$ in the direction of I
2. $\frac{1}{2}MI^2\left[\dfrac{x}{Ml}+\dfrac{1}{M+m}\right]^2 + \frac{1}{2}mI^2\left[\dfrac{x}{ml}-\dfrac{1}{M+m}\right]^2$
   where x is the distance of I from the C of M. At the C of M of the particles
3. $I(1+3x^2/a^2)/m$; $x=0$, i.e. the blow is given at the C of M.
4. $\pi a/3$
5. $\omega' = \omega/4$; $ma^2\omega^2/4$
6. $11V/15a$
7. $\frac{1}{4}mv^2/25$

EXERCISES 20

1. $2u/3a$
2. $t = 7a\omega/(5g\sin\alpha)$, $d = 7a^2\omega^2/(10g\sin\alpha)$,
   No. of revs $= 7a\omega^2/(20\pi g\sin\alpha)$
3. $AX = 4a/3$
5. $\cos^{-1}\frac{10}{17}$
6. $12(\sqrt{2}-1)g/5a$, $3(\sqrt{2}-1)g/2a$
7. $\omega' = k^2\omega/(k^2+h^2)$
8. $4I/5m$, $I/5m$
9. $\dot\theta^2 = 6g(\sin\alpha - \sin\theta)/a(1+3\sin^2\theta)$
13. $-3:1$
14. $mu(4a-3x)/4a$ (a) $\frac{4}{7}$ (b) $\frac{4}{7}$
17. $n=6$, $v^2/2\mu g$
18. $5mv/16$, $v = \sqrt{(32ga/9)}$
19. $a(1+m/M)$
20. $\tan^{-1}\left(\frac{71}{187}\right)$
21. $P = Mg\sin 2\alpha/3$, angular acceleration $= 2g\sin\alpha/3$
23. $\omega_{AB} = 7I/4ma$, $\omega_{BC} = 0$, $\omega_{CA} = I/4ma$, $R_B = 5I/12$, $R_C = I/12$
24. $3I/[2m\sqrt{(a^2+b^2)}]$
25. $\omega = 4x/(k^2+x^2)$;
    loss of K.E. $= \frac{1}{2}mu^2k^2/(k^2+x^2)$

Exercises 21a

1. 1 kg wt., along the rod $2\sqrt{3}$ kg wt., perpendicular to the rod, 1 kg wt.
2. 10 kg wt., 30 kg wt.
3. Each one is 4 kg wt., $4\sqrt{17}$ kg wt. at $\tan^{-1}4$ to the horizontal
4. $33W_1/32$
5. 30 kg wt.
6. $T = W\tan\theta/2$. Reactions $W\tan\theta/2$ and W
7. $\sin^{-1}(8\mu/3)$
8. $T = 5W/\sqrt{3}$, $R = W\sqrt{21}/3$ at $\tan^{-1}(\sqrt{3/5})$ to AB
9. $45°$
10. $\frac{2}{5}$, 17·5 cm from C along CD

12. $T = F = W\sin\alpha/(1 + \cos\theta)$,
    $N = W[\cos(\theta - \alpha) + \cos\alpha]/(1 + \cos\theta)$
13. $N = W$, $F = W/\sqrt{3}$
14. At C $(89{\cdot}28 - 28{\cdot}8x)$ kg wt., at D $(6{\cdot}72 + 28{\cdot}8x)$ kg wt., all of the plank

Exercises 21b

1. $\tan^{-1}(\tfrac{3}{8})$, $\tan^{-1}(\tfrac{1}{2})$
2. $\tan^{-1}(\tfrac{12}{41})$
3. $2a\sqrt{3}$
4. $\tan^{-1}(3a/h)$

Exercises 21c

1. $50\sqrt{17}$ kg wt.
2. a, $W\sqrt{2}$, W
3. $30°$
4. $6°\,35'$
5. $\mu \geqslant 4\sin\theta/3\pi$
6. $T = W$, $R = 3W/2$
7. $\tan^{-1}[(1 - \mu_1\mu_2)/2\mu_2]$
8. $T = 5W/8$, $N = 39W/40$
9. $c\tan\theta/a$

Exercises 21d

1. $3W/\sqrt{2}$
4. $\tfrac{1}{9}, \tfrac{1}{9}, \tfrac{1}{3}$
5. $W\tan\theta/2$ horizontally
6. $3W/2$, $W/2$ vertically
7. At A: $W/\sqrt{3}$ horizontally, $3W/2$ vertically
   At B, C: $W/\sqrt{3}$ horizontally, $W/2$ vertically
8. $3W$, W, $2W/\sqrt{3}$; $W\sqrt{(7/12)}$
9. AB thrust $W\sec\theta$, BC tension $W\tan\theta$
10. AB thrust $W/2$, BC thrust $W\sqrt{3}/2$, AC tension $W\sqrt{3}/4$

Exercises 21e

1. $T = 6$ kg wt. Horizontally $4{\cdot}8$ kg wt., Vertically $16{\cdot}4$ kg wt.
2. $P = [\tfrac{1}{2}\tan\alpha + a/h]$, $N = W[\tfrac{1}{2}\sin\alpha\tan\alpha + a\sin\alpha/h + \cos\alpha]$,
   $F = W[a\cos\alpha/h - \tfrac{1}{2}\sin\alpha]$
4. $\mu > \tfrac{1}{8}$
5. $3\sqrt{3}/5$
6. $2W - Wx/2a$; $2W + Wx/2a$; $5a/4$
7. $W\sqrt{2}$ at walls, W at the floor
8. W at walls, $2W$ at the floor
9. W
10. $T_1 = T_2 = 7W/8\operatorname{cosec}\theta$, $X = 7W\cot\theta/8$, $Y = 17W/8$

EXERCISES 21

1. $W\sin\beta/\sin(\alpha + \beta)$, $W\sin\alpha/\sin(\alpha + \beta)$
2. $3W/2\sqrt{2}$; $W/2\sqrt{2}$
4. $35°\,21'$

EXERCISES

8. $\sqrt{5}W$ at $\tan^{-1} 2$ to the horizontal
9. $F = W/2 \tan \theta$
11. $F = W/\sqrt{3}$, $W\sqrt{21}/6$ at $\tan^{-1}(2/\sqrt{3})$ with the horizontal
13. $\mu \geqslant (\pi + 2)/4$
15. $(\tan \theta + 2 \tan \alpha)/(2 - \tan \theta \tan \alpha)$
16. *At A*: horizontally 84 g wt., vertically 187 g wt.
    At C: horizontally 84 g wt., vertically 163 g wt.
19. $W/2$, $W/2$ at 60° to the vertical
21. $W/2$, bW/a, $W(\frac{1}{2} - b/a)$ 23. $2r\sqrt{2}$, $W/\sqrt{2}$
24. $5\sqrt{2}a/4$ 27. $2\frac{8}{11}a$
28. At C $W/2\sqrt{3}$ horizontally. Thrust $2W/\sqrt{3}$
29. $W/2\sqrt{3}$, $W\sqrt{3}/2$ 32. $W(23 + 2\sqrt{2})/20$
37. $16a/5$, $4a/5$ 38. $b < 4a$
39. at l_2, $\mu \geqslant 0.5$ 42. $\sqrt{[W(W + \omega)/2]}$
43. $W\pi/(\pi - 2\sqrt{2})$ 46. $\frac{3}{4}$, 1, 1 lb wt., $\sin^{-1}(\sqrt{3}/4)$
48. $(3\sqrt{2} - 4)a$
49. $W[8a^2/(a^2 + b^2) \tan^{-1}(a/b) - 1]$

INDEX

INDEX

Acceleration,
    angular, 366
    definition, 85
    gravity, 121
    linked accelerations, 143
    relation to force, 116
    S.I. units of, 4, 85
    variable, 94
Angular momentum,
    conservation of, 322
    definition, 320
    S.I. units of, 320
Angular speed, 73 *et seq.*
    velocity, 76
Associative rule, 19, 21
Astroid, 52
Average speed, 57

Basic quantities, 3
Bolas, 317

Centre of gravity,
    composite bodies, 356 *et seq.*
    rigid body, 347
    some standard bodies, 348 *et seq.*
    standard results, 355
    system of particles, 347
Centre of mass of a system of particles,
    definition, 315
    motion of, 317
    motion relative to, 323
Centre of percussion, 397
Centroids, 352, 355
Circular motion,
    acceleration of a particle, 295, 298

Circular motion, (*contd.*)
    forces acting, 297
    in a vertical circle, 301
Coefficient of restitution, 250
Commutative law, 19, 20, 37
Compound pendulum, 381, 383
Connected particles, 136, 139, 143, 145
Conservation of energy principle, 228
Conservation of momentum, 241, 250, 255
Conservative force, 226
Cotangent rule for a triangle, 436
Couple,
    definition, 330
    work done by, 392
Cross product, *see* Vector product

Derived quantities, 4
Directed segment, 6
Displacement, 8, 87
Distributive law, 19, 24, 37, 40
Dot product, *see* Scalar product

Energy methods for solution of problems, 390 *et seq.*
Equilibrium,
    coplanar forces, 449 *et seq.*
    general conditions, 425
    jointed rods, 442
    particle, 188 *et seq.*
    several bodies in contact, 440
    three force problems, 434
    two force problems, 433
Equimomental systems, 399
Equivalent sets of forces, 329

INDEX

Force,
  centre of parallel forces, 346
  conservative, 226
  coplanar systems, 188, 331
  definition, 116
  equivalent systems, 329, 331
  relation to,
    acceleration, 116
    momentum, 116
  resultant, 336
  S.I. units of, 4, 116
Friction,
  angle of, 124
  coefficient of sliding friction, 124
    static friction, 124
  definition, 123
Fundamental quantities, 3

Gravitation,
  acceleration, due to, 121
  attraction, 121
  local constant, 121
  universal constant, 121

Hooke's law, 122, 298

Impact,
  elastic bodies,
    direct, 249
    oblique, 254
  inelastic, 240
  Newton's experimental law, 250, 255
Impulse,
  applied to rigid bodies, 415
  constant force, 236
  impulsive motion about a fixed axis, 394
  impulsive tension in strings, 244
  relation to momentum, 236
  S.I. units of, 236
  variable, 236
Impulsive tension, 244
Instantaneous speed, definition of, 58

Jointed rods, 442

Kinetic energy,
  definition, 222
  relation to work, 222
  rigid body, 411
  system of particles, 324
Kilogramme, 3, 188

Lami's theorem, 194 (Ex. 10)
Light framework, 446
Line integral, 211
Linked acceleration, 143
Localized vector, 42

Mass, 3
Metre, 3
Modules of elasticity, 122
Moment of a vector, 42, 43, *see also* Torque
Momentum,
  conservation of, 241, 319
  definition 115
  moment of, 320
  rate of change of, 116
  relation to impulse, 238
  S.I. units of, 115
Moment of inertia,
  calculation of, 367 *et seq.*
  definition, 366
  list of standard results, 375
  parallel axis theorem, 371
  perpendicular axis theorem, 374
Motion in a straight line, 85 *et seq.*
  under constant forces, 134
  under variable forces, 175
Movable pulleys, 143

Newton's laws of motion, 115, 121
Null vector, 7

Parallelogram rule, 9
Particle, definition of, 2

INDEX

Physical laws, 121 *et seq.*
Polygon rule, 13
Position vector, 30, 46
Potential energy, 227
Power,
    definition, 213
    relation to velocity and force, 214
    S.I. unit of, 213
Principle of conservation of energy, 228
Principle of conservation of momentum, 241, 250, 255
Principle of work,
    particles, 222, 390
    rigid body moving in two dimensions, 412
Projectile,
    equation of path, 275
    impact, 282
    maximum range,
        horizontal plane, 274
        inclined plane, 280
    range,
        horizontal plane, 274
        inclined plane, 279
    time of flight, 273
Pulley systems, 143, 145

Radius of gyration, 368
Reaction—total, normal, 123
Relative velocity, 64 *et seq.*
Right-handed set of axes, 27
Rigid body,
    definition, 313
    equations of motion,
        general, 405
        rotating about a fixed axis, 365
    kinetic energy,
        general, 411
        rotating about a fixed axis, 366
    principle of work, 342
    reaction at axis of rotation, 387
    uniform angular acceleration, 376
Routh's rule, 375

Scalar product, 35
Scalar quantity, 6
Second, 3
Sideslip of a car, 406
Simple equivalent pendulum, 384
Simple harmonic motion,
    amplitude, 100
    centre, 156
    formulae, 101
    maximum speed, 100
    motion under forces causing S.H.M., 155 *et seq.*
    period, 101
    relation to motion in a circle, 105 (Ex. 4)
Simple pendulum, 159 (Ex. 2)
Speed,
    definition, 58
    S.I. units, 4, 58
Speed–time graphs, 91
Springs, 122, 211
Systems of connected particles, 195
Systems of units, 3

Tension in a light rod, 446
Thrust in a light rod, 446
Time, 3
Tonne, 4
Toppling and sliding, 451
Torque,
    definition, 320
    relation to angular momentum, 321
    S.I. unit, 320
Triangle rule, 9

Uniform acceleration, 86
Unit tangent to a curve, 31
Unit vector, 7

Vectors,
    addition, 20
    angle between, 7
    collinear, 25
    components, 11, 26

INDEX

Vectors, (*contd.*)
  compounding, 8
  coplanar, 25
  definition, 6
  differentiation, 30
  differentiation of unit vectors, 34
  division by a scalar, 24
  equation of a straight line, 48
  integration, 30
  modulus, 7
  multiplication by a scalar, 23
  product, 40
  rectangular resolution, 27
  resolved part, 37
  resultant, 13, 18
  subtraction, 20
  unit, 7
  variable, 30
  zero, 7
Velocity,
  definition, 62
  S.I. unit, *see* Speed

Weight, 121
Work,
  relation to kinetic energy, 221
  S.I. unit, 206
Work done by,
  constant force, 206
  couple, 392
  extending an elastic thread or spring, 211
  variable force, 210